# Mycoagroecology

During the 20th century, agriculture underwent many unsustainable changes for the sake of greater food production. Today, the effects of climate change are becoming ever more apparent and the global population continues to grow, placing additional pressures on agricultural systems. For this reason, it is vital to turn international agriculture towards a sustainable future capable of providing healthy, bountiful foods by using methods that preserve and reconstruct the balance of natural ecosystems.

Fungi are an underappreciated, underutilized group of organisms with massive potential to aid in the production of healthy food and other products while also increasing the sustainability of agricultural systems. *Mycoagroecology: Integrating Fungi into Agroecosystems* lays the foundations for integrated fungal-agricultural understanding and management, the proposed practice of "mycoagroecology". Suitable for students and professionals of multiple disciplines, this text includes nine introductory chapters that create a firm foundation in ecosystem functioning, evolution and population dynamics, fungal biology, principles of crop breeding and pest management, basic economics of agriculture, and the history of agricultural development during the 20th century. The latter half of the text is application-oriented, integrating the knowledge from the introductory chapters to help readers understand more deeply the various roles of fungi in natural and agricultural systems:

PARTNERS: This text explores known benefits of wild plant-fungal mutualisms, and how to foster and maintain these relationships in a productive agricultural setting.

PESTS AND PEST CONTROL AGENTS: This text acknowledges the historical and continuing role of agriculturally significant fungal pathogens, surveying modern chemical, biotechnological, and cultural methods of controlling them and other pests. However, this book also emphasizes the strong potential of beneficial fungi to biologically control fungal, insect, and other pests.

PRODUCTS: This text covers not just isolated production of mushrooms on specialized farms but also the potential for co-cropping mushrooms in existing plant-based farms, making farm systems more self-sustaining while adding valuable and nutritious new products. An extensive chapter is also devoted to the many historical and forward-facing uses of fungi in food preservation and processing.

# Mycology

*Series Editor:*
Donald O. Natvig

## PUBLISHED TITLES

For more information about this series, please visit: https://www.crcpress.com/Mycology/book-series/CRCMYCOLOGY

# Mycoagroecology
## Integrating Fungi into Agroecosystems

Edited by
### Elizabeth "Izzie" Gall
Morel Dilemma

### Noureddine Benkeblia
University of the West Indies

CRC Press
Taylor & Francis Group
Boca Raton  London  New York

CRC Press is an imprint of the
Taylor & Francis Group, an **informa** business

First edition published 2023
by CRC Press
6000 Broken Sound Parkway NW, Suite 300, Boca Raton, FL 33487-2742

and by CRC Press
4 Park Square, Milton Park, Abingdon, Oxon, OX14 4RN

*CRC Press is an imprint of Taylor & Francis Group, LLC*

© 2023 Taylor & Francis Group, LLC

*Library of Congress Cataloging-in-Publication Data*
Names: Gall, Elizabeth, editor. | Benkeblia, Noureddine, editor.
Title: Mycoagroecology : integrating fungi into agroecosystems / Elizabeth
Gall, Noureddine Benkeblia.
Description: First edition. | Boca Raton : CRC Press, 2023. |
Includes bibliographical references. |
Summary: "Fungi serve as partners of plants that can bolster disease resistance and improve crop yields while reducing the need for artificial inputs; as pests that can cause significant diseases of major crops; and as valuable products with nutrient content and low input requirements. This new textbook on mycoagroecology introduces basic concepts of ecology, mycology, and systems of agriculture. It combines this information into an understanding of mycoagroecology, highlighting various roles of fungi in agriculture. This textbook is for students in plants sciences and agriculture departments and contains beautiful color illustrations throughout"—Provided by publisher.
Identifiers: LCCN 2022019507 (print) | LCCN 2022019508 (ebook) |
ISBN 9780367335243 (hardback) | ISBN 9780429320415 (ebook)
Subjects: LCSH: Fungi in agriculture. | Fungi as biological pest control
agents. | Agricultural ecology.
Classification: LCC SB733 .M93 2023 (print) | LCC SB733 (ebook) |
DDC 632/.4—dc23/eng/20220829
LC record available at https://lccn.loc.gov/2022019507
LC ebook record available at https://lccn.loc.gov/2022019508

ISBN: 9780367335243 (hbk)
ISBN: 9781032365701 (pbk)
ISBN: 9780429320415 (ebk)

DOI: 10.1201/9780429320415

Typeset in Times
by codeMantra

# Dedication

---

*I dedicate this book to every scholar who thinks*
*learning has to be done in a classroom.*

*It doesn't.*

**Elizabeth "Izzie" Gall**

*In memory of Sonia (B) Kheitmi and Françoise Varoquaux*

**Noureddine Benkeblia**

# Contents

# SECTION I   Partners

# SECTION II   Pests and Pest Control Agents

# SECTION III   Products

# Preface

Fungi are everywhere. They are degraders of wood and skin, flowers and hair, leaves and bones. Fungi live on our hands and in our lungs, on animal dung and on stone, on land and in the water. They kill and strengthen crops, curdle milk and ferment grapes into wine, cause famines and turn waste into edible protein.

Fungi are inescapable, key members of Earth's natural balance, and yet most of them are so small as to be invisible. Indeed, they are largely forgotten members of our world. Introductory biology courses at the high school and undergraduate levels typically cover bacterial, plant, and animal biology but only mention fungi in passing or omit them entirely. Further, many courses that do include the fungi focus on their role as pests either in animal medicine or in plant pathology. Many people with urban or suburban lifestyles do not know or remember that fungi even exist until a mushroom suddenly asserts itself in an apartment, in a refrigerator, or on a grassy lawn, and then it is not greeted with wonder, but with fear. Fungi, however, are vital members of natural ecosystems and their exclusion, whether accidental or purposeful, contributes to the unsustainability of the current industrial agricultural paradigm. To move agriculture toward a more sustainable future, it is vital to incorporate fungi when teaching agricultural or agroecological science and when managing agricultural systems.

I had two major goals in creating this text: first, to help readers of various backgrounds recognize the importance and wonderful diversity of fungi and, second, to encourage readers to incorporate fungi into agricultural systems. This book outlines all the basic concepts needed for readers with only basic biological experience to understand the cyclic functioning of agroecosystems, the significance of genetic inheritance in plant and fungal variety selection, the role of fungi in agroecosystems, and the potential management practices that unite mycology and agroecology into integrated **mycoagroecology**.

## PREREQUISITES AND SCOPE

This book is written for the undergraduate or graduate student of any subject who may have a high school level of biology background. The introductory chapters (1–4) provide the information necessary to benefit from the remainder of the book. Familiarity with taxonomic systems (mainly Kingdom, Phylum, Genus, and Species), the scientific naming convention (*Genus species* or *Genus* spp.), and basic cell biology will be helpful.

This book is directed at adding fungi and mycological understanding to existing agricultural systems, so it does not include all the information required to start a business or farm (mushroom-based or otherwise). To start an operation from scratch, readers may need additional plant biology, agricultural management, and business management courses or texts.

## ORGANIZATION

The first half of the book lays out introductory concepts key to agricultural management. Chapters 1–3 cover nutrient cycling, evolutionary development, and population dynamics, the mechanics that govern all terrestrial ecosystems. Chapter 4 helps close the mycological gap in general biology education with a detailed introduction to the Kingdom Fungi: its diversity, modes of reproduction within major taxonomic groups, and the variety of niches filled by fungi. Chapters 5 and 6 introduce major limiting factors in agriculture, including the stresses that these limits place on plants and on agricultural production. Chapter 6 explores the three photosynthetic systems in land plants as an example of the evolutionary limitations to selective breeding, a concept which applies to all of Earth's organisms. Chapters 7 and 8 begin the focus on agricultural management rather than underlying theory, covering fundamentals of pest management and economic considerations, respectively. Finally, Chapter 9 provides important historical, social, and economic factors that have led to the current agricultural paradigm of high inputs and the use of marginal lands at the expense of surrounding ecosystems.

Chapter 10 provides the philosophical basis for the latter half of the book, which integrates concepts from the introduction to reveal the vital roles fungi play in both natural and agricultural ecosystems. The Partners section (Chapters 11 and 12) details the mechanisms fungi use to partner with plants in mutualistic relationships and the ways that farm managers can encourage these mutualisms in recovering agricultural soils. In Pests and Pest Control Agents (Chapters 13–15), readers will learn about significant fungal diseases in agriculture and silviculture (forestry), many of which use similar mechanisms to those described in the Partners section. Chapters 13 and 14 also detail several management methods for pathogenic fungi, including the use of beneficial biocontrol fungi. Chapter 15 reveals the important evolutionary relationships between fungi and pests such as insects and nematodes, demonstrating that many of these relationships are beneficial to agriculture. Finally, the Products section (Chapters 16–18) outlines how fungi are currently cultivated and used to create foods, food additives, and other agricultural products. Chapter 16 details how fungi are cultivated on devoted mushroom farms, while Chapter 17 illustrates how fungi can be incorporated into existing farm systems to add healthy, valuable products while closing farm waste loops. The book concludes with Chapter 18, a tour of the many traditional, modern, and upcoming products that make use of fungi to the benefit of human cuisine.

## TEACHING/READING ORDER

Undergraduate courses (or undergraduate-level readers) can proceed through the book in the order it is presented. Graduate-level readers are encouraged to begin with Chapter 4 and proceed from Chapter 9 onward, treating the other introductory chapters as supplemental or refreshers in the topics they present. I do not recommend skipping Chapter 4, as most readers will lack the mycology background presented there.

The latter chapters (10–18) are organized with the Pest role of fungi between the positive roles of Partners and Products. The Partners section should always be read first because the mycorrhizal partnerships taught there are important to both the Pests and Pest Control Agents and Products sections. The Partners section also helps establish the locations on plants where fungi can live and their methods of interaction with plants, which is important background for the Pests and Pest Control Agents section. However, the Pests and Pest Control Agents and Products sections can be read in either order.

## CONCLUDING REMARKS

By no means does this text include every aspect of fungal biology, the performance of fungi in natural or agricultural systems, or their potential to improve agricultural production and quality. Readers are highly encouraged to read the papers and textbooks referenced throughout this volume. I hope this book not only acts as a foundation to help readers incorporate fungi into agroecological systems but also inspires further research and future innovative uses of fungi.

**Elizabeth "Izzie" Gall**

# Acknowledgments from Elizabeth "Izzie" Gall

I am deeply grateful to Alice Oven for reaching out in September 2017 and offering to let a pod-caster write a textbook. I am also indebted to Randy Brehm for her extensive guidance throughout this process.

Thanks very much to my amazing husband, Nathaniel Neligh, and my wonderful mother, Barbara Weil Laff, for contributing their endless support and their professional expertise. They and Megan France-Peterson provided invaluable help by responding to three years of late-night read-ability requests and my (hopefully infrequent) episodes of editor panic.

I also gratefully acknowledge my contributing authors, in the order they joined the project: Barbara Weil Laff, Steven Stephenson, Noureddine Benkeblia, Jason Slot, Guillermo E. Valero David, Juan Francisco Barrera Gaytán, Denita Hadziabdic-Guerry, Aaron Onufrak, and Romina Gazis-Seregina. Thank you for responding to a first-time textbook editor and coming onto this proj-ect. I have learned an incredible amount from each of you and I thank you very much for your time and efforts. I hope this project has been as rewarding for you as it has been for me.

To every listener of *Morel Dilemma*: Thank you for your support, for helping me find my voice, and for letting me chase my passion wherever it led. I never expected it to lead here.

Finally, I wish to extend a profound thanks to Dr. George Ellmore of Tufts University, who taught fungal biology in *Plants and Humanity* in Spring 2012 and set me on this journey. His boundless enthusiasm for his research and his students' interests always inspires me. GSE, I will never forget having tea in a hurricane with you or the way you muttered "They'll never believe it" before every lecture. If I ever become a professor, I want to be like you. Thank you from the bottom of my heart.

# Editors

**Elizabeth "Izzie" Gall** graduated from Tufts University in 2015 with a plant biology degree and a fierce passion for mycology. As an intern at the New York Botanical Garden, she founded *Morel Dilemma*, a podcast intended to impart her love of fungi to listeners of diverse backgrounds. After the internship ended, Izzie worked as a laboratory technician at New York University but continued the podcast, researching and producing episodes about mushroom cultivation, mycorrhizal relationships, foraging practices and safety, and fungi in the fossil record, among other topics. In 2017, she attended graduate school in mycology and plant pathology at Michigan State University before returning to outreach. She is delighted to bring readers of all levels a deeper understanding of fungi and their utility in natural and agricultural spaces. She currently lives in Knoxville, Tennessee, with her husband Nate.

**Noureddine Benkeblia** is a Professor of Crop Science, Department of Life Sciences, Faculty of Science and Technology, University of the West Indies Mona Campus, Jamaica. He is also Head of the Laboratory of Crop Science (Life Sciences Department) and Head of the Laboratory of Tree Fruits and Aromatic Crops (Biotechnology Centre) and Coordinator of the Graduate Agriculture program. His main research areas focus on the environmental stresses on the physiology and biochemistry of crops including pre- and postharvest metabolism, and the metabolome. Prof. Benkeblia has extensive research (more than 30 years) experience in horticultural sciences, and experience teaching at undergraduate, graduate, and postgraduate levels in many countries. Prof. Benkeblia has published over 230 publications including books, chapters, and research papers. He is a member of many international scientific societies and NGOs, and a recipient of many awards. He is also Associate Editor of several scientific journals, including the *Canadian Journal of Plant Science* and *PLoS Climate*, among others.

# Contributors

**Juan F. Barrera**
El Colegio de la Frontera Sur
Chiapas, Mexico

**Romina Gazis**
Department of Plant Pathology
University of Florida
Gainesville, Florida

**Denita Hadziabdic**
Department of Entomology and Plant
    Pathology
The University of Tennessee
Knoxville, Tennessee

**Barbara Weil Laff, Esq.**
Ireland Stapleton Pryor & Pascoe P.C.
Denver, Colorado

**Aaron Onufrak**
Department of Entomology and Plant
    Pathology
The University of Tennessee
Knoxville, Tennessee

**Jason C. Slot**
College of Food, Agricultural, and
    Environmental Sciences
The Ohio State University
Columbus, Ohio

**Steven Stephenson**
Department of Biological Sciences
University of Arkansas
Fayetteville, Arkansas

**Guillermo E. Valero David**
The Ohio State University
Columbus, Ohio

# 1 Nutrient Cycling and Trophic Lifestyles

*Elizabeth "Izzie" Gall and Noureddine Benkeblia*

## CONTENTS

## 1.1 INTRODUCTION

At its core, **agriculture** is the act of manipulating a biological system (**ecosystem**) so that the system produces more resources, or different resources, than it would produce on its own. Farms, gardens, nurseries, plantations, and greenhouses are **agroecosystems**, spaces that humans have made distinct from the surrounding "natural" environment to create a steady supply of food, fiber, construction materials, and other natural products. Agroecosystems are clearly separated from their surroundings by the addition of desirable organisms (crops or livestock) and the removal of unwanted organisms (weeds or pests). Farmers can also add water, fertilizer, and other **abiotic** (nonliving) material to further improve their farms' products. Unfortunately, placing too high of a demand on the ecosystem can cause serious problems for human consumers as well as the native life around a farm (see Chapter 9 in this volume).

With the global human population continuing to increase, we need to practice agricultural methods that will deliver high-quality products consistently over the long term. The importance of environmental stewardship has also never been clearer as problems tied with pollution, like changing weather patterns and habitat destruction, become increasingly common. The goal of **sustainable agriculture** includes both environmental responsibility and dependable production of agricultural materials over time.

To make agricultural systems more sustainable, we must first explore:

- The processes that govern both natural and agricultural ecosystems (Chapters 1–3)
- The role of decomposers and resource gatherers in ecosystems (Chapter 4)
- The processes that must occur in a farm to maintain production and profit (Chapters 5–8)
- Some of the historical, cultural, and economic factors that have led modern agriculture to be unsustainable (Chapter 9)

Agricultural practices cannot be made truly sustainable if any of these factors is ignored. Full sustainability requires that the agricultural and "natural" ecosystems be treated with equal care and that farmers avoid the damaging pitfalls of the past. This chapter reviews how energy becomes stored in **terrestrial** (land-based) ecosystems and how both energy and nutrients move between organisms.

DOI: 10.1201/9780429320415-1

## 1.2  PHOTOSYNTHESIS: PRIMARY CALORIE PRODUCTION

Every living being on Earth shares one pool of resources. Many originate from the Earth itself: liquid water, atmospheric gases like oxygen and carbon dioxide, and solid minerals like copper and zinc. To gather and use these resources, organisms need a source of energy. Most of the energy stored in and cycled through Earth's ecosystems originally comes from the sun.

**Photosynthesis** is the process by which certain organisms convert solar energy into chemical (biologically usable) energy. Solar energy is made up of **photons**, highly energetic particles that can cause major damage to living cells if not handled carefully. Each photon contains enough energy to split one water molecule into two excited electrons, two hydrogen atoms, and one oxygen atom. Photosynthetic organisms use the power from the electrons to combine hydrogen atoms with **carbon dioxide** ($CO_2$) from the air, ultimately storing the energy in molecules like **glucose** sugar. For every two water molecules split in this way, one molecule of oxygen gas ($O_2$) is released (Figure 1.1).

The glucose and other products created during photosynthesis provide materials for molecules with diverse biological functions, including but not limited to long-term energy storage (starch and fats), information storage (DNA and RNA backbones), and structural support (proteins and membrane components). You are probably familiar with the idea of **calories**, which measure the amount of biologically available energy in food. The storage of solar energy in molecules like glucose is so effective and important that most of the calories in terrestrial ecosystems are produced with photosynthesis. Therefore, photosynthetic organisms are often called **primary producers**.

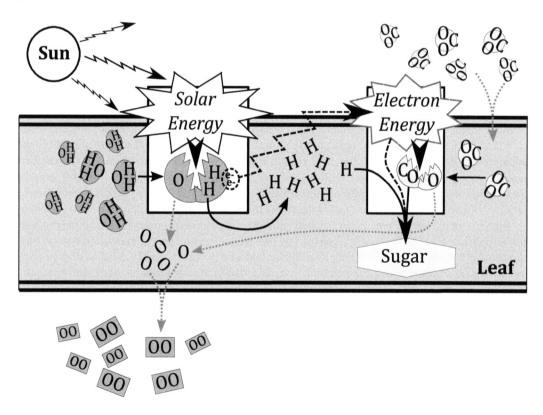

**FIGURE 1.1**  The basic process of photosynthesis. Sunlight reaches the membrane proteins of a photosynthetic organism; the proteins use the solar energy to split cellular water into oxygen, hydrogen, and two electrons. Different proteins use the electron energy to affix hydrogen onto a $CO_2$ molecule, the first of many steps leading to the formation of sugar and the release of another oxygen atom. Oxygen gas is released to the atmosphere; the sugar is retained and used by the cell or transported elsewhere in the organism.

## 1.3   PRIMARY AND SECONDARY METABOLITES

All living organisms are constantly conducting a stream of complex, tightly controlled chemical reactions. These reactions enable movement, communication, growth, and other essential functions of life. These overlapping chemical processes are known as the **metabolism** (from the Greek *metaballein*, "to change"). The molecules that move through the pathways and undergo cell-directed changes are called **metabolites**.

**Primary metabolites (PMs)** are molecules involved in normal growth, development, and reproduction of living organisms. All other compounds an organism produces, like color or defense molecules, are termed **secondary metabolites (SMs)**. While secondary metabolites may play a role in growth and development, they also have powerful ecological functions, strongly affecting organism interactions (as discussed below). Many secondary metabolites are found only in specific organisms or cells (Table 1.1) (Erb and Kliebenstein 2020).

Organisms that can assemble all the metabolites they need using raw materials in their environment are called **autotrophs** (from the Greek *auto* meaning "self" and *troph*, "nourishment"). Because photosynthetic plants derive the energy for metabolite assembly from photons, they are referred to as **photoautotrophs**. There are other types of autotrophs, such as deep-sea bacteria that harness geothermal energy, but they rarely appear in human agricultural systems!

In contrast with autotrophs, **heterotrophs** (from the Greek *hetero*, "other") lack the ability to generate their own biological molecules from scratch. Instead, they must take the molecules, or their components (polymers), from autotrophs. Several heterotrophic lifestyles are discussed in Section 1.6.

In living organisms, including animals, plants, and microorganisms, a multitude of metabolites are found. The vast diversity of the **metabolome** (total grouping of metabolites) is governed by specific catalyzing enzymes (specialized proteins) and the availability of their specific substrates, although a few enzymes can be involved in multiple pathways (Pott et al. 2019). The distinction between primary and secondary metabolism is not always clear-cut. Secondary metabolites are expensive, requiring that cells gather and spend more energy than what is needed for basic function. To justify the cost in energy and resources, such molecules usually provide an important advantage to the organism that makes them. For example, the pigment molecule **melanin** makes cells opaque, but also protects cells from water loss and absorbs energetic photons, deflecting damage from sensitive genetic material (Jennings and Lysek 1999). **Alkaloids** like caffeine are defensive compounds that also act as metabolic storage (Jennings and Lysek 1999). While neither melanin nor caffeine are required for cell growth or reproduction, they can certainly extend an organism's life by reducing the destructive impacts of dehydration, radiation, predation, and starvation. Thus, while PMs allow

---

**TABLE 1.1**
**Roles and Functions of Primary and Secondary Metabolites**

| Primary Metabolites | Secondary Metabolites |
| --- | --- |
| The most fundamental molecules used for organism functioning | Derivatives of primary metabolites |
| Physiological functions (directly involved in growth, development, and reproduction) | Ecological functions (involved in defense, pigmentation, etc.) |
| Formed during the growth stage due to energy metabolism | Formed near or during stationary stage of growth |
| Produced in large quantities; extraction is easy | Produced in small quantities; extraction is difficult |
| Same in all organisms | Might be unique to different organisms, species, or tissues |
| A part of the basic molecular structure of an organism e.g., Ethanol, lactic acid, and pyruvic acid | Not a part of the basic molecular structure of an organism e.g., Phenolics and terpenes |

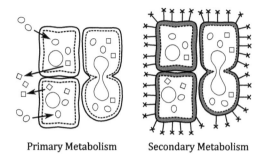

Primary Metabolism        Secondary Metabolism

**FIGURE 1.2**  Primary metabolism relates to basic cell functions like bringing in nutrients (ovals), expelling waste (squares), growing, and reproducing. Secondary metabolism includes molecules with ecological functions like pigmentation and defense (shown here as stylized spikes).

an organism to live at the most basic level, "secondary" metabolites may have ecological functions that are necessary for an organism to survive within the wider context of its ecosystem (Figure 1.2).

Secondary metabolites can also act as important defense mechanisms. Fungi are some of the most prolific producers of defensive SMs, which can strongly affect other organisms for better or for worse; for example, fungal antibacterial molecules like penicillin are known as potent drugs that strongly benefit humans, while **mycotoxins** can poison humans and other animals that try to ingest them (Keller et al. 2005). In fungi, SMs are classified into four categories, namely **polyketides** (e.g., aflatoxins and fumonisins, which are also toxic), **non-ribosomal peptides** (e.g., the toxin sirodesmin, the anti-insect grazing molecule peramine, and siderophores[1] such as ferricrocin), **terpenes** (e.g., T-2 toxin, deoxynivalenol mycotoxin [DON]), and **indole terpenes** (e.g., the mycotoxins paxilline and lolitrems) (see Figure 1.3) (Chen and Liu 2017; Fox and Howlett 2008; Keller et al. 2005; Mousa and Raizada 2013).

Analytical chemistry has seen tremendous progress in the identification and production of secondary metabolites through the study of filamentous fungi. Many such compounds, including terpenoids (Figures 1.4 and 1.5), steroids (Figure 1.6), alkaloids (Figure 1.7), phenylpropanoids (Figure 1.8), aliphatic compounds (Figure 1.9), polyketides (Figure 1.10), and peptides (Figure 1.11), have shown anti-microbial activities (Mousa and Raizada 2013).

While the roles of many SMs in fungi and other organisms are not well known or reported yet (Fox and Howlett 2008; Keller 2019; Mousa and Raizada 2013), different studies have demonstrated that some fungal SMs exhibit biological activities with potential uses in medicine, agriculture, and food processing. The availability of fungal genome sequences has led to an enhanced effort at identifying the biosynthetic genes for these useful molecules. However, some fungal SMs are toxic to humans, such as the mycotoxins known to contaminate cereals (such as wheat and rye parasitized by Ergot, *Claviceps purpurea*) and oil seeds (such as peanuts parasitized by *Aspergillus flavus*). Although SMs exhibit many types of structure (see Figures 1.4–1.11), the biosynthesis pathways for fungal SMs share numerous key characteristics such as precursors and intermediary metabolites (Bills and Gloer 2021; Nielsen and Nielsen 2017).

The plethora of bioactive natural products produced by fungi display a broad range of bioactivities for pharmaceutical and agricultural purposes. The development of *-omics* technologies, particularly genomics and metabolomics, has increased our understanding of fungi and fungal metabolism to the point where we can develop further biosynthetic pathways to produce important or novel SMs. There is still massive potential for fungal genetics and genomics to promote SM research, in general, and in the pharmaceutical sciences, in particular (Brakhage and Schroeckh 2011; Hautbergue et al. 2018; Misiek and Hoffmeister 2007).

---

[1] Iron-attracting compounds.

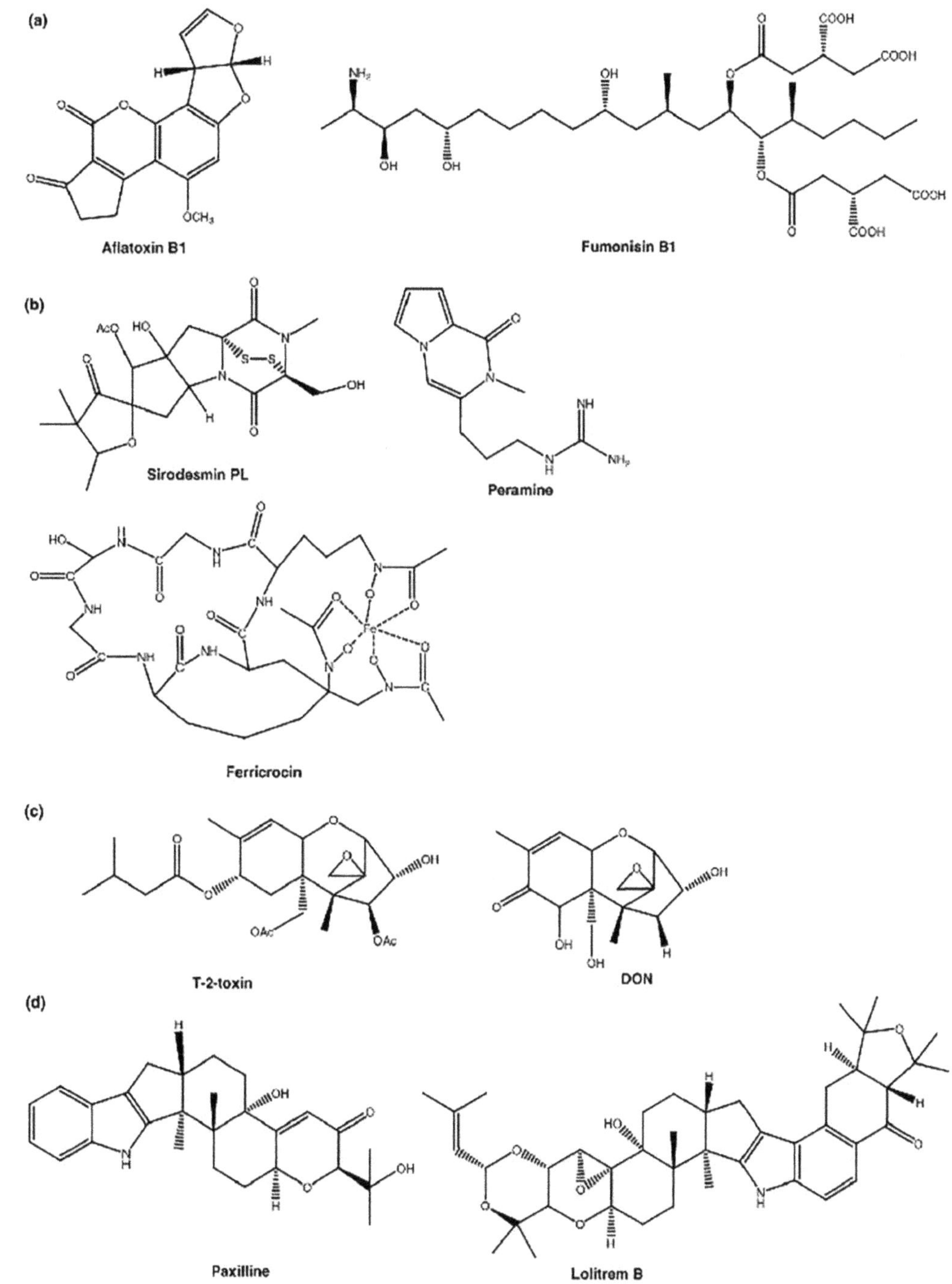

**FIGURE 1.3** The four classes of fungal secondary metabolites. (a) Polyketides: Aflatoxin B1, produced by *Aspergillus flavus* and *A. parasiticus*, and Fumonisin B1, produced by *Fusarium verticillioides*. (b) Non-ribosomal peptides: Sirodesmin PL, produced by *Leptosphaeria maculans*, Peramine produced by *Epichloe/Neotyphodium* spp., and Ferricrocin, produced by *Cochliobolus heterostrophus*. (c) Terpenes: T-2 toxin, produced by *Fusarium sporotrichioides* and Deoxynivalenol (DON), produced by *Fusarium graminearum*. (d) Indole terpenes, for example Paxilline, produced by *Penicillium paxilli* and *lolitrem* B, produced by *Epichloe/Neotyphodium* spp. (Reprinted with permission from Elsevier: Fox, E. M., and B. J. Howlett. 2008. Secondary metabolism: regulation and role in fungal biology. *Current Opinion in Microbiology* 11:481–7.)

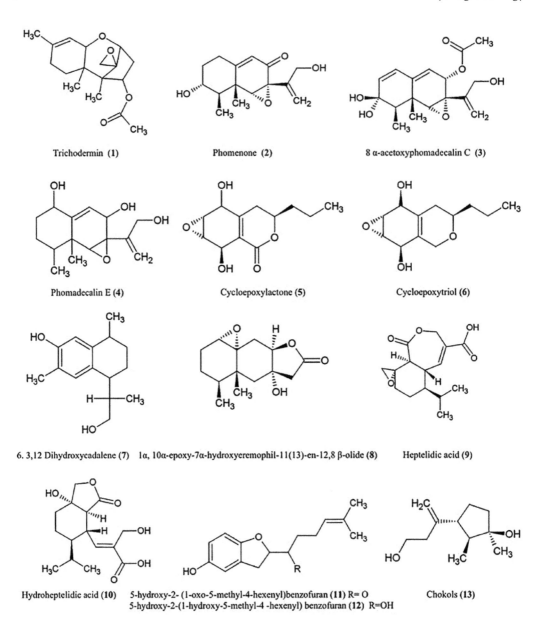

**FIGURE 1.4** Structures of sesquiterpene derivatives of fungal endophyte origin. (Reprinted from Mousa, W. K., and M. N. Raizada. 2013. The diversity of anti-microbial secondary metabolites produced by fungal endophytes: an interdisciplinary perspective. *Frontiers in Microbiology* 4:65; article published under the terms of the Creative Commons Attribution License, with free use permission.)

In fungi as well as other living beings, the complex networks of metabolites store calories and energy in a delicate **biotic** (living) web. **Ecology** is the study of how calories and other essential biological compounds move through ecosystems and throughout the world. In healthy ecosystems, all the factors essential to life cycle between autotrophs and heterotrophs and between biotic and abiotic systems. What is left behind by one organism can be used by another. Almost no materials are wasted in a healthy ecosystem.

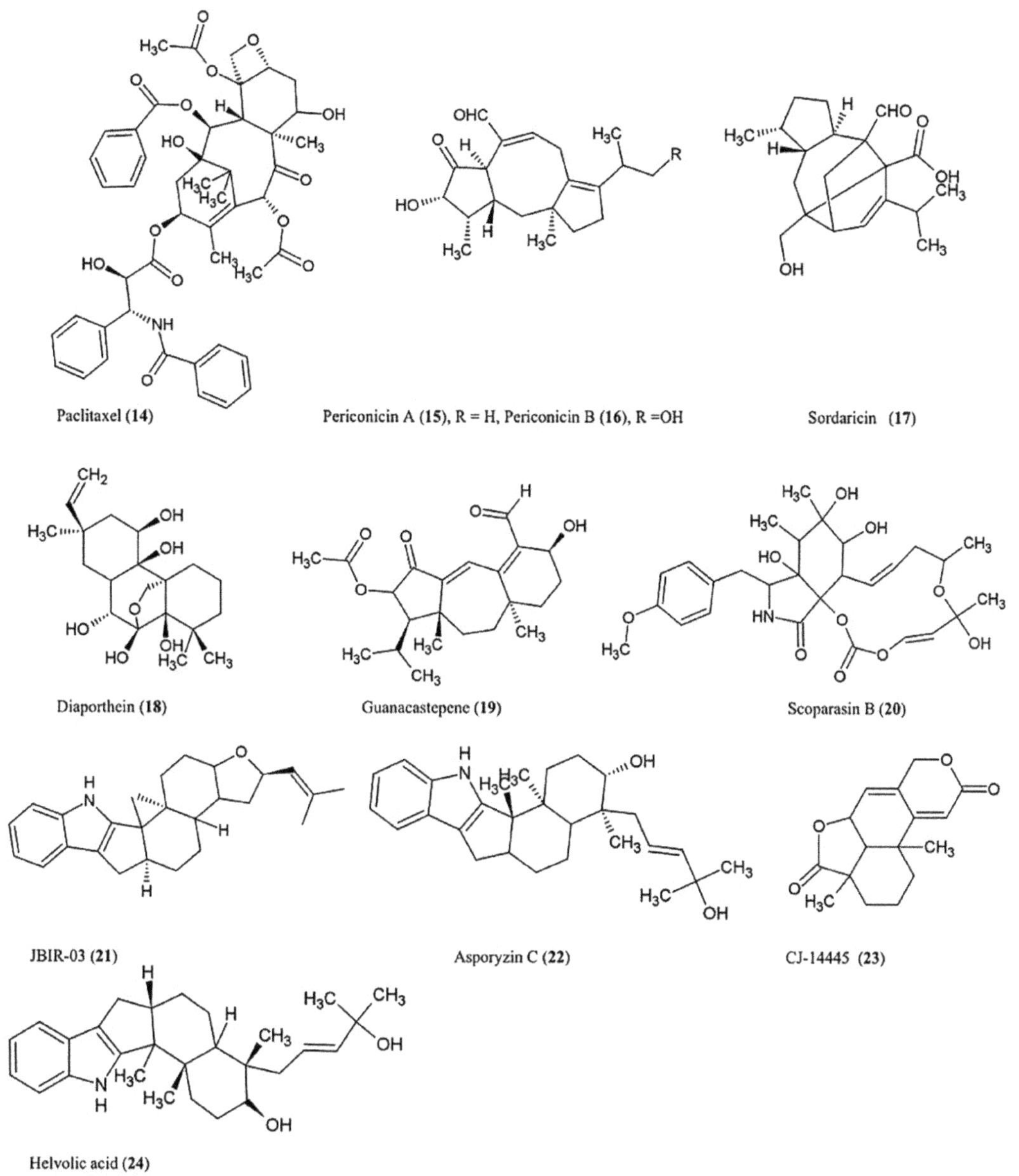

Paclitaxel (14)  Periconicin A (15), R = H, Periconicin B (16), R =OH  Sordaricin (17)

Diaporthein (18)  Guanacastepene (19)  Scoparasin B (20)

JBIR-03 (21)  Asporyzin C (22)  CJ-14445 (23)

Helvolic acid (24)

**FIGURE 1.5** Structures of diterpene and triterpene derivatives of fungal endophyte origin. (From Mousa and Raizada 2013; article published under the terms of the Creative Commons Attribution License, with free use permission.)

## 1.4 THE CARBON CYCLE

The element carbon (elemental symbol C) is the chemical backbone of all life on Earth, is the basic building block of life, and helps form the bodies of living organisms. Most carbon on Earth is stored in rocks and sediments, while the rest resides in the atmosphere, oceans, and living organisms. Biologically stored carbon is released back into the atmosphere through cellular respiration (metabolism) and through the mineralization (decomposition) of animals and plants (Figure 1.12). Carbon is then taken back up from the environment by organisms as they build or repair themselves.

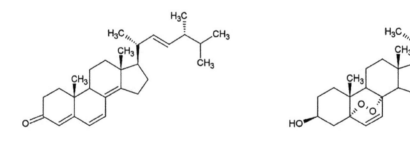

Penicisteroid A (25)                3β,5α-dihydroxy-6β-acetoxy-ergosta-7,22 –diene (26)

3β, 5α-dihydroxy-6β-phenylacetyloxy-ergosta-7,22 –diene (27)                3β-hydroxy-ergosta-5-ene (28)

3-oxo-ergosta-4,6,8(14),22-tetraene (29)                3β-hydroxy- 5α 8α-epidioxyergosta-6,22-diene (30)

**FIGURE 1.6** Structures of additional steroid derivatives of fungal endophyte origin. (From Mousa and Raizada 2013; article published under the terms of the Creative Commons Attribution License, with free use permission.)

Fungi play an essential role in the global carbon cycle because they are the main degraders of plant biomass, especially lignin, in nature (Benocci et al. 2017; Mäkelä et al. 2014; Zak et al. 2019; Zeikus 1981; Zhu and Miller 2003; see Chapter 17 in this volume). Abundant research has shown that fungi make considerable contributions to carbon and energy cycling (Harlery 1971); however, the amount of carbon sequestration or release varies by fungal species and lifestyle (Gao et al. 2007). Fungi that partner with plants are also key to the movement of carbon from plants to the

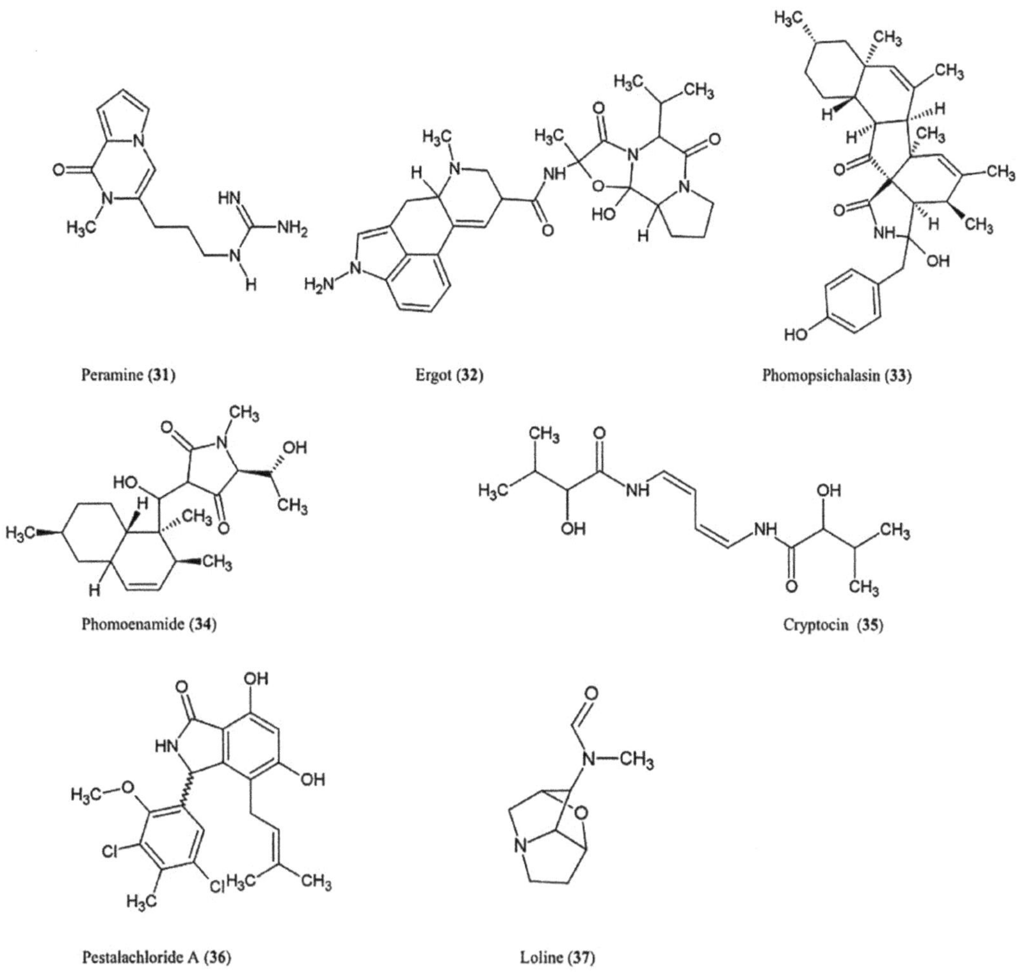

**FIGURE 1.7** Structures of alkaloid derivatives of fungal endophyte origin. (From Mousa and Raizada 2013; article published under the terms of the Creative Commons Attribution License, with free use permission.)

soil (see Chapter 4) and can encourage carbon sequestration in the soil by enhancing plant growth (Figure 1.13) (Rosenzweig et al. 2018; Six et al. 2006). More generally, fungi have been found to play an important role in different terrestrial ecosystem processes including soil carbon sequestration, litter decomposition, and weathering (de Graaff et al. 2015; Hoffland et al. 2004; Landeweert et al. 2001; Quirk et al. 2014; Zhong et al. 2018). Indeed, based on the respiration from their massive amounts of hyphal material, fungi are considered one of the driving forces in the biological end of the terrestrial carbon cycle (Barron 2003; Johnson 2008).

## 1.5 THE NITROGEN CYCLE

Although nitrogen accounts for 79% of the atmosphere, its availability to living organisms is limited because it must be incorporated into complex forms to be utilized by animals and plants (Delwiche 1970). Nitrogen availability plays a key role in the species composition and functioning of many terrestrial, freshwater, and marine ecosystems (Bebber et al. 2011; Chen et al. 2014; Gullis et al. 1996; Pajares and Bohanna 2016; Vitousek et al. 1997). Through **biological nitrogen fixation (BNF)**, unreactive molecular nitrogen in the atmosphere is reduced ("fixed") to ammonium compounds

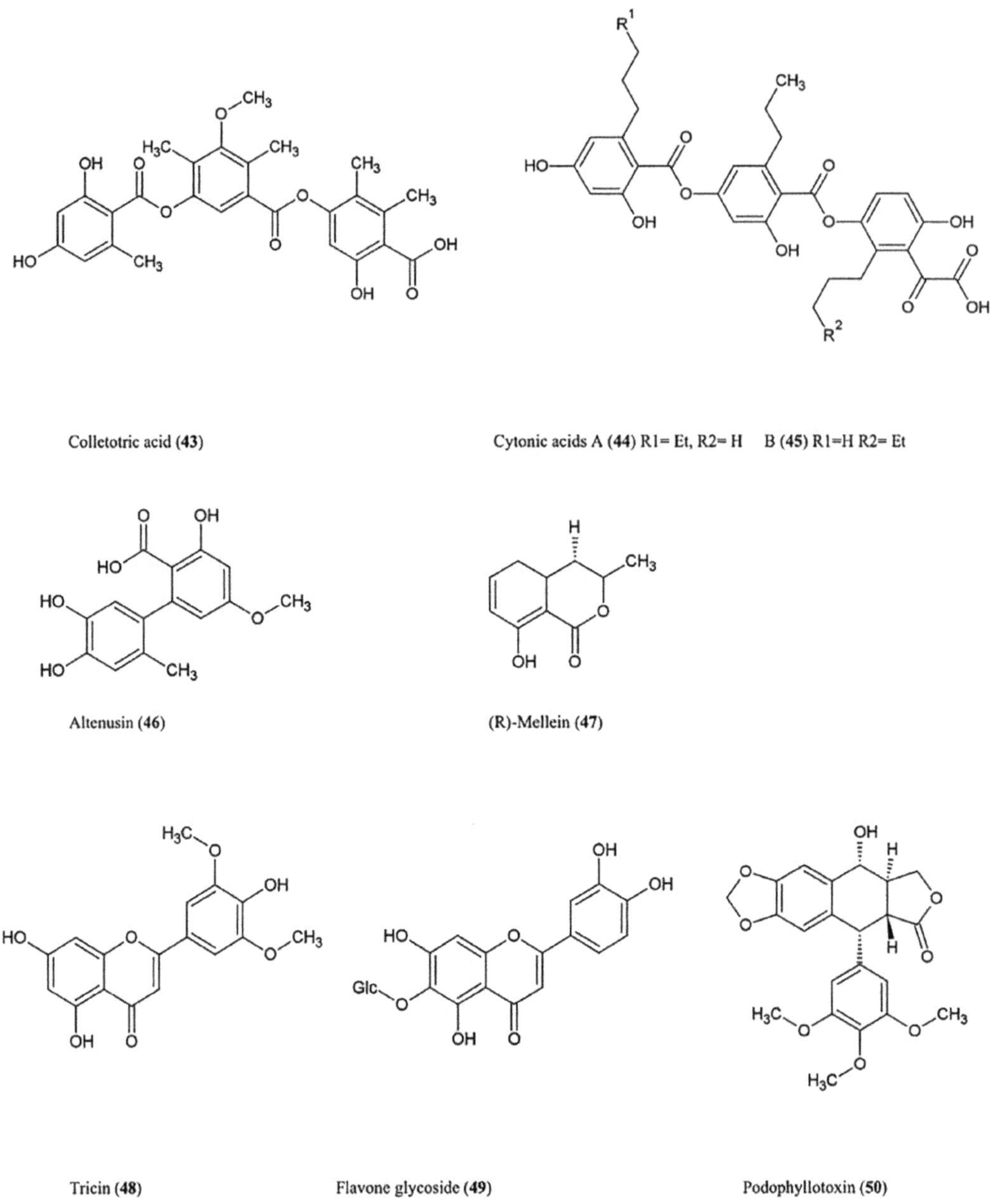

Colletotric acid (43)          Cytonic acids A (44) R1= Et, R2= H    B (45) R1=H R2= Et

Altenusin (46)                              (R)-Mellein (47)

Tricin (48)                    Flavone glycoside (49)              Podophyllotoxin (50)

**FIGURE 1.8** Structures of phenolic compounds of fungal endophyte origin. (From Mousa and Raizada 2013; article published under the terms of the Creative Commons Attribution License, with free use permission.)

which are available to living organisms (Figure 1.14). This fixed nitrogen is subsequently transformed to numerous organic compounds by living organisms and finally returned to the atmosphere as molecular nitrogen through microbial denitrification in soils, fresh and marine waters, and sediments (Fowler et al. 2013; Galloway et al. 2004; Stein and Klotz 2016). However, the massive industrial use of nitrogen fertilizers to increase food productivity and yields has accelerated the nitrogen cycle, causing serious environmental problems such as water acidification and **eutrophication** (too much algal growth resulting in a lack of dissolved oxygen and leading to mass death of animals) (Galloway 1998; Gruber and Galloway 2008; see Chapter 9 in this volume).

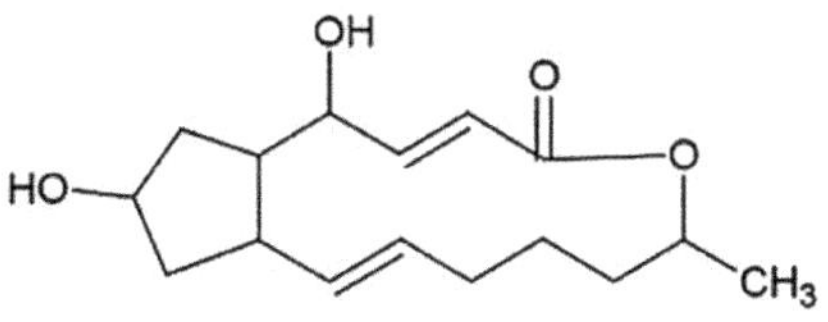

Brefeldin A (51)

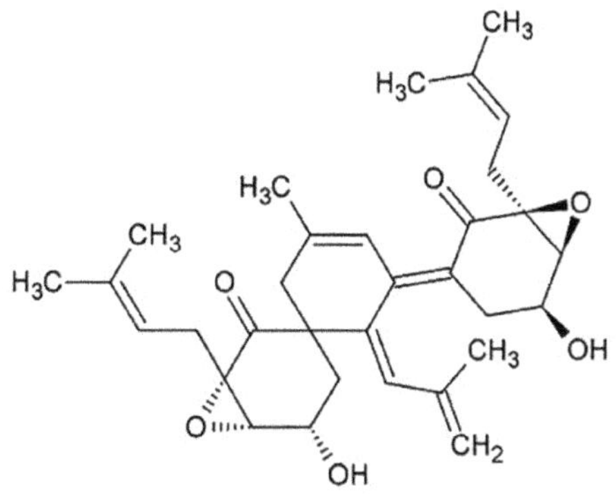

Pestalofone C (52)

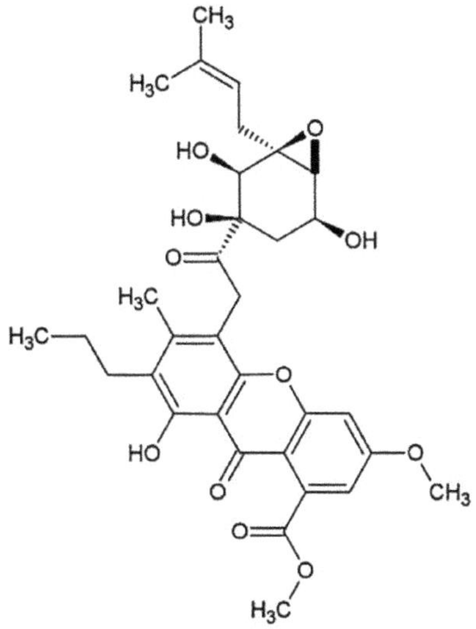

Pestalofone E (53)

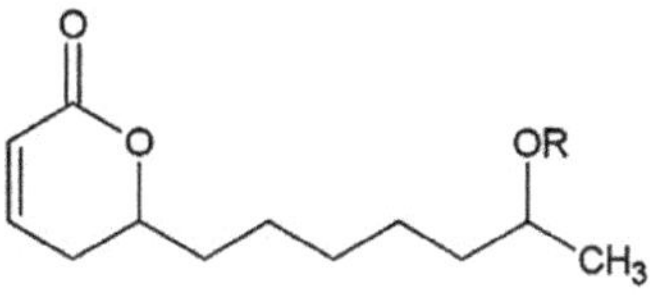

Gamahonolides
A (54) R=H    B (55) R= COCH₂CH₂COOCH₂CH

**FIGURE 1.9** Structures of aliphatic derivatives of fungal endophyte origin. (From Mousa and Raizada 2013; article published under the terms of the Creative Commons Attribution License, with free use permission.)

In 2008, a review was published by Jetten (2008) highlighting interesting discoveries in the microbial nitrogen cycle, including the diversity of nitrogen-fixing bacteria, the regulation of metabolism in nitrifying organisms, and the molecular diversity of denitrifying microorganisms and their enzymes. In addition to bacteria, fungi have been shown to play important roles in the nitrogen cycles of various ecosystems (McClaugherty et al. 1985; Chen et al. 2014), notably in denitrification and the production of $N_2O$ (nitrous oxide, a potent greenhouse gas) and $N_2$ (gaseous nitrogen) (Hayatsu et al. 2008; Wrage-Mönnig et al. 2018). Soil is one of the major sources of $N_2O$ (Thomson et al. 2012) via a complex process which is still not fully known (Butterbach-Bahl et al. 2013). Understanding the fungal production of $N_2O$ is of great import, as any step we can take to mitigate greenhouse gas production could go a long way in reducing the effects of global climate change. Mothapo et al. (2015) reported a total of 119 $N_2O$-producing fungal species representing 60 genera, with 90% belonging to the phylum Ascomycota, 7% to Basidiomycota, and 3% to Zygomycota (see Chapter 4 in this volume). Lab work by Shoun (2006) further clarified the biochemical mechanisms and the physiological roles of fungal denitrification, while studies in soils also showed the

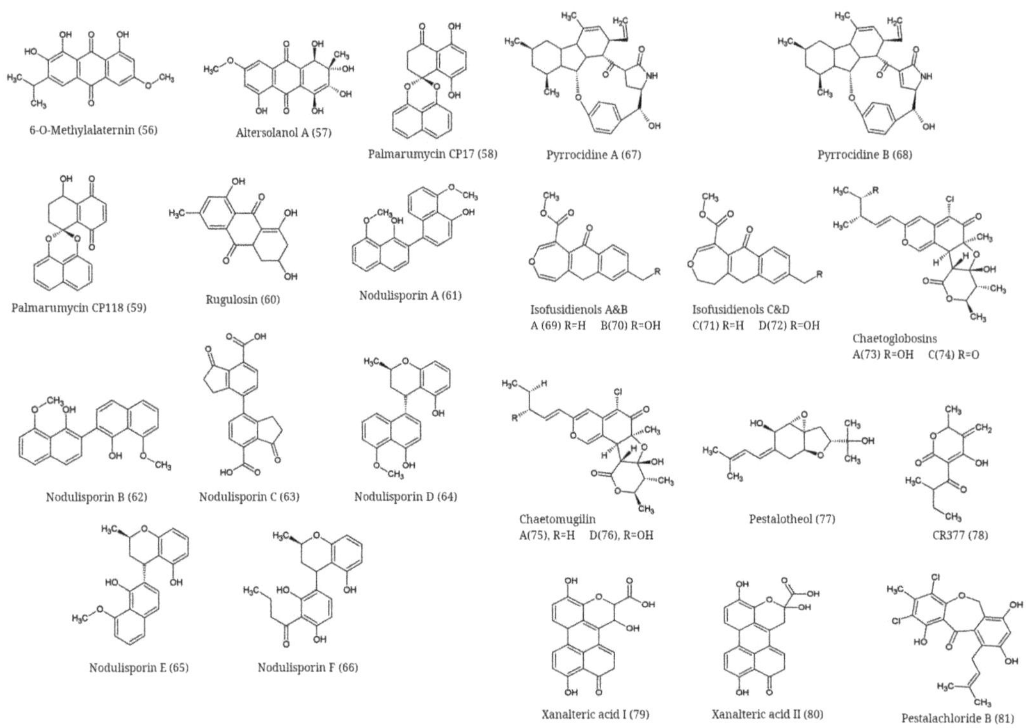

**FIGURE 1.10** Structures of polyketide derivatives of fungal endophyte origin. (Adapted from Mousa and Raizada 2013; article published under the terms of the Creative Commons Attribution License, with free use permission.)

denitrification potential of fungi *in vivo* and their role in $N_2O$ production under both aerobic and anaerobic conditions (Mothapo et al. 2015). In desert and semi-arid grassland soils of different water levels, as well as in acidic soils, fungi are major producers of $N_2O$ (Huang et al. 2017; Marusenko et al. 2013). However, some studies report net movement of $N_2O$ from the atmosphere to the soils, suggesting that soil and the soil microbiome might be sinks rather than, or as well as, sources of $N_2O$ (Chapui-Lardy et al. 2007).

## 1.6   HETEROTROPH LIFESTYLES

Both carbon and nitrogen are major elements required for biological functions and their cycling is complex, involving both abiotic and biotic processes. Autotrophs sequester and use the carbon and nitrogen in the atmosphere or in mineral forms, creating the biological molecules they need from scratch.

Heterotrophs require many of the same biological materials as autotrophs but cannot assemble them from the basic molecules in the environment. Instead, heterotrophs need to harness the biological materials made by autotrophs (or, sometimes, other heterotrophs; see Chapter 3 in this volume). We can very generally refer to the heterotroph that successfully harvests biological molecules as a 'winner' of resources; the source organism is a victim, the 'loser' of both the resources and the metabolic effort that went into making them. Understandably, many autotrophs have developed defenses that deter heterotrophs from stealing these hard-won resources; however, some autotrophs form intimate, mutually beneficial relationships with the heterotrophs around them (see Chapter 2 in this volume).

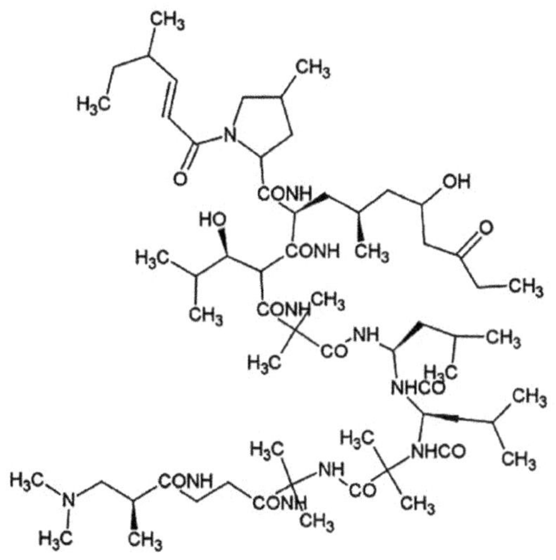

Leucinostatin A (82)

Echinocandin A (83)

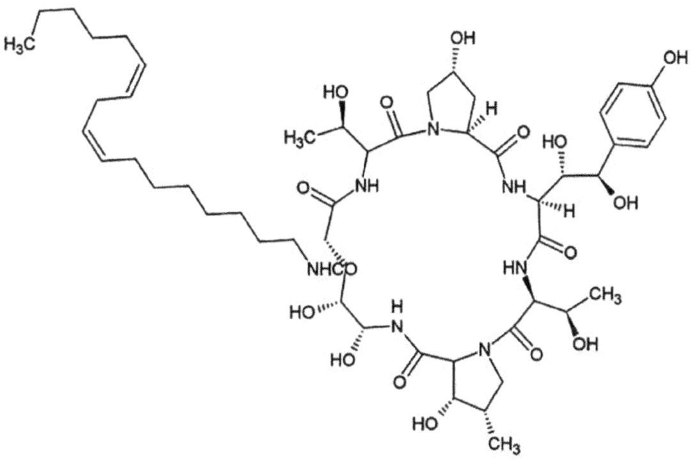

Cryptocandin (84)

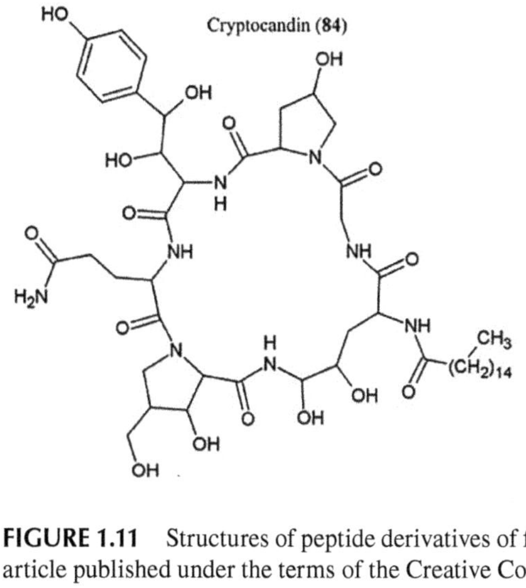

**FIGURE 1.11** Structures of peptide derivatives of fungal endophyte origin. (From Mousa and Raizada 2013; article published under the terms of the Creative Commons Attribution License, with free use permission.)

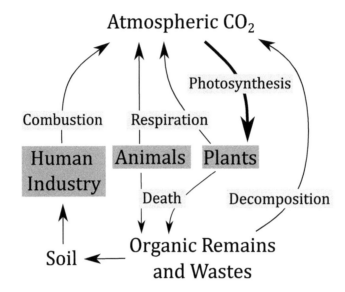

**FIGURE 1.12**  The carbon cycle represents the movement of carbon between living organisms, the soil, and the atmosphere. Arrows indicate the direction of carbon flow.

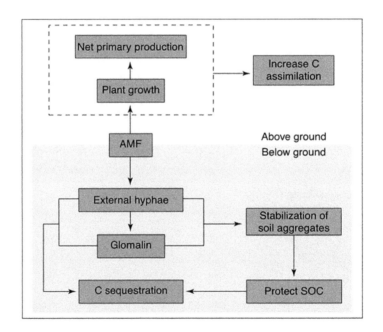

**FIGURE 1.13**  Role played by arbuscular mycorrhizal fungi (AMF; see Chapter 4 in this volume) in regulating carbon fluxes between the biosphere and the atmosphere. *Abbreviation: SOC*, soil organic carbon. (Reprinted with permission from Elsevier: Zhu, Y. G., and R. M. Miller. 2003. Carbon cycling by arbuscular mycorrhizal fungi in soil–plant systems. *Trends in Plant Science* 8:408–10.)

The categories listed below were created by human scientists to help understand and compare the lifestyles of various organisms. However, as ever more organisms are observed in ever greater detail, we are discovering that some of the distinctions can be blurry. As you continue studying biology, you may find ways that these categories overlap or fall short of real observations. Generalizing allows us to recognize broad biological patterns and understand how organisms or ecosystems function, but there are exceptions to almost all definitions in biology.

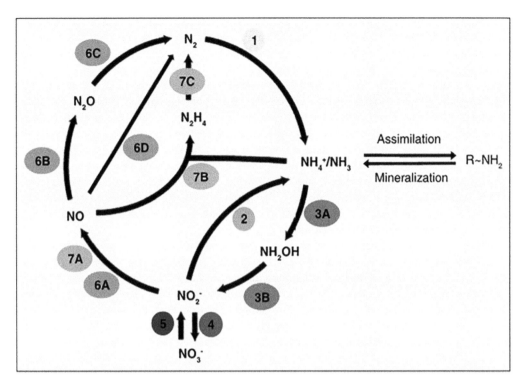

**FIGURE 1.14** Major processes of the nitrogen cycle. Reactions that comprise the seven major processes of the nitrogen cycle are represented by the numbered circles. Ammonification (creation of ammonium, $NH_4^+$) may be accomplished either by process 1, reduction of molecular nitrogen (also referred to as 'nitrogen fixation'), or by process 2, dissimilatory nitrite ($NO_2^-$) reduction to ammonium (DNRA). Nitrification is composed of process 3, oxidation of ammonia to nitrite, and process 4, oxidation of nitrite to nitrate ($NO_3^-$). Process 5, reduction of nitrate to nitrite, can be coupled to processes 2, 6, or 7 in a population or a community. Denitrification is shown as process 6, which returns molecular nitrogen to the atmosphere. Anammox, also referred to as "coupled nitrification–denitrification", is shown as process 7. (Reprinted with permission from Stein, L. Y., and M. G. Klotz. 2016. The nitrogen cycle. *Current Biology* 26:R94–8; caption modified.)

### 1.6.1 INGESTIVE HETEROTROPHS

If you are human, the form of heterotrophy most familiar to you is probably **ingestive feeding** (from the Latin *ingest*, "bring in"). Ingestive heterotrophs bring food inside of their bodies before digesting it. Once inside the organism, the food is mixed with digestive enzymes that break the food down into small biological molecules. The digested metabolites are often referred to as biological "building blocks" because they are used to assemble whatever larger molecules the heterotrophic organism needs.

**Herbivores** are ingestive heterotrophs that only consume molecules made by autotrophs. The autotroph might be killed by herbivory, such as when yam and potato plants are uprooted for their tubers. However, herbivory can also leave much of the autotroph intact, such as when a squirrel eats a nut dropped by a living tree or grazing cows crop the top few inches off a clump of grass. The difference usually comes down to how much of the victim organism's **tissue** is consumed or destroyed during herbivory. (In biology, tissue refers to a group of similar cells that do not make up a whole organ.) When a turnip moth caterpillar consumes some leaf tissue from a young Eucalyptus tree, the rest of the tree will probably be able to continue living. However, if a swarm of locusts were to descend, the loss of leaf tissue could be severe enough to kill the sapling (Agriculture Victoria 2017).

Sometimes, heterotrophs need to ingest compounds that are not created by autotrophs. For example, most multicellular organisms *can* create the amino acid **arginine**, which is required to remove waste products from cells – an ability that domestic cats (*Felis catus*) have lost. To survive, domestic cats require large amounts of arginine in their diets, which means they must eat the flesh of other animals. Usually, we refer to heterotrophs that consume other heterotrophs as **carnivores** (from the Latin *caro*, "meat"). Carnivores usually kill their victim organisms, such as when a lioness brings down an antelope or a mouse eats a grasshopper. However, it is possible to eat heterotroph tissues without killing the source animal. For example, some lizards can cleanly release their tails when under attack, giving a carnivorous predator muscle tissue to eat while the lizard escapes. Note that organisms that consume only unicellular heterotrophs are not considered carnivorous. Virtually no multi-celled organisms can produce vitamin B12, which is necessary for proper DNA replication and neurological function (Office of Dietary Supplements 2020). Large organisms need to acquire vitamin B12 from the bacteria, algae, and fungi that produce it. A vegan human who takes B12 supplements is not considered a carnivore!

Very few organisms are purely herbivorous or purely carnivorous. For example, deer and cows are known to eat fish, birds, and other small animals (Landers 2012); fearsome gray wolves often enjoy fresh fruits like berries and pears (Fuller 2019). However, the major portion of a cow's diet is herbivorous and gray wolves are mostly carnivorous. There is a special term for ingestive feeders that require a balanced mixture of autotroph and heterotroph tissues. Such **omnivores** (from the Latin *omnis*, "all") may be able to survive on an herbivorous diet but gain distinct advantages from consuming meat. Non-vegan humans tend to be omnivores; this group also includes chickens, pigs, and domesticated dogs.

### 1.6.2  ABSORPTIVE HETEROTROPHS

The other type of heterotrophy is **absorptive** feeding, where food is digested before it is taken in. Absorptive heterotrophs release their digestive enzymes into the local environment; then, once the enzymes contact a food source, they break it down into biological building blocks. This process can also be called **lysotrophic** nutrition (from the Greek *lysis*, "release") because the small metabolites are released from the food source before being ingested. Absorptive nutrition is the main feeding style of many microorganisms, including the fungi. An absorptive heterotroph is further categorized based on the state of its victims' tissues when they are consumed.

**Biotrophs** digest and eat the tissues of living victims, such as when a ringworm fungus (*Tinea* spp.) eats human skin and causes a rash. The Greek prefix *bio-* is used to indicate that these heterotrophs can only complete their life cycles on living victims that supply a steady stream of nutrients. Therefore, a biotrophic infection can cause problems (disease) for the victim but is usually not **lethal** (deadly). Some biotrophic microbes are actually beneficial to plants (these are known as endophytes; see Chapter 11 in this volume). Biotrophy is also a lifestyle of some ingestive animals; fleas, ticks, aphids, tapeworms, lampreys, and vampire bats are multicellular biotrophic **parasites** (see Chapter 2 in this volume).

**Necrotrophs** eat freshly dead tissue, usually by infecting and then killing the victim's cells (Greek *nekros* = "corpse"). The dead tissue is then digested and taken in by the necrotroph. Because necrotrophs do not rely on a steady stream of fresh resources from their host victim, they can complete their life cycles even after the host dies. Necrotrophic infections are often lethal. Dutch elm disease, which wiped out most of the mature elm trees in Europe and North America during the 20th century, is caused by a necrotrophic fungus (*Ophiostoma* spp.) partnered with a boring beetle (see Chapters 14 and 15 in this volume). Several necrotrophic microorganisms can cause gangrene in humans and other animals. Necrotrophy is a category that only applies to absorptive feeders; an ingestive parallel would be lethal carnivory, such as a lion killing a gazelle to ingest its meat.

Finally, **saprotrophs** eat dead tissue from the environment (Greek *sapros*, "rotten"). Where necrotrophs actively work to kill their victims, saprotrophs usually encounter cells that have already died or may already be partially digested. Fungi that grow on heaps of cow dung are saprobes, feeding on the partially digested plant matter that was killed and consumed by the cow. Some absorptive heterotrophs can "switch" between biotrophy and saprotrophy, living inoffensively inside a host's living cells but digesting dead tissues once the host has been killed by other means. Saprotrophy is also a category exclusive to absorptive feeders. An ingestive parallel would be scavenging, such as a buzzard feeding on the carcass of a large animal.

## 1.7 AGROECOLOGICAL APPLICATIONS

In any ecosystem, the availability of nutrients and other inputs limits biological growth and production. By understanding what organisms need and learning how they gather materials in the wild, we can more effectively deliver those materials to crop plants and livestock. If we observe that a farm location is lacking in some necessary factor, we can make informed changes to the landscape that enable or improve agricultural production. The issue of dry soil can be temporarily overcome with irrigation; frost-sensitive plants can be cultivated in cold climates if they are protected in greenhouses. Over the course of human history, farmers have studied the ecosystems around them, conducted tests, and identified countless techniques to increase agricultural quality and yield. In the late 20th and early 21st centuries, farmers are also directing their attention back to the health of wild ecosystems (see Chapter 10 in this volume). As local, national, and international organizations work to improve crop species, protect plants from diseases, and maintain the beauty and diversity of the natural world, all the policies and techniques must work around the essential movement of energy and nutrients covered in this chapter.

## REFERENCES

Agriculture Victoria. 2017. *Fact Sheet: Forestry and Plantations*. Victoria State Government. http://agriculture. vic.gov.au/agriculture/pests-diseases-and-weeds/pest-insects-and-mites/plague-locusts/fact-sheet-forestry-and-plantations (accessed April 3, 2020).

Barron, C. L. 2003. Predatory fungi, wood decay, and the carbon cycle. *Biodiversity* 4:3–9.

Bebber, D. P., Watkinson, S. C., Boddy, L., and P. R. Darrah. 2011. Simulated nitrogen deposition affects wood decomposition by cord-forming fungi. *Oecologia* 167:1177–84.

Benocci, T., Aguilar-Pontes, M. V., Zhou, M., Seiboth, B., and R. P. de Vries. 2017. Regulators of plant biomass degradation in ascomycetous fungi. *Biotechnology for Biofuels* 10:152. https://doi.org/10.1186/s13068-017-0841-x.

Bills, G. F., and J. B. Gloer. 2021. Biologically active secondary metabolites from the fungi. *Microbiology Spectrum* 4(6). https://doi.org/10.1128/microbiolspec.FUNK-0009-2016.

Brakhage A. A., and V. Schroeckh. 2011. Fungal secondary metabolites – strategies to activate silent gene clusters. *Fungal Genetics and Biology* 48:15–22.

Butterbach-Bahl, K., Baggs, E. M., Dannenmann, M., Kiese, R., and S. Zechmeister-Boltenstern. 2013. Nitrous oxide emissions from soils: how well do we understand the processes and their controls? *Philosophical Transactions of the Royal Society B: Biological Sciences* 368:20130122. https://doi.org/10.1098/rstb.2013.0122.

Chapui-Lardy, L., Wrage, N., Metay, A., Chotte, J. L., and M. Bernoux. 2007. Soils, a sink for $N_2O$? A review. *Global Change Biology* 13:1–17.

Chen, H., Mothapo, N. V., and W. Shi. 2014. The significant contribution of fungi to soil $N_2O$ production across diverse ecosystems. *Applied Soil Ecology* 73:70–7.

Chen, H. P., and J. K. Liu. 2017. Secondary metabolites from higher fungi. In *Progress in the Chemistry of Organic Natural Products*, eds. A. Kinghorn, H. Falk, S. Gibbons, and J. Kobayashi, 1–201. Cham: Springer.

de Graaff, M. A., Adkins, J., Kardol, P., and H. L. Throop. 2015. A meta-analysis of soil biodiversity impacts on the carbon cycle. *Soil* 1:257–71.

Delwiche, C. C. 1970. The nitrogen cycle. *Scientific American* 223:136–47.

Erb, M., and D. J. Kliebenstein. 2020. Plant secondary metabolites as defenses, regulators, and primary metabolites: the blurred functional trichotomy. *Plant Physiology* 184:39–52.

Fowler, D., Coyle, M., Skiba, U., et al. 2013. The global nitrogen cycle in the twenty-first century. *Philosophical Transactions of the Royal Society B* 368:20130164. http://doi.org/10.1098/rstb.2013.0164.

Fox, E. M., and B. J. Howlett. 2008. Secondary metabolism: regulation and role in fungal biology. *Current Opinion in Microbiology* 11:481–7.

Fuller, T. K. 2019. *Wolves: Spirit of the Wild*. London: Chartwell Books Publisher.

Galloway, J. N. 1998. The global nitrogen cycle: changes and consequences. *Environmental Pollution* 102:15–24.

Galloway, J. N., Dentener, F. J., Capone, D. G., et al. 2004. Nitrogen cycles: past, present, and future. *Biogeochemistry* 70:153–226.

Gao, L., Sun, M. H., Liu, X. Z., and Y. S. Che. 2007. Effects of carbon concentration and carbon to nitrogen ratio on the growth and sporulation of several biocontrol fungi. *Mycological Research* 111:87–92.

Gruber, N., and J. Galloway. 2008. An earth-system perspective of the global nitrogen cycle. *Nature* 451:293–6.

Gulis, V., Kuehn, K., and K. Suberkropp. 2006. The role of fungi in carbon and nitrogen cycles in fresh water ecosystems. In *Fungi and Biochemical Cycles*, ed. G. M. Gadd, 404–35. Cambridge: Cambridge University Press.

Harley, J. L. 1971. Fungi in ecosystems. *Journal of Ecology* 59:653–68.

Hautbergue, T., Jamin, E. L., Debrauwer, L., Puel, O., and I. P. Oswald. 2018. From genomics to metabolomics, moving toward an integrated strategy for the discovery of fungal secondary metabolites. *Natural Product Reports* 35:147–73.

Hayatsu, M., Tago, K., and M. Saito. 2008. Various players in the nitrogen cycle: diversity and functions of the microorganisms involved in nitrification and denitrification. *Soil Science and Plant Nutrition* 54:33–45.

Hoffland, E., Kuyper, T. W., Wallander, H., et al. 2004. The role of fungi in weathering. *Frontiers in Ecology and the Environment* 2:258–64.

Huang, Y., Xiao, X., and X. Long. 2017. Fungal denitrification contributes significantly to $N_2O$ production in a highly acidic tea soil. *Journal of Soils and Sediments* 17:1599–606.

Jennings, D. H., and G. Lysek. 1999. *Fungal Biology: Understanding the Fungal Lifestyle*. 2nd ed. Oxford: BIOS Scientific Publishers Ltd.

Jetten, M. S. M. 2008. The microbial nitrogen cycle. *Environmental Microbiology* 10:2903–9.

Johnson, D. 2008. Resolving uncertainty in the carbon economy of mycorrhizal fungi. *New Phytologist* 180:3–5.

Keller, N. P. 2019. Fungal secondary metabolism: regulation, function and drug discovery. *Nature Reviews Microbiology* 17:167–180.

Keller, N. P., Turner, G., Bennett, J. W. 2005. Fungal secondary metabolism 1/m from biochemistry to genomics. *Nature Reviews. Microbiology* 3:937–947. PMID 16322742 DOI: 10.1038/Nrmicro1286.

Landers, J. 2012. *Bambi Ate Thumper. Why Herbivores Sometimes Eat Meat*. Slate Group. https://slate.com/technology/2012/11/deer-eat-meat-herbivores-and-carnivores-are-not-so-clearly-divided.html (accessed April 2, 2020).

Landeweert, R., Hoffland, E., Finlay, R.D., Kuyper, T. W., and N. van Breemen. 2001. Linking plants to rocks: ectomycorrhizal fungi mobilize nutrients from minerals. *Trends in Ecology & Evolution* 16:248–54.

Mäkelä, M. R., Donofrio, M., and R. P. de Vries. 2014. Plant biomass degradation by fungi. *Fungal Genetics and Biology* 72:2–9.

Marusenko, Y., Huber, D. P., and S. H. Hall. 2013. Fungi mediate nitrous oxide production but not ammonia oxidation in aridland soils of the southwestern US. *Soil Biology and Biochemistry* 63:24–36.

McClaugherty, C. A., Pastor, J., Aber, J. D., and Melillo, J. M. 1985. Forest litter decomposition in relation to soil nitrogen dynamics and litter quality. *Ecology* 66:266–75.

Misiek, M., and D. Hoffmeister. 2007. Fungal genetics, genomics, and secondary metabolites in pharmaceutical sciences. *Planta Medica* 73:103–15.

Mothapo, N., Chen, H., Cubeta, M. A., Grossman, J. M., Fuller, F., and W. Shi. 2015. Phylogenetic, taxonomic and functional diversity of fungal denitrifiers and associated $N_2O$ production efficacy. *Soil Biology and Biochemistry* 83:160–175.

Mousa, W. K., and M. N. Raizada. 2013. The diversity of anti-microbial secondary metabolites produced by fungal endophytes: an interdisciplinary perspective. *Frontiers in Microbiology* 4:65. https://doi.org/10.3389/fmicb.2013.00065.

Nielsen, J. C., and J. Nielsen. 2017. Development of fungal cell factories for the production of secondary metabolites: linking genomics and metabolism. *Synthetic and Systems Biotechnology* 2:5e12. https://doi.org/10.1016/j.synbio.2017.02.002.

Office of Dietary Supplements. 2020. *Vitamin B12 Fact Sheet for Health Professionals*. https://ods.od.nih.gov/factsheets/VitaminB12-HealthProfessional/ (accessed April 2, 2020).

Pajares, P., and. B. J. M. Bohanna. 2016. Ecology of nitrogen fixing, nitrifying, and denitrifying microorganisms in tropical forest soils. *Frontiers in Microbiology* 7:1045. https://doi.org/10.3389/fmicb.2016.01045.

Pott, D. M., Osorio, S., and J. G. Vallarino. 2019. From central to specialized metabolism: an overview of some secondary compounds derived from the primary metabolism for their role in conferring nutritional and organoleptic characteristics to fruit. *Frontiers in Plant Science* 10:835. https://doi.org/10.3389/fpls.2019.00835.

Quirk, J., Andrews, M. Y., Leake, J. R., Banwart, S.A., and D. J. Beerling. 2014. Ectomycorrhizal fungi and past high $CO_2$ atmospheres enhance mineral weathering through increased below-ground carbon-energy fluxes. *Biology Letters* 10:20140375. http://doi.org/10.1098/rsbl.2014.0375.

Rosenzweig, S. T., Fonte, S. J., and M. E. Schipanski. 2018. Intensifying rotations increases soil carbon, fungi, and aggregation in semi-arid agroecosystems. *Agriculture, Ecosystems & Environment* 258:14–22.

Shoun, H. 2006. Denitrification and anaerobic energy producing mechanisms by fungi. *Tanpakushitsu Kakusan Koso* 51:419–49.

Six, J., Frey, S. D., Thiet, R. K, and K. M. Batten. 2006. Bacterial and fungal contributions to carbon sequestration in agroecosystems. *Soil Science Society of America Journal* 70:555–69.

Stein, L. Y., and M. G. Klotz. 2016. The nitrogen cycle. *Current Biology* 26:R94–8.

Thomson, A. J., Giannopoulos, G., Pretty, J., Baggs, E. M., and D. J. Richardson. 2012. Biological sources and sinks of nitrous oxide and strategies to mitigate emissions. *Philosophical Transactions of the Royal Society B* 367:1157–68.

Vitousek, P. M., Aber, J. D., Howarth, R. W., et al. 1997. Human alteration of the global nitrogen cycle: sources and consequences. *Ecological Applications* 7:737–50.

Wrage-Mönnig, N., Horn, M. A., Well, R., Müller, C., Velthof, G., and O. Oenema. 2018. The role of nitrifier denitrification in the production of nitrous oxide revisited. *Soil Biology and Biochemistry* 123:A3–16.

Zak, D. R., Pellitier, P. T., Argiroff, W., et al. 2019. Exploring the role of ectomycorrhizal fungi in soil carbon dynamics. *New Phytologist* 223:33–9.

Zeikus, J. G. 1981. Lignin metabolism and the carbon cycle. In *Advances in Microbial Ecology*, ed. M. Alexander, 211–43. Boston: Springer.

Zhong, Y., Yan, W., Wang, R., Wang, W., and Z. Shangguan. 2018. Decreased occurrence of carbon cycle functions in microbial communities along with long-term secondary succession. *Soil Biology and Biochemistry* 123:207–17.

Zhu, Y. G., and R. M. Miller. 2003. Carbon cycling by arbuscular mycorrhizal fungi in soil–plant systems. *Trends in Plant Science* 8:408–10.

# 2 Evolution and Symbiosis

*Elizabeth "Izzie" Gall*

## CONTENTS

## 2.1 INTRODUCTION

All organisms must compete for the limited resources in the environment. Due to underlying differences in their genetic codes, each organism has a slightly different chance of accessing and using the resources it needs while avoiding environmental dangers. The most successful organisms reproduce, increasing the proportion of favorable traits in a population. This leads to slow shifts in the average traits of a species, refining toward those traits that best suit the environment. The incredible variety of life forms on Earth results from uncountable mutations arising in breeding populations and being subjected to natural selection. While undergoing this process, species are constantly interacting with many others and may develop symbiotic relationships that help each other survive – or antagonistic relationships that increase one species' chance of success at the cost of another's survival.

## 2.2 FITNESS AND GENES

The organisms in any ecosystem are constantly competing for limited space, nutrients, water, and energy. Poor competitors may receive too little of these necessary components and starve; they could also die from disease or be killed and eaten by predators. Strong competitors are better at gathering and using resources, resisting diseases, and evading predators, so they have a better chance of reaching adulthood and reproducing.

An organism's ability to survive to reproductive age and produce **viable** (living) offspring is called its **fitness**. Most of an organism's fitness is based in its genetic code, or **genome**, a collection of long strands of DNA that helps determine how the organism will develop, look, and behave. Every protein, molecule, membrane, and structure made by an organism originates with instructions in its genome. Each section of DNA that includes a command or encodes a product is called a **gene**.

When genes work together to produce a certain trait, they are known collectively as the trait's **genotype**. If the trait affects how the organism appears or functions, the appearance or function

DOI: 10.1201/9780429320415-2

is known as a **phenotype** (from the Greek *phainein*="to show"). For example, if a rabbit has two genes that both control the level of pigment produced in its fur, then those genes together form its fur color genotype. The actual color of the rabbit's fur is the corresponding phenotype. The genetic code is **redundant**, meaning that individual genes can influence multiple phenotypes and most phenotypes are controlled by multiple genes.

It is important to note that not all phenotypes are genetically based. Some aspects of behavior, development, or appearance are the result of an organism's interaction with its environment. Muscle weakness due to Duchenne muscular dystrophy is genetically based, but muscle weakness due to prolonged heavy metal exposure is not. Low body weight due to poor digestion is genetically based, but low body weight due to famine is not. However, the way an organism interacts with its surroundings is strongly influenced by the patterns encoded in the genome.

Some genes may give an organism a higher risk of getting certain diseases if specific environmental or behavioral conditions are also met. We can also say, equivalently, that a certain environmental or behavioral condition has a greater chance to cause disease in organisms with specific genes (Ottman 1996). This is known as a "gene-environment interaction" or **GxE**. In cases of GxE, a disease is caused by a combination of environmental and genetic factors – one alone is not enough to cause the disease state. For example, human asthma is a disease resulting from GxE. Risk factors of childhood asthma include behavioral and environmental factors (such as obesity of the mother during pregnancy, exposure to secondhand tobacco smoke, or amount of time spent in polluted urban air rather than green spaces) as well as genetic factors (such as a "high risk" variation in the EPHX1 gene, which has been linked to a 50% increase in the risk of developing asthma) (Miller et al. 2019).

The difference between genomes of individuals of the same species is known as **genetic variation**. All genetic variation on Earth derives from **mutations**: small, accidental changes in the genetic code. Mutations come from rare mistakes in normal DNA replication; modern human cells have about one mutated base pair out of every two billion replicated base pairs (Moorjani et al. 2016). Mutations can also arise when chemicals or radiation produce holes in DNA, resulting in information gaps or inaccurate copying of the genome during cell replication. Some mutations are catastrophic, resulting in the death of a cell or organism. However, due to the redundant nature of the genome, some mutations may have no phenotypic effect or may remain phenotypically invisible until additional mutations accumulate. New phenotypes arise when mutations make an organism's growth, behavior, or appearance different from the traits of closely related organisms who lack those mutations.

An unusual phenotype can affect an organism's fitness. For example, a mutation that leads to lower pigmentation in a tundra rabbit's fur could make camouflage against the snow easier, creating a more difficult target for predators and increasing the likelihood that the rabbit will survive and reproduce (an increase in fitness). By contrast, a mutation that decreases the pigment in a flower's petals could reduce its visibility to pollinators, resulting in a lower chance that the plant will be able to reproduce (a decrease in fitness). Because their underlying genotypes are part of the physical DNA in each cell, the phenotypes that increase or maintain an organism's fitness will be passed along to its offspring. Genotypes that encode lower-fitness phenotypes are less likely to be passed along.

## 2.3   SPECIES

Every **species** is a group of organisms that are genetically similar enough to mate and produce viable offspring together (**interbreed**). The word "species" can be either singular or plural. As new phenotypes arise within a species, they can affect the ability of members to interbreed. If two groups that begin with similar genomes accumulate enough changes in different directions, it may become impossible for the groups to interbreed. At that point, the groups will have become separate species. For example, as humans bred thousands of generations of *Zea mays* (maize) individuals with large kernels and other desirable phenotypes, the underlying genome changed so much that modern sweet corn cannot interbreed with its ancestor, teosinte (see Chapter 6 in this volume).

Sometimes, members of the same species develop pronounced genetic distinctions that do not prevent interbreeding. In that case, we say the species has several **subspecies** or **varieties**. More than a dozen distinct varieties of *Zea mays* are grown specifically for use as popcorn, but they could still interbreed with the varieties of *Z. mays* grown for human consumption, animal feed, or ethanol production. A **population** is a group of organisms belonging to one species (or variety) that occupies a certain space at the same time. One corn field growing in Iowa, USA, and another in Uttar Pradesh, India, represent two populations of the corn species *Zea mays*. Regardless of the number of varieties present, a group of organisms from the same species that are present in the same space, present at the same time, and able to interbreed creates a **breeding population**.

When members of a breeding population undergo sexual reproduction, bits of their genomes are carefully and randomly shuffled in a process called **meiosis** before being re-combined and passed to offspring. This process creates offspring with a mixture of existing and new genotypes and phenotypes. Due to the slow, steady introduction of random mutations and the routine recombination of genotypes in each sexual generation, every breeding population is made up of organisms of varying fitness.

## 2.4   EVOLUTION AND EVOLUTIONARY PRESSURES

Each generation of a species includes individuals with varied genomes, phenotypes, and fitness levels. Individuals with phenotypes that improve their ability to find and utilize resources, avoid disease, and evade predators have better fitness than individuals that lack those traits. Therefore, resource availability, disease, and predator activity are some of the environmental factors that mark the differences between high-fitness and low-fitness phenotypes. When organisms with low fitness die before reproducing, their genomes are removed from the breeding population; we can say that by process of elimination, the environment "selects" the surviving, high-fitness organisms to reproduce. This **natural selection** increases the proportion of individuals with favorable traits, and the frequency of the underlying genotypes, in a population.

As natural selection proceeds within a population, the frequency of different genotypes shifts in a process known as **evolution**. Evolutionary shifts drive diversification and the development of new species. Any phenotype that has a genetic basis can be passed on to offspring and is susceptible to evolution by natural selection. Because natural selection leads to evolution, the environmental factors which "select" for individuals with certain traits can also be called **evolutionary pressure**. Over time, persistent evolutionary pressure can lead to the development of **adaptations**, phenotypes that help a population thrive in their preferred environment. For example, some evergreen trees have adapted to cold winter temperatures by developing specialized leaves (such as the needles of pine, *Pinus* spp.) that do not freeze. Remember that phenotypes without a genetic basis do not contribute to the process of evolution.

Cold winter temperature is an example of evolutionary pressure from a nonliving, or **abiotic**, source. Seasonal flooding, daily temperature variations, the salt content of a lake, and the amount of light reaching a field are all examples of abiotic environmental pressures that drive the development of certain adaptations. **Biotic** pressure comes from organisms in the environment. For example, predator selection may increase the frequency of genotypes conveying effective camouflage, defensive chemicals, or evasive behavior phenotypes in the prey. In turn, those prey adaptations place evolutionary pressure on predators to develop better sight, chemical tolerance, and increased speed, respectively. When multiple species influence each other's fitness on an evolutionary scale, causing genotype frequency shifts in each other's populations, we say that those species are **coevolving**.

## 2.5   COEVOLUTION AND SYMBIOSIS

When two or more species coevolve for long enough, they may enter a stable relationship known as **symbiosis** (from the Greek *syn*, "together"). Each member of the symbiosis is known as a **symbiont**.

While the word "symbiosis" is often used to imply a mutually beneficial relationship, it simply means that multiple species are interacting in a predictable or recurring way. Mutually beneficial symbioses do exist, but there are many types of symbiosis in which not all participants benefit.

### 2.5.1 MUTUALISM

In a **mutualism**, each participating species in an interaction receives an increase in fitness. For example, flowers pollinated by insects have an easier time completing their sexual life cycle, while the insects receive a rich, stable source of nutrients. Eighty percent of known land plants have partnerships with soil fungi. These fungi, which are discussed extensively in Chapter 11, receive a reliable source of photosynthetic carbon from the plant in exchange for nutrients that most plants cannot extract from soil on their own, such as insoluble phosphorus.

### 2.5.2 PREDATORS AND PREY

Predator-prey and herbivore-plant relationships are common examples of coevolution, with the prey species developing increasingly sophisticated defenses and the predator (or herbivore) species evolving continuous countermeasures. If one partner gains a fitness advantage, the other must adapt to compensate or risk extinction. This cycle of developing new, genetically based defenses and attacks is commonly known as the **evolutionary arms race**.

### 2.5.3 COMPETITION

Multiple species that try to live in the same part of the environment and use the same resources (species living in the same **niche**) are living in another non-beneficial symbiosis. **Competition** for space, water, and nutrition exerts pressure on all organisms in the niche, driving additional adaptations like territorial behavior or novel foraging techniques to develop in different species.

### 2.5.4 COMMENSALISM AND PARASITISM

In some cases, one species may be able to increase its fitness by relying throughout its life on another species that does not benefit from the interaction. We will refer to the species that benefits from this symbiotic relationship as the *instigator*, since its partner has no reason to start the relationship. We will call the other species the *passive partner*, though it may actually be able to defend against the instigator.

If the instigator benefits from its passive partner, but the passive partner is not affected, the relationship is called a **commensalism**. The name comes from the Latin root words *com*, "together", and *mensa*, "table". We can imagine that the two species are having dinner and the instigator claims resources that the passive partner does not need, such as its unfinished salad. For example, if a vine climbs a non-photosynthetic tree trunk, does not restrict the tree's growth or cover its leaves, and gains access to better sunlight, the vine and tree are in a commensalism. In this example, the vine (instigator) receives better access to light than it would have at ground level and its leaves are protected from ground-level herbivores. The tree is not denied any of the resources it needs but also receives no benefit from the vine's presence (making it the passive partner). **Decomposition** is another example of commensalism; bacteria and fungi that colonize dead organisms benefit from a rich source of nutrients, but the fitness of the deceased organism is unaffected. The decomposer bacteria and fungi are known as **saprotrophic** organisms (see Chapter 1 in this volume).

In contrast, **parasitism** describes a symbiosis in which the instigator benefits directly from harming its passive partner. This word comes from the Latin *parasitus*, "one who eats at another's table". In this case, we might imagine that the instigator has burst into the dining room uninvited and snatched the resources right out of the passive partner's hands. Ringworm fungus, hookworms, and tapeworms are familiar parasites of mammals; peach leaf curl, potato blight, and root-knot nematodes are examples that feed on plants. Parasite species rely on their passive partner, known in this case as the **host**, for nutrients and the means to reproduce. Many parasites rely entirely on their hosts to survive, siphoning away resources to such an extent that the host may die. **Biotrophic** parasites require a living host (Greek *bios* = "life"). Parasites that live from biotrophy are highly specialized to certain host species and often have highly complex life cycles. However, **necrotrophic** ("dead-eating") parasites can feed on a variety of freshly killed hosts as well as the dead portions of living organisms.

## 2.6   CONTEXT DEPENDENCE OF SYMBIOSIS

Categories in biology are created for human study, helping scientists understand and discuss complex natural patterns. However, as we gather more evidence and observe the world in increasing detail, we may discover that our artificial categories do not describe meaningful biological distinctions. By distilling a complex interaction between organisms into a single word, we are obscuring many factors that can influence the nature of a symbiotic relationship.

Any interaction between species is context dependent. The balance in a mutualistic relationship may shift depending on the abiotic environment and interactions with other species outside the defined symbiosis (see Chapter 3 in this volume). For example, mycorrhizal fungi exchange nutrients trapped in the soil, such as phosphorus, for sugar their partner plants generate through photosynthesis. However, when fertilized with nitrogen, plants do not rely as heavily on mycorrhizae and may opt to keep their sugar for themselves. Due to a shift in an abiotic factor (nitrogen availability), the plant could terminate a partnership that we normally classify as mutualism.

Context can significantly blur the boundary between biotrophy, saprotrophy, and necrotrophy, especially in the fungi. Some biotrophic commensal fungi living inside of plants (**endophytes**) can become necrotrophic if the plant is invaded by a disease, feeding on the cells killed by the invader. Those same, normally biotrophic fungi may continue to utilize the plant host's resources after its death, technically becoming saprotrophs (Koide et al. 2008). While the categories of symbiosis can be useful to begin describing species interactions, the line between different lifestyles can become quite blurry.

It is also important to remember that our understanding of symbiotic relationships is limited by our observations. Despite our best efforts, limitations in time, technology, and biological understanding may lead us to misinterpret natural interactions. For more than a hundred years, **lichens** were considered a classic example of mutualism, with one fungus and one photosynthetic organism partnering to create a very successful colonial organism totally unlike either individual. However, in the 1980s research began to suggest that the photosynthetic member of a lichen is often treated as a host for the parasitic, biotrophic fungus (Wrzosek et al. 2017). In 2004, it was discovered that some lichenizing fungal species can live independently as saprotrophs if they germinate on certain substrates (Wedin et al. 2004), and it turns out that lichens require not one, but two types of fungus to form (Spribille et al. 2016). Such upsets in established biological paradigms can be frustrating and require significant changes to systems of naming and categorization – but they are also exciting reminders of all the work yet to be done in biology!

## 2.7   AGROECOLOGICAL APPLICATIONS

The principles of evolution and symbiosis apply just as well to a controlled field as any wild ecosystem. From the moment the first early humans selected one plant to cultivate and one to exclude, we have been an important part of the environment these organisms inhabit. By adding fertilizers, irrigating fields, and removing shade we substantially change the evolutionary pressures acting on crop species. The "natural" evolutionary pressures like limited nutrient availability, drought, and shade have been replaced by "artificial", **anthropogenic** (human-driven) pressure such as demand for bright coloration and high protein content. These artificial pressures are discussed in detail in Chapter 6.

For example, consumers often prefer seedless fruits for their texture and convenience, but fruits that complete their sexual life cycles produce seeds. Therefore, many modern crops are grown without completing their sexual cycles. For instance, banana trees across the world are grown by removing side stalks from existing banana trees. Because the "next" generation of trees does not arise from sexual processes and come from the same individual, they are **clones** and, barring mutations, each banana tree has effectively the same genome. This results in large populations that have the same favorable phenotypes such as short ripening time, good shelf stability, strong peel color, and high sugar content. Unfortunately, it also means that all the trees have identical defense mechanisms, so the entire growing population is susceptible to the same diseases. The Gros Michel banana variety, previously the top cultivated banana worldwide, was effectively wiped out by Panama disease (fungal species *Fusarium oxysporum*) in the 1960s. It was replaced by clonal cultivation of the current world favorite, the Cavendish banana, which is currently under threat from Black Sigatoka (fungus *Mycosphaerella musicola*) (J.P. 2014).

Non-clonal **monocultures**, large areas planted with only one species, also present an issue. While there is some genetic variation reducing the chance that a single disease will kill every individual, the sheer number of similar crops means that if an effective disease does hit, it can cause considerable damage. In more diverse environments, a disease might encounter plants it cannot infect and be unable to spread further. In monocultures, every neighboring plant is the same species, so the population is highly susceptible to disease. Similarly, when an annual crop is cultivated on the same land year after year, diseases and parasites can establish a strong presence and significantly impact crop yield and quality. Monocultures are further discussed in Chapter 9.

## REFERENCES

J.P. 2014. We have no bananas today. *Economist*. https://www.economist.com/feast-and-famine/2014/02/27/we-have-no-bananas-today (accessed 15 Oct 2019).

Koide, R.T., Sharda, J.N., Herr, J.R., and G.M. Malcolm. 2008. Ectomycorrhizal fungi and the biotrophy-saprotrophy continuum. *New Phytologist* 178(2):230–233. https://www.jstor.org/stable/30147674.

Miller, M., Schettler, T., Tencza, B., and M. Valenti. 2019. *A Story of Health*. Agency for Toxic Substances and Disease Registry, Commonweal, Science and Environmental Health Network, Western States PEHSU. https://www.healthandenvironment.org/docs/SoH_Brett_asthma_update%202019.pdf (accessed March 3, 2022).

Moorjani, P., Gao, Z., and M. Przeworski. 2016. Human germline mutation and the erratic evolutionary clock. *PLoS Biology* 14(10):e2000744. DOI: 10.1371/journal.pbio.2000744.

Ottman, R. 1996. Gene-environment interaction: definitions and study designs. *Preventive Medicine* 25(6): 764–770. DOI: 10.1006/pmed.1996.0117.

Spribille, T., Tuovinen, V., Resl, P., et al. 2016. Basidiomycete yeasts in the cortex of ascomycete macrolichens. *Science*. DOI: 10.1126/science.aaf8287.

Wedin, M., Döring, H., and G. Gilenstam. 2004. Saprotrophy and lichenization as options for the same fungal species on different substrata: environmental plasticity and fungal lifestyles in the *Stictis–Conotrema* complex. *New Phytologist* 164:459–465. DOI: 10.1111/j.1469-8137.2004.01198.x.

Wrzosek, M., Riszkiewicz-Michalska, M., Sikora, K., Damszel, M., and Z. Sierota. 2017. The plasticity of fungal interactions. *Mycological Progress* 16:101–108. DOI: 10.1007/s11557-016-1257-x.

# 3 Population Dynamics

*Elizabeth "Izzie" Gall*

## CONTENTS

## 3.1 INTRODUCTION

When species in an ecosystem have coevolved, their complex relationships impact how energy and resources flow through the system. Herbivores have access to some of the energy stored in plants; higher consumers access some of the energy stored in the tissues of their prey. Diagrams like food chains and food webs can help illustrate the movement of resources to each species within and through an ecosystem.

The complex networks observed in nature develop slowly, as discussed in Chapter 2. Newly established populations of any species follow a characteristic growth pattern, which can be visualized as a growth curve. Once a population nears the carrying capacity of its environment, we may be able to observe specific factors that impact its size. Seasonal changes, abiotic shifts, and relationships with other species in the area can all influence the size of a local population.

## 3.2 ENERGY FLOW IN ECOSYSTEMS

Solar radiation is at the base of all terrestrial ecosystems. Plants harness photons to generate energy-rich molecules that fuel their maintenance, growth, reproduction, and nutrient storage. Because photosynthetic plants are the most abundant terrestrial autotrophs, the heterotrophs in a given ecosystem must compete for the resources stored in the communal "pool" of local plant life.

A **food chain** is a visual representation of how energy and nutrients flow through organisms at different levels of an ecosystem. **Primary producers** of sugars, such as photosynthetic plants, are usually at the base of the diagram. All other points on the food chain are **trophic levels**, showing organisms that obtain resources by consuming the organisms in the levels below. The first trophic level above the primary producers holds the **herbivores** that directly consume plant material. Next come the **primary consumers**, carnivores and omnivorous predators that prey on the herbivores. **Secondary consumers** prey on primary consumers, such as when a hawk eats a snake. Arrows are used to show the direction of nutrient flow from primary producers through the final level of consumers, which are known as **apex** (top) **predators** (Figure 3.1).

Ecosystems are extremely complex and often include many more interactions than can be represented by a single linear food chain. For example, the chain seems to end with apex predators, but they are susceptible to consumption by parasites and diseases; upon death, even apex predators can be eaten by scavengers and will be broken down by decomposers. It is also extremely uncommon for one species to be exclusively hunted by one other species; multiple higher-level predators may share a preferred prey species and herbivores tend to have varied diets. **Food webs** join overlapping

DOI: 10.1201/9780429320415-3

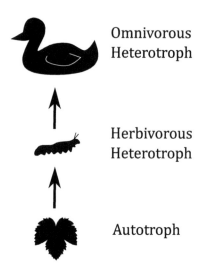

Omnivorous
Heterotroph

Herbivorous
Heterotroph

Autotroph

**FIGURE 3.1**   One example of a food chain: grape leaves (primary producers), the caterpillars that consume the leaves (primary consumers), and ducks that consume the caterpillars (apex predators).

food chains together, mapping nutrient flow through an ecosystem in greater detail. Some food webs incorporate decomposers and detritivores, which recycle certain nutrients so they are again available to primary producers (see Chapter 1 in this volume) (Figure 3.2).

Even though food webs are more complex diagrams, they can still omit a lot of important details about how resources move within, into, and out of an environment. For example, Chinook salmon are born in inland streams but only inhabit freshwater for a few months, migrating into the sea as they age. Adults may live in the ocean for up to eight years, eating and growing, before swimming back to their native streams to reproduce and die (U.S. National Park Service 2019). The fittest (or luckiest) salmon consume nutrients in the marine habitat, then die and decompose in the mountain freshwater habitat, depositing ocean-sourced nutrients in the temperate forest ecosystem.

Nutrients can also move between ecosystems abiotically. When dead leaves and wood fall into a mountain freshwater river, they are washed downstream, removing resources from mountain decomposers. Heavy windstorms can remove topsoil from fertile areas and drop dust or sand as they move, changing soil composition along the storm path (see Chapter 9 in this volume). Rain erosion can release minerals from rocks, locally increasing the concentration of some nutrients; however, heavy rains can also wash nutrients out of soil and into nearby or underground bodies of water.

In addition to the resources that leave systems through animal migration or abiotic processes, ecosystems also suffer energy loss at each trophic phase. All living beings generate **waste** products, substances that are not useful to them and may cause problems if not expelled. For example, cell proteins do not work forever. Old or damaged proteins are often digested so that their building blocks can be reused. Unfortunately, protein digestion generates ammonium, which at high concentrations is toxic to both animals and plants. In addition to molecular byproducts, waste also includes the undigestible portions of a consumer's diet, such as insoluble fiber, spores, fur, or bones. The energy and nutrients locked inside those compounds are lost to the organism that **excretes** (releases) them. Some organisms also abandon resources at certain life stages, such as when birds and insects molt, deer shed their antlers, and deciduous trees drop their leaves. Most excreted and shed materials will eventually be recycled by consumers and decomposers, though materials like spores, fur, or bones are usually broken down by specialized enzymes and can take much longer to digest than other tissues. For example, gut bacteria begin breaking down guts and muscles in a mammal within three days of death, but mammalian hair decomposition takes between 50 days and one year (Australian Museum 2020). Finally, every time a cell uses energy – whether to create or read DNA,

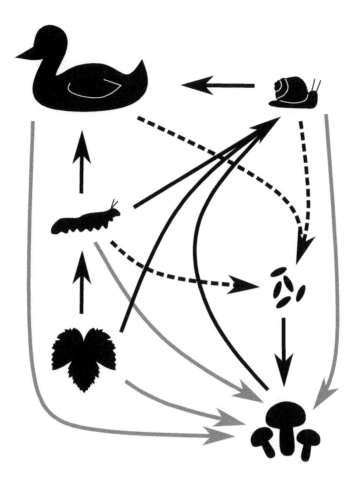

**FIGURE 3.2**   A food web expanding on the chain from Figure 3.1. Snails are omnivores that prey on caterpillars and grape leaves and are eaten by ducks. The animals all produce feces (dotted arrows) which are consumed by detritivores like fungi. On death (gray arrows), all the organisms in the web are digested by decomposers, represented again by the fungi. For visual simplicity, some arrows are not included; for example, snails also feed on fungi, ducks may feed on grape leaves, and nutrients return from detritivores to primary producers.

communicate with other cells, or help move a muscle – the molecules release some of the energy as heat, which cannot be recovered. Due to the consistent loss of energy along the food web, fewer organisms are supported at each higher trophic level.

## 3.3   GROWTH CURVES

Overall, food webs represent established networks of organisms that have coevolved within an ecosystem. Herbivores have evolved to utilize not just plant biomolecules, but the nutrients that the eaten plants originally pulled from the soil. If the plants have defense mechanisms like synthesizing defensive toxins or accumulating high levels of toxic metals, or both, herbivores need to cope with these materials as well. Species at higher trophic levels must be adapted to overcome the defenses of lower trophic level organisms, while species within a trophic level must be adapted to competition. Scavengers and detritivores have adapted to use resources given up by other members of the food web.

In order to coevolve into a stable network, the species in a food web must be present together for dozens or hundreds of generations, responding to the evolutionary pressures created by each other and by the environment. During this extended coevolutionary period, each population grows or shrinks depending on the advantages or pressures they face. Starting from its initial members, the size of any given population can be charted over time on a graph known as a **growth curve**. The following growth curve (Figure 3.3) does not reference a real population; instead, it is an illustration of the general way population sizes have been observed to grow and change.

When a new compound, food source, or livable space becomes available, it represents an untapped resource free from immediate competition. Since resources are always at a premium, species that can quickly take advantage of new environmental features have increased fitness over species that continue to vie over more established resources. Also, a population with a suddenly expanded food source (for example) will have less internal competition, so more of its members can survive to reproductive age. Therefore, the individuals that take advantage of a new environmental feature will be able to establish a new population or expand an existing population.

A **local** species may adapt to utilize the new feature. If the adaptations are substantial, this can result in the formation of new genetic varieties, subspecies, or even species. A non-local species adapted to the feature could also **immigrate** into the newly favorable local ecosystem, establishing a new **invasive** population.

At first, the population will grow slowly (segment A of the curve in Figure 3.3). This may be because of the slow addition of adaptations that allow organisms to use a novel resource or because of limited numbers of breeding members present (or both). This initial growth phase is shown as a gently upward-sloping curve on the left of the graph.

If availability of the novel feature stays high relative to the size of the population using it, then the adapted population can grow quickly through reproduction, continued immigration, or both. This is the population's **exponential growth phase**, reflected as the steep upward slope on the growth curve (Figure 3.3, segment B). However, exponential growth cannot continue forever. Abiotic factors like water availability and biotic factors like the presence of predators determine the maximum possible population size for a given species within an ecosystem. This practical limit is known as the environment's **carrying capacity** for the population (represented mathematically as $k$ and shown as the dotted line in Figure 3.3).

As the population size approaches $k$, fewer resources are available to each individual and competition will occur. When less-fit members of the population begin to die (or start failing to reproduce), the population growth rate declines and the growth curve begins to level out (Figure 3.3C). Once carrying capacity is reached, the population can remain approximately the same size, utilizing a sustainable amount of the available resources and contributing to the normal cycling of nutrients

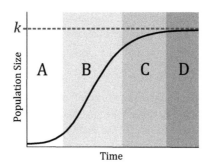

**FIGURE 3.3** An idealized population growth curve. At first, a population of newly adapted or immigrating organisms will be small and increase slowly (segment A). If resources remain plentiful, the population can grow rapidly (B). As the population approaches carrying capacity ($k$), the growth rate declines (C). Ultimately, the population will remain approximately the same size, hovering near the environment's carrying capacity for the population (D).

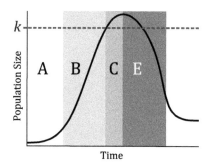

**FIGURE 3.4** A population growth curve illustrating a crash. If a population grows too rapidly during the exponential growth phase (B), the population size will exceed $k$. Population growth will slow down (C), but if the population remains too high above $k$, the unsustainable population size leads to a population crash (E) rather than the more sustainable plateau seen in Figure 3.3D.

through the ecosystem. This stable phase, where the growth rate is approximately zero, is represented by the growth curve leveling off on the right side of the graph (Figure 3.3D).

This very general pattern of a population's establishment and growth creates the S-shaped, **logistic** growth curve shown. However, because actual populations are exposed to different environmental factors and random chance, there are several ways a population's growth can divert from this common pattern.

During the exponential phase, a population can grow rapidly enough to exceed the carrying capacity of the environment. The carrying capacity may also rapidly decrease due to a catastrophic change, such as a storm or drought that eliminates the primary producers from an area. As the ratio of niche resources to niche organisms decreases, the population will rapidly **crash** (Figure 3.4E). The reduction in population may be due to a high death rate, such as from starvation due to low availability of prey. However, it may also be due to **emigration**, movement of individuals out of their current environment into a more favorable location. A population crash may not lead to local extinction of a population; it may result in a small population that can later grow again, as alluded to by the right side of the graph in Figure 3.4.

## 3.4 CARRYING CAPACITY AND POPULATION CYCLES

Because carrying capacity ($k$) depends on both biotic and abiotic factors, it is not a fixed value. Carrying capacity can change quickly or slowly, with sudden, unexpected events or in a predictable cycle. One of the most common unexpected events that influence carrying capacity is disease. When a prey species is struck by a disease outbreak, the availability of edible prey will decrease, lowering the environmental carrying capacity for certain predator populations. Such a change in carrying capacity is reflected by a population decrease, as illustrated in Figure 3.5a. Predictable changes can occur with seasonal shifts. For example, the amount of vegetation supported by a grassland is increased during the rainy season. This increase in primary production may raise the carrying capacity for herbivores, and through them the carrying capacity of higher trophic levels, relative to the carrying capacity of the same environment during the dry season, as illustrated in Figure 3.5b. Therefore, rather than staying at some perfect size, a population will fluctuate around the ever-changing, theoretical carrying capacity of its environment. Because of all the conflicting and compounding variables affecting each species, it is most useful to consider the carrying capacity for each population individually rather than trying to consider the total number of organisms an ecosystem can support.

Graphing a population's size through several environmental changes can give us an estimate of the population's carrying capacity. The same mechanisms acting during the population's establishment

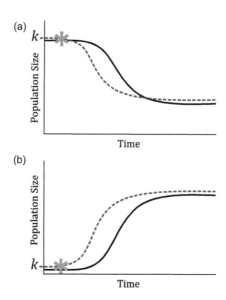

**FIGURE 3.5** Population growth curves illustrating how events (gray star) can lead to population crashes or booms. If an event like food scarcity decreases $k$, the graphed population decreases until a new equilibrium is reached (a). If an event like seasonally increased rain increases $k$, the graphed population increases until the new equilibrium is reached (b). These are idealized curves.

are still in effect but are happening on a smaller overall scale as the population remains fairly stable. As competition between organisms is low, the population growth rate is able to increase gently. As the number of individuals increases, the carrying capacity may be exceeded slightly, and competition increases once more, leading to a gentle population decrease. If these population shifts occur regularly, they are known as **population cycles** (Figure 3.6).

By observing when and how an established population changes size, we may discover certain key factors that affect the environment's carrying capacity for that population. We may even be able to predict when the population will next change in size or how it will respond to different environmental shifts. Combining observations from different populations of the same species can reveal general patterns. If the population responds to aspects of the environment that humans can change, this information may be useful in controlling pests or increasing the populations of threatened species.

Organisms in symbiotic relationships frequently have linked population cycles. In a mutualism, the cycles are synchronized because each species relies on and contributes to the success of the other. For relationships like parasitism and predation, the instigator population grows or shrinks in response to availability of its food source. In times of plenty, consuming populations can reproduce quickly, increasing population size within even one breeding season; however, when resources become lean, it may take some time for emigration or death rates to bring the population back down. Therefore, the instigator's population cycle will lag behind the respondent's cycle. The population cycles of competing species may be similar, since they respond to the availability of similar resources. However, the success of one species is achieved at the cost of the others in competition, so their population cycles may be roughly inverted. In all these cases, each population in a symbiotic relationship influences the environmental carrying capacity for its symbiont.

## 3.5 AGROECOLOGICAL APPLICATIONS

Because carrying capacity deals with the number of organisms in a population rather than biomass or other productivity measures, it does not translate directly to crop yield. For example, an Iowa

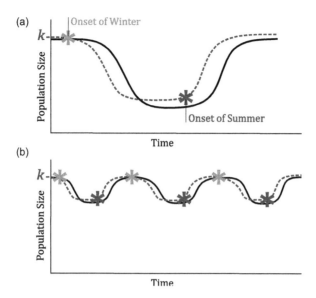

(a) Onset of Winter

$k$

Population Size

Onset of Summer

Time

(b)

$k$

Population Size

Time

**FIGURE 3.6** If a population responds to regular changes in the environment, such as seasonal shifts (a), the changes in population size may form a predictable pattern or "cycle" (b).

University study achieved the same yield from fields planted with 125,000 or 175,000 soy seeds per acre (Pedersen and De Bruin 2007). Since they did not contribute to farm income or production, the additional 50,000 planted seeds in the latter trial wasted up-front inputs like seed stock, sowing labor, and sowing time.

In some cases, however, increased carrying capacity *can* increase yields, leading to greater production from the same amount of space. Between 1960 and 2020, corn harvests in Iowa increased from 63.5 to 184 bushels per acre (National Agricultural Statistics Service 2020). The technological improvements during that time focused on delivery of irrigation, addition of fertilizer, and pest control (see Chapter 9 in this volume), all of which raise the carrying capacity of farmland. As a result of those additions, corn seeding rates in Iowa increased from 16,000 per acre in 1960 to more than 30,000 plants per acre in 2020 (Elmore and Abendroth 2006; National Agricultural Statistics Service 2020). While more productive corn varieties were also introduced during that time, the changes in planting density were the main force behind higher yields (Elmore and Abendroth 2006).

Meanwhile, a thorough understanding of population cycles can help farmers improve crop quality and use space efficiently without additional inputs. By comparing the population cycles and life histories of different crop species, we can identify plants that would grow well together in an intercropping system, using field space and resources more effectively (see Chapter 6 in this volume). Crops with alternating life cycles can be used in crop rotation systems, helping soil structure and nutrition recover between each planting and reducing the chances of long-term disease outbreaks. Comparing the population cycles of crop species and their pests can reveal the best seasonal timing for events like seed sowing, pest deterrent application, and harvest. Even more advanced methods of pest protection, like encouraging soil symbionts and native insect predators, require even greater understanding of the ecological web surrounding agricultural land.

Finally, agricultural managers must acknowledge that the carrying capacity of indigenous species does not equate to agricultural capacity. When native grasslands or forests are cleared to make way for farmland, a richly balanced ecosystem is sacrificed. Long-lived trees and other established plants are a type of **carbon sink**, since they remove $CO_2$ from the atmosphere and store (**sequester**) it in solid molecules. Carbon sinks help counter greenhouse and climate change effects. When native plants and other complex cycling systems in the soil are removed, the carbon sequestration rate drops considerably. Crop plants also sequester some $CO_2$, but since the modern agricultural ecosystem is often monocropped (i.e., trees are not cultivated with field crops), the density of carbon sinks is much lower. Agricultural waste products or cover crops are typically removed at least once a year, their breakdown by saprotrophic organisms releasing $CO_2$ back into the atmosphere regularly. Furthermore, with natural nutrient cycles disturbed and compost material moved off-site, soil-based carbon and other resources run out quickly. As a result, an industrialized field cannot support crops in nearly the same density as it could support native plant species. Attempting to achieve high yields on converted land with low crop carrying capacity has caused devastating environmental and agricultural problems in the past (see Chapter 9 in this volume).

## REFERENCES

Australian Museum. 2020. Stages of decomposition. *Australian Museum*. https://australian.museum/learn/science/stages-of-decomposition/ (accessed March 3, 2022).

Elmore, R. and L. Abendroth. 2006. Seeding rate relative to seed cost. *Iowa State University Extension and Outreach*. IC-496(6):82–83. https://crops.extension.iastate.edu/encyclopedia/seeding-rate-relative-seed-cost (accessed November 20, 2020).

National Agricultural Statistics Service. 2020. *Quick Stats*. United States Department of Agriculture. https://quickstats.nass.usda.gov/ (accessed November 20, 2020).

National Park Service. 2019. *The Salmon Life Cycle*. NPS.gov. https://www.nps.gov/olym/learn/nature/the-salmon-life-cycle.htm (accessed October 17, 2019).

Pedersen, P. and J. De Bruin. 2007. How many seeds does it really take to get 100,000 plants per acre at harvest? *Iowa State University Extension and Outreach. Iowa State University*. IC-498(5): 107–108. https://crops.extension.iastate.edu/encyclopedia/how-many-seeds-does-it-really-take-get-100000-plants-acre-harvest#:~:text=The%20recommendation%20is%20to%20seedfield%2C%20and%20farmer%20to%20farmer (accessed November 18, 2020).

# 4 The Kingdom Fungi

*Steven Stephenson and Elizabeth "Izzie" Gall*

## CONTENTS

## 4.1 INTRODUCTION

It is difficult for the average person not to be aware of the larger plants and animals that share the world with us, but most people know very little about the widely distributed, ecologically very important, and exceedingly abundant organisms known as **fungi** (singular: fungus). Although found almost everywhere in nature, fungi are often overlooked, usually underappreciated and underutilized, and frequently misunderstood. The objective of this chapter is first to introduce the group and then to outline some of the many ways in which they are important, especially in an agricultural context.

For a very long time, fungi were considered to be members of the plant kingdom. In some ways, fungi are similar to plants, since the fruiting body increases in size as it matures and a multicellular fungus has both an underground, sometimes rootlike structure and an aerial, stem-like portion. However, there are some major fundamental differences. Perhaps the most important of these is that fungi lack the green pigment chlorophyll. As a result, fungi cannot produce their own food through the process of photosynthesis. Instead, fungi obtain their food by breaking down dead organic matter or, in some instances, by deriving it from living plants, animals, or even other fungi. The formal study of fungi is called **mycology**, and the people who consider at least some aspect of the biology, ecology, or economic importance of these organisms are known as **mycologists**.

DOI: 10.1201/9780429320415-4

## 4.2   BASIC STRUCTURES OF FUNGI

The "body" of all but the simplest fungi consists of a system of very finely branched, microscopic, threadlike structures called **hyphae** (singular: hypha). Unlike plant roots, fungal hyphae are just one cell thick. The cells may be separated from each other with walls called **septa** (singular: septum) or they may grow as one long, **coenocytic** unit, without walls. The system of hyphae making up a single fungus is referred to as its **mycelium** (plural: mycelia) and is often densely branched. Mycelia grow throughout **substrates** like soil, leaf litter, and decaying wood, where individual hyphae obtain the nutrients and water the fungus needs to grow (Stephenson 2010). The growth and branching of mycelia usually occur during the **trophic**, or feeding, phase of the life cycle. The reproductive phase involves the growth of fruiting bodies and usually the release of spores, as discussed below (Figure 4.1).

From a practical point of view, there are two major groups of fungi. By definition, **macrofungi** produce fruiting bodies or other structures that are large enough to be observed directly with the naked eye. **Microfungi** are too small to be observed individually without a hand lens or even a microscope, although the clustered fruiting structures of a large colony or the mycelium of a particular microfungus can be clearly visible. More technically, fungi are named and grouped based on their physical characteristics, life cycles, and ancestral relationships, if they are known. Their formal classification is discussed below.

## 4.3   BASIC CLASSIFICATION OF FUNGI AND FUNGUS-LIKE ORGANISMS

The system of classification used for the kingdom Fungi is constantly evolving as observational tools and study methods keep improving. Traditionally, four major phyla of "true fungi" have been recognized: the Chytridiomycota, the Zygomycota, the Ascomycota, and the Basidiomycota (Alexopoulos et al. 1996). Recent treatments have separated some members of the Zygomycota into the new phylum Glomeromycota, which contains the fungi involved in endomycorrhizal relationships (see Section 4.8.2 below).

In addition to these "true" fungi, there are several other groups of organisms that have been studied almost exclusively by mycologists (Stephenson 2010), since they are similar to the fungi in at least some respects. The best example is the group known as water molds, which have long been considered as the fungal phylum Oomycota. These organisms have a vegetative (trophic) stage consisting of **filameners** that superficially resemble fungal hyphae, and they obtain their food in the same manner as fungi (Stephenson 2010). However, water molds actually belong to a different kingdom (the Stramenopila) and are only distantly related to the fungi.

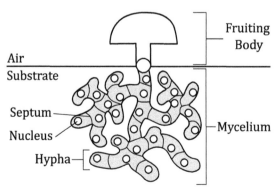

**FIGURE 4.1**   Anatomy of a generalized fungus with septate hyphae. In a coenocytic species, septa would be absent.

### 4.3.1 SLIME MOLDS

Another group usually considered along with the fungi is made up of organisms commonly referred to as slime molds, the **myxomycetes** (Gr. *myxa* = "slime") and the **dictyostelids** (Gr. *dicty-* = "net"; *stelid* = "pillar").

Myxomycetes, also called plasmodial slime molds or myxogastrids, are commonly associated with soil and various types of decaying plant material in virtually every type of terrestrial ecosystem examined to date (Martin and Alexopoulos 1969). Approximately 1,000 species are known worldwide (Lado 2005–2019). The life cycle of a myxomycete encompasses two vegetative trophic stages and a reproductive stage when fruiting bodies are produced. The first of the two trophic stages is microscopic, but the second trophic stage (termed a **plasmodium**) and the fruiting body for a number of species are large enough to be observed with the naked eye. Some of the fruiting bodies are miniature objects of considerable beauty (Stephenson and Stempen 1994) (Figure 4.2).

Dictyostelids, also called cellular slime molds or social amoebae, feed primarily on bacteria. They are common and sometimes abundant inhabitants of forest soil and leaf litter (Raper 1984), the soil of grasslands (Rollins and Stephenson 2013) and agricultural fields (Stephenson and Rajguru 2010), and animal dung (Stephenson and Landolt 1992). Both the vegetative and reproductive stages in the dictyostelids are microscopic, observable only under laboratory conditions (Figure 4.3).

## 4.4 TAXONOMY OF MACROFUNGI

"Macrofungi" is an artificial, size-based category that does not constitute a distinct taxonomic group, but conveniently, all macrofungi occur within the two phyla Ascomycota and Basidiomycota.

Some of the most highly regarded edible macrofungi, the truffles and morels, belong to phylum Ascomycota. Other than these notable exceptions, most macrofungi in this phylum are characteristically cup- or saucer-shaped, which is why they are often referred to as the "cup" fungi. However, the majority of the Ascomycota (commonly referred to as **ascomycetes**) would be classified as microfungi (see Section 4.5) (Figure 4.4).

In contrast with the Ascomycetes, most members of the Basidiomycota (commonly referred to as **basidiomycetes**) do produce macroscopic fruiting bodies. Basidiomycetes are among the most charismatic of all fungi, since the phylum contains such familiar and easily recognized examples as the oyster, shiitake, button, and porcini mushrooms, as well as the more exotic and colorful puffballs, chanterelles, jelly fungi, and some polypores. All of these are edible and widely consumed by mankind. In most instances, a particular basidiomycete can be assigned to one of a number of morphological groups based on the overall shape of the fruiting body and where the spores are produced. For example, the **agarics** (or gilled mushrooms) are basidiomycetes with fruiting bodies characterized by the presence of thin, blade-like gills upon which the spores are produced. The button mushroom is a very common and typical example of an agaric. In contrast, the porcini mushroom is a member of the **boletes**, in which the spores are produced on the inner walls of a series of tubes that make up the fertile portion of the fruiting body. The basidiomycetes also include the coral fungi, stinkhorns, bird's nest fungi, earthstars, corticoid fungi, and tooth fungi in addition to two groups of parasitic microfungi, the rusts and smuts (Figure 4.5).

## 4.5 TAXONOMY OF MICROFUNGI

The ascomycetes are the largest taxonomic assemblage of fungi, and the microfungal members include numerous saprotrophs, plant pathogens, and parasites along with most of the often-unicellular fungi known as **yeasts**. Most people are aware that yeasts, through the process of fermentation, are important for brewing beer, baking bread, and making wine and other preserved foods (see Chapter 18 in this volume). Beyond their importance in human cuisine, yeasts are the most abundant fungi on Earth and are the second-most abundant of all living organisms (after bacteria). Yeasts are present on virtually all plant surfaces, including the bark of trees, fruits and flowers, and even roots, and are common in most soils. There are some basidiomycete yeasts of note, such as the human

**FIGURE 4.2** A selection of myxomycetes. The still developing plasmodium (a) and mature fruiting body (b) of *Ceratiomyca fructiculosa* var. *porioides*. The plasmodium of a species of *Physarium* (c) and a fruiting body (d) of *Physarium viride*. The plasmodium (e) and fruiting bodies (f) of *Badhamia utricularis*. [Images (c, d) provided by Dr. Stephenson. (a, b, e, f) reproduced from Shutterstock with permission, then cropped: (a) Henri Koskinen. (b) NK-55. (e, f) Gertjan Hooijer.]

pathogen *Cryptococcus* and the basidiomycete yeasts that occur in the thallus of a lichen. However, most yeast species of ecological or economic importance are ascomycetes.

The remaining three phyla of fungi (the Chytridiomycota, the Zygomycota, and the Glomeromycota) are best considered as microfungi. The Chytridiomycota (chytrids) are unique among the true fungi in that they produce flagellated, self-propelling cells called zoospores.

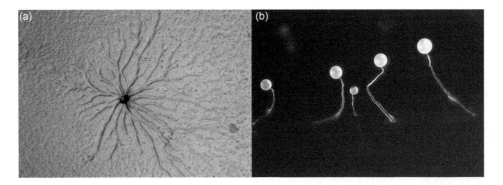

**FIGURE 4.3**  Vegetative stage (a) and fruiting bodies (b) of a species of *Dictyostelium*. (Images provided by Dr. Stephenson.)

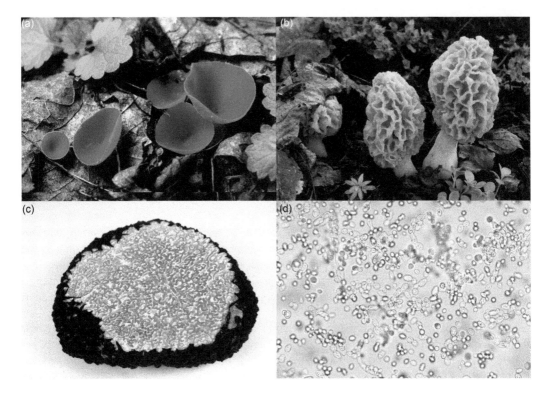

**FIGURE 4.4**  A selection of ascomycetes. (a) A "fairy cup", *Sarcoscypha occidentalis*. (b) A morel, *Morchella esculenta*. (c) A truffle, *Tuber* sp. (d) The bread and brewing yeast, *Saccharomyces cerevisiae*. [Images (a, b) provided by Dr. Stephenson. (c, d) Reproduced from Shutterstock with permission, then cropped: (c) Stephen Farhall. (d) Rattiya Thongdumhyu.]

Although chytrids are regarded as primarily aquatic, these organisms also occur in soils. Because most species are essentially microscopic, chytrids represent an understudied group of fungi; relatively little is known about their distribution and ecology, especially for those species found in soil. The majority of species are probably saprotrophic, but at least a few are parasitic or pathogenic. For example, Synchytrium endoboticum causes a condition in potatoes known as black wart disease.

The Zygomycota (zygomycetes) are a diverse assemblage of terrestrial microfungi typically associated with a wide range of organic materials including dead insects, dung, and macrofungi as well as stored grain and various other food products consumed by humans. In fact, one of the most widely

**FIGURE 4.5** A selection of basidiomycetes. (a) A "coral fungus", *Ramaria formosa*. (b) Oyster mushroom, *Pleurotus ostreatus*. (c) King bolete, *Boletus edulis*. (d) Lion's mane mushroom, a species of *Hericium*. (Images provided by Dr. Stephenson.)

distributed and well-known species is *Rhizopus stolonifer*, the black bread mold. Zygomycetes are rather ubiquitous in all kinds of terrestrial habitats and can be exceedingly abundant on compost piles and accumulations of various types of organic waste. A few species produce a mycelium extensive enough to be seen with the naked eye. The distinguishing features of the zygomycetes are coenocytic hyphae and the fact that many members of the phylum produce a special type of spore called a zygospore (Alexopoulos et al. 1996).

The Glomeromycota are exclusively plant symbionts known as **mycorrhizae**. By growing bundles of hyphae into the interior of root cells, glomeromycetes are able to exchange nutrients and water from the soil for a steady supply of plant sugars. This relationship is explored in detail in Section 4.8.2.

## 4.6  SEXUAL REPRODUCTION

Sexual reproduction involves fusion of two diploid parent nuclei, reshuffling of corresponding genetic information, and meiosis to create unique, haploid offspring nuclei. In the fungi, this process typically takes place within a fruiting body.

### 4.6.1  BASIDIOMYCETES

The single most consistent feature of the basidiomycetes is the production of sexual spores (basidiospores) attached externally to a special club-shaped hypha called a **basidium** (L. "pedestal"; plural: basidia). Typically, each basidium will produce four haploid, uninucleate basidiospores, but there is considerable variation between different species (Alexopoulos et al. 1996). The life cycle of a basidiomycete begins with the germination of a haploid ($n$) spore that has been dispersed from a fruiting body. The spore germinates to yield a haploid hypha that branches outward, seeking both nutrients and a mate – the hypha derived from another spore of the same species but representing a different, compatible mating type. When two compatible hyphae meet, their membranes fuse, initiating the **dikaryotic** ($n+n$) phase of growth (Figure 4.6b). The term dikaryotic (literally, "having two nuclei") refers to the fact that although the cell has two full complements of DNA, the parent nuclei do not fuse. The dikaryotic hyphae of basidiomycetes are septate, divided into distinct cell-like compartments by incomplete walls. The central pore in each septum helps ensure that new hyphal cells receive both parental nuclei, maintaining the dikaryotic state (Alexopoulos et al. 1996). Dikaryotic mycelial growth makes up the majority of a basidiomycete's life cycle.

Once the dikaryotic mycelium has undergone a period of growth, a basidiomycete can theoretically produce a fruiting body at any time that conditions are favorable (Jennings and Lysek 1999). Environmental conditions suitable for fruiting are much more specific than the conditions under which trophic (mycelial) growth can occur. Once the fruiting conditions are met, specialized hyphae begin to grow above the surface of the substrate and differentiate, eventually creating the fruiting body (Figure 4.6d and e). When the essential structure of the fruiting body is in place, the cell walls of the aerial hyphae weaken and expand, allowing the volume of the entire structure to increase considerably (Jennings and Lysek 1999). The aerial hyphae terminate in a fertile layer called the **hymenium**, which bears the basidia (as gills in agarics and tubes in boletes; Figure 4.6f). The two parental nuclei finally fuse inside each basidium, which becomes briefly diploid ($2n$) before generating haploid ($n$) basidiospores (Figure 4.6g and h). When the spores are mature, the basidia release them to complete the sexual life cycle.

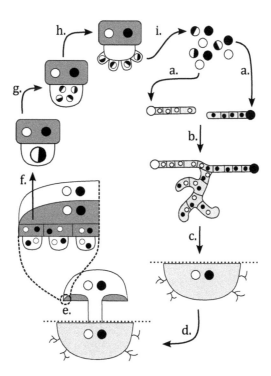

**FIGURE 4.6** The basidiomycete sexual life cycle. (a) Haploid basidiospores germinate to form haploid hyphae. (b) Compatible haploid hyphae fuse to form dikaryotic hyphae, which spread throughout the substrate. (c) The mycelium reaches the edge of its substrate (dotted line indicates air interface). (d) Aerial hyphae form dikaryotic fruiting bodies. (e) The terminal cells of the fruiting body (the hymenium) give rise to basidia. (f) In each basidium, the parent nuclei fuse. (g) The fused nuclei undergo meiosis. (h) The resulting haploid nuclei migrate to the outside of the basidium, forming mature spores that can be expelled. (i) Mature spores are actively expelled.

Many basidiomycetes can also produce a type of asexual spore called a **conidium** (plural: conidia) to rapidly increase their numbers or to take advantage of additional hosts. For example, the wheat rust *Puccinia graminis* uses one type of conidium to spread through wheat hosts during summer and autumn and uses a different type of asexual spore when living on its alternate host, barberry.

Some basidiomycetes produce specialized asexual structures to wait out unfavorable conditions. Chlamydospores are thick-walled resting cells produced from vegetative hyphae which can germinate into new mycelia when favorable conditions return (Burnett 2003). **Sclerotia** (singular: sclerotium) are masses of highly branched, asexual hyphae packed with oils, proteins, and carbohydrates. In some species of basidiomycetes, sclerotia can produce viable hyphae after decades of dormancy (Burnett 2003). However, while asexual spores and structures have their advantages, the sexual cycle remains the major method for introducing variation into a population and is necessary for the continued evolution of a species (see Chapter 2 in this volume).

### 4.6.2 Ascomycetes

The ascomycetes share some features with the basidiomycetes, such as septate hyphae with incomplete walls, a dikaryotic portion of the life cycle, production of both sexual and asexual spores, and a wide range of morphological variation. However, there are a number of distinguishing features. Most importantly, whereas basidiomycetes have a limited haploid state, the haploid mycelium produced by an ascomycete spore persists for a greater portion of the life cycle. When two compatible mycelia of the same species come into contact, some of their hyphae fuse to form dikaryotic **ascogenous** ("sac-creating") hyphae, which are typically very short-lived, usually lasting no more than a few days (Figure 4.7b and c). As in basidiomycetes, the fruiting body of an ascomycete consists of

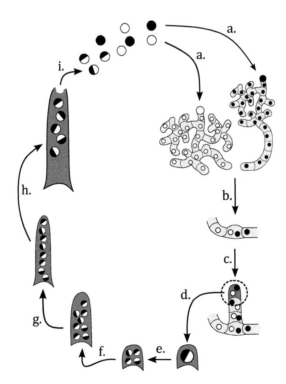

**FIGURE 4.7** The ascomycete sexual life cycle. (a) Haploid ascospores germinate to form haploid hyphae, which grow extensively throughout the substrate. (b) Compatible haploid hyphae fuse to form a small number of dikaryotic hyphae. (c) Dikaryotic hyphae rapidly generate ascogenous hyphae (dark gray). (d) Inside the ascogenous cells, the parent nuclei fuse. (e) Parent nuclei undergo meiosis, (f) then mitosis. (g) The eight resulting haploid nuclei align within the ascus. (h) Mature spores are released from the ascus.

both a sterile portion of vegetative hyphae and the fertile portion, in this case the ascogenous hyphae rather than a hymenium (Stephenson 2010). In a large fruiting body, the system of ascogenous hyphae can be rather extensive. Some of the ascogenous hyphae ultimately produce **asci** (singular: ascus), the specialized reproductive structures unique to ascomycetes. In the formation of an ascus, the nuclei in the terminal ascogenous cell fuse, giving rise to a diploid ($2n$) nucleus (Figure 4.7d). The nucleus immediately undergoes meiosis to produce four haploid ($n$) nuclei (Figure 4.7e). In the majority of ascomycetes for which the life cycle has been studied in detail, the four haploid nuclei then undergo mitosis to yield eight nuclei, which develop into ascospores. However, in some cases, multiple mitotic divisions occur, resulting in asci with hundreds or even thousands of ascospores! In contrast to externally borne basidiospores, ascospores are located within the ascus, often lined up and having the general appearance of peas in a pod (Figure 4.7g). The ascus is the single defining characteristic of the phylum Ascomycota.

Although many ascomycetes are capable of reproducing sexually, asexual reproduction is exceedingly common, more so than in the basidiomycetes. In fact, some ascomycetes rely almost exclusively upon asexual spores to produce new individuals. These spores are called conidiospores (or **conidia**), the same term used for basidiomycetes (Alexopoulos et al. 1996). Conidia are often produced at the tip of an elevated hypha called a conidiophore. Traditionally, the overall morphology of the conidiophore and the way in which multiple conidiophores arise from a particular mycelium have provided the basis for the classification used for these fungi (Carmichael et al. 1980), but with the advent of modern molecular biology, modern mycologists identify different taxa on the basis of DNA sequences.

For ascomycete yeasts, asexual reproduction takes place by means of a parent cell budding off an initially smaller version of itself. The bud increases in size while still attached to the parent cell but eventually breaks off and forms a new individual cell. This process can take place over a relatively short period of time, allowing yeasts to greatly increase in number rather rapidly.

## 4.7   ASEXUAL AND SEXUAL STAGES

In asexual reproduction, no nuclei fuse, meiosis does not occur, and no specialized reproductive structures are involved. Instead, new spores (usually conidia) are produced by mitosis. As noted above, asexual reproduction is much more common in the ascomycetes than in the basidiomycetes, although many species in both phyla have clearly distinct sexual and asexual stages. Mycologists have adapted a system of terminology to refer to this situation. The term **teleomorph** is used for the sexual stage (Gr. *téleios* = "complete"), whereas the term **anamorph** (Gr. *ana-* ≈ "repeated") is used for the asexual stage (Alexopoulos et al. 1996). The biological entity or species that incorporates both stages (the "whole" fungus) is referred to as the **holomorph** (Gr. *holos* = "whole"). More recently, it has been proposed that the terms meiosporic fungi (for the teleomorph) and mitosporic fungi (for the anamorph) should be used, since they better reflect the type of nuclear division involved – meiosis in the first instance and mitosis in the second (Figure 4.8).

Based on observations and laboratory studies, some fungi are apparently not capable of sexual reproduction. Presumably, the complex trait has been lost over the course of their evolution. This poses a problem, since the classification of fungi is based largely on the type of sexual spores they produce, and such structures are lacking in those forms characterized only by asexual reproduction. Mycologists traditionally have placed these forms in the artificial phylum Deuteromycota (or deuteromycetes). However, with the advent of modern molecular techniques, it is now possible to extract and sequence the DNA of these fungi, revealing that in many instances what had previously been considered two different deuteromycete species actually represent the teleomorph and anamorph of a single species. Although it seems apparent that many deuteromycetes do indeed have corresponding sexual stages, there are some examples that truly reproduce only vegetatively by the production of new hyphae. These *mycelia sterilia* (or sterile fungi) do not produce any known spores, either sexual or asexual.

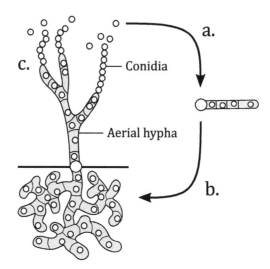

**FIGURE 4.8** The asexual life cycle for a generalized ascomycete. (a) Haploid conidia disperse and germinate into hyphae. (b) After growing hyphae throughout its substrate, the vegetative fungus gives rise to aerial hyphae. (c) More conidia are produced at the tips of the aerial hyphae through mitosis. Note that, barring mutations, there is no change or exchange in genetic information.

## 4.8  FUNDAMENTAL NICHES OF FUNGI

At least some fungi are found in every habitat on Earth, and some groups of fungi are among the most abundant of all living organisms (Stephenson 2010). Although the relatively large fruiting bodies produced by some species of fungi are quite conspicuous, most such fruiting bodies are ephemeral, disappearing within days, and during some seasons of the year they may be limited in number or entirely absent. The rest of the fungal organism, the mycelium, is almost always completely out of sight in substrates like decaying wood, soil, or the trunk of a living tree, making fungi very easy to overlook. Moreover, since most people know relatively little about fungi, occasional cases of mushroom poisoning or black mold infestations lead to widespread fear and often overshadow the positive services fungi perform. However, this world would be a very different place if there were no fungi! The various roles that fungi play in nature are absolutely essential for the proper functioning of ecosystems.

### 4.8.1  Saprotrophic Fungi

As already noted, fungi must meet their nutritional needs by exploiting other organisms, either directly or indirectly. Saprotrophic fungi depend upon dead organic matter to meet their energy needs, and their ecological impact is considerable. Between 32.2% and 34% of a tree's new yearly growth falls to the ground as leaf litter. This is true of both broadleaf and coniferous (evergreen) trees and for both temperate and tropical climates (Neumann et al. 2018). These leaves do not build up year after year because various saprotrophic fungi break them down. Some of these fungi are specific to the type (or types) of leaves they digest, forming distinct communities around particular plant species. Different assemblages of fungi tend to be associated with each of the other components that make up the litter layer on the forest floor, with certain fungal species associated with woody twigs, fruits, seeds, and flowers.

Each fungus produces various digestive enzymes to release into its immediate environment. Different fungi produce different enzymes and consequently can degrade different substances, so fungi as a group can break down almost every type of organic substance. All fungi can produce the enzymes required to break down relatively repetitive organic molecules such as starch. Many can also degrade and utilize cellulose, a complex carbohydrate which forms about one-third of all plant matter and is the most abundant organic substance on the Earth. Far fewer fungi (mostly basidiomycetes, along with a few ascomycetes) have the capability of decomposing lignin, which is the second-most abundant organic substance on the Earth. Both cellulose and lignin are major components of wood (see Chapter 17 in this volume).

In a technical sense, **wood** refers to the secondary **xylem**[1] of vascular plants that have secondary growth, such as the gymnosperms (e.g., conifers) and certain angiosperms (e.g., oak and beech trees). Nonwoody or **herbaceous** plants generally do not undergo secondary growth. With its regularly repeating, unbranched structure, cellulose lends wood rigidity and tensile strength but is relatively easy to degrade. However, the lignin in wood is an irregular polymer which acts as a glue to hold cellulose fibers together. Due to its complex and highly polymerized molecular structure, lignin digestion requires specialized enzymes.

Wood-decay fungi can be classified into two groups based on the component of wood they specialize at breaking down. "Brown-rot" fungi digest cellulose but leave lignin, which has a brown coloration, intact. The destruction of cellulose results in cubical fracture of the wood, creating richly colored brown cubes of wood that mix into soil and humus. The sulfur shelf mushroom, *Laetiporus sulphureus*, is an edible brown-rot fungus which can either colonize dead broadleaf trees or weakly parasitize mature ones. "White-rot" fungi specialize at digesting lignin, leaving the remaining wood pale and fibrous. Several wild and cultivated mushrooms fit into this category, including saprotrophic shiitake and oyster mushrooms as well as the parasitic honey mushrooms of the genus *Armillaria*. Some wood-decay fungi can colonize living trees, since most of the xylem cells (and some of the phloem cells) produced through secondary growth are dead and thus subject to being exploited. In some instances, the entire center of the trunk can be decomposed, resulting in a living but hollow tree (Figure 4.9).

Some saprotrophic fungi decompose the dung of herbivores. Dung actually consists of undigested plant material to which animal waste products have been added. The exact nature of the dung depends upon the type of animal from which it was derived, and the original type of plant material consumed. Once deposited, dung decomposes (is utilized by microbes) rapidly because the organic material inside is already broken up to some extent by the animal's digestive enzymes and because the dung has a relatively high nitrogen and moisture content. The fungi associated with dung are

**FIGURE 4.9** Characteristic types of wood rot. (a) White-rot fungi leave pale, fibrous wood behind. (b) Brown-rot fungi leave darker wood that breaks into cubic formations.

---

[1] Pronounced "*zye-lem*" and derived from the Greek *xulon* = "wood". Primary xylem is the water-conducting tissue at the center of a woody plant. It is deposited during primary growth, such as extension of the shoot or roots. Secondary xylem is generated during the growing season each year, expanding the width of the stem and creating clearly visible rings.

taxonomically very diverse and include representatives of all of the major phyla. Some of these **coprophilous** (dung inhabiting) fungi are highly specialized for surviving on dung. A notable example is *Pilobolus*, a member of the Zygomycota that has evolved the capability of dispersing its spores yards (meters) away from the dung on which it grows. This increases the chance of having its spores reach an untouched, richly nutritious dung pile.

### 4.8.2 FUNGI IN SYMBIOSES

Many fungi can establish a symbiotic relationship with the roots of trees and other plants. This relationship, known as a **mycorrhizal** association, is mutually beneficial to both the plant and the fungus. The roots of most plants are relatively inefficient at nutrient uptake; the thin hyphae of fungi are much better at absorbing nutrients from the soil and can scavenge water from further away than plant roots can reach. In brief, when a fungus is allowed to colonize a plant root, the fungus enables the plant to access nutrients and water that would otherwise be unavailable. In return, the plant provides the fungus sugars that are produced by photosynthesis. The vast majority of plants worldwide are involved in mycorrhizal associations (see Chapter 11 in this volume). Many of the more obvious and best-known examples of mycorrhizal fungi are basidiomycete macrofungi, including members of such widely known genera as *Amanita*, *Boletus*, and *Russula*. The charismatic fruiting body we typically see above ground is just the reproductive portion of a mycorrhizal fungus; the bulk of its biomass is belowground in the intricately branched mycelial network. Though the threadlike hyphae are only one cell wide, they can be very dense in healthy soils. There have been reports of several kilometers' worth of these hyphae in just a few grams of soil, leading to the belief that some basidiomycetes are among the largest living organisms in the world. This network of mycelia can connect plants within the same genus, and sometimes also within different genera.

There are two fundamentally different types of mycorrhizal associations, ectomycorrhizal and endomycorrhizal. The former is the better known since the fungus involved is usually a basidiomycete, such as members of *Amanita*, *Boletus*, *Cortinarius*, *Lactarius*, and *Russula*. These genera all produce large, often beautiful fruiting bodies, including examples that range from highly edible to deadly poisonous. An **ectomycorrhizal** association (Gr. *ecto* = "outside") begins when the hyphae of an ectomycorrhiza-forming fungus encounter the root system of a potential host plant. These hyphae appear to be attracted to actively growing root tips, possibly in response to the sugary substances being exuded by the plant (for more on these materials, see Chapter 12 in this volume). Once in contact with the root, the hyphae branch considerably and give rise to a covering that extends over the outside of the root, which is known as a sheath or **mantle** (see Figure 4.10). In most examples, the mantle is differentiated into two layers. The outermost layer consists of relatively thick-walled hyphae that are tightly packed, while the inner layer is composed of thin-walled hyphae that are more loosely bundled. This inner layer gives rise to hyphae that penetrate the root, growing down among the cells in the root's outermost layer (the cortex) to create what is known as the **Hartig net**. Other hyphae, produced by the outmost layer of the mantle, extend out into the soil to retrieve resources and water. Only about 3% of plant species form ectomycorrhizal associations, but prominent examples include eucalyptus and southern beech trees.

In contrast, **endomycorrhizal** association is known to occur in more than 75% of all plant families (Wang and Qiu 2006), including most agricultural crops and ornamental species. The vast majority of endomycorrhiza-forming fungi belong to the phylum Glomeromycota. Endomycorrhizal associations do not involve the formation of a sheath around the root of the host plant. In fact, the presence of an endomycorrhizal-forming fungus is only indicated by a loose, very sparse network of coenocytic, usually thick-walled hyphae in the soil near the root. When the hyphae of an endomycorrhiza-forming fungus first contact the root of a potential host plant, they penetrate the outmost layer of root cells and then form numerous branches that grow not just between, but into individual cells of the cortex (Gr. *endo* = "inside"). Eventually, the hyphae of the fungus may

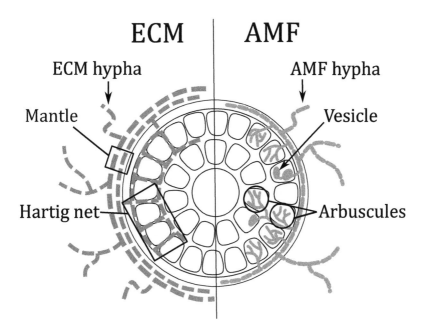

**FIGURE 4.10** Generalized ectomycorrhizae and endomycorrhizae on a section of plant root. Ectomycorrhizae (ECM), left, grow a thick layer, known as the mantle or sheath, around the plant root. The inner layer of the sheath grows between plant cells to form the Hartig net, where nutrient exchange takes place. Arbuscular mycorrhizal fungi (AMF), right, grow into the plant root cells, where they generate storage vesicles and the arbuscules that are their sites of nutrient exchange. Both ECM and AMF also grow hyphae into the soil surrounding the plant root.

proliferate throughout the entire cortex, but they do not disrupt the vascular tissue (e.g., xylem) at the center of the root. Once inside an individual root cell, an endomycorrhizal hypha generates very finely branched, treelike structures called **arbuscules**, often filling the interior of the plant cell. Hence, endomycorrhizae are also known as arbuscular mycorrhizae (or AM). The branches of the arbuscules represent the primary sites of the exchanges of water, nutrients, and organic molecules that occur between the fungus and the plant.

In many endomycorrhiza-forming fungi, a second type of structure, called a **vesicle**, is produced within the cells of the root. Vesicles develop at the tips of individual hyphae. They are thin-walled, ovoid or spherical in shape, and appear to serve a storage function for the fungus. Those endomycorrhiza-forming fungi that produce these two types of structures are often referred to as vesicular-arbuscular mycorrhizae (or VAM) (Figure 4.11).

Not all fungal symbioses are beneficial to the fungus' host. For example, many fungi can feed upon living plant or animal tissues. For those situations in which the host is harmed but not killed, the fungus is considered a parasite. If the presence of the fungus produces a condition (called a disease) that has the potential of resulting in the death of the host, it is known as a **pathogen**. The distinction between parasite and pathogen is not necessarily absolute, and a parasite may become a pathogen over time or under a different set of circumstances (see Chapter 2 in this volume).

## 4.9 SECONDARY METABOLITES OF INTEREST

Like plants, many fungi create secondary metabolic compounds – chemicals not used directly for self-maintenance, growth, or reproduction. These include pigment molecules, such as those that give portabella mushrooms their earthy brown color or give certain *Amanita* species their bright orange and red coloration. Many fungi synthesize flavor compounds that make them particularly attractive

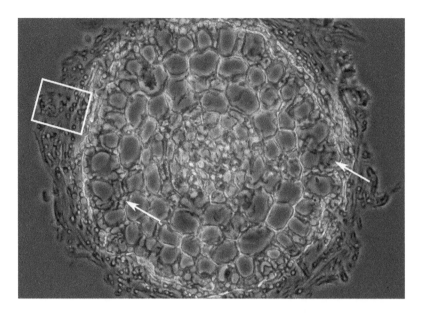

**FIGURE 4.11** Ectomycorrhizae on a section of plant root. White arrows indicate dense regions of the Hartig net; box draws attention to the hyphal mantle. (Image provided by Dr. Stephenson.)

to humans or other animals. Some compounds of particular interest are those with harmful, antibiotic, or toxic properties, which should be handled with caution, if ever (see Chapter 1 in this volume for examples). However, there is great potential to use other fungal metabolites for medicinal purposes. This section explores some of these interesting secondary metabolites.

### 4.9.1 CULINARY USE

As mentioned earlier in this chapter, some fungi are collected for human consumption, and a few are well-known as delicious edibles. The vast majority of fungi are not poisonous, but fungi are such a large and diverse group of organisms that there are no general rules that can be used to determine whether a particular specimen is edible. The best rule to follow is to be exceedingly careful and not only learn to identify the edible species but also know the other species with which they could be confused. Having one or more good field guides available is exceedingly useful when foraging. Only a very few species of fungi, including the button mushroom, are safely consumed raw. Even certain widely known edibles, such as morels, can be dangerous if eaten raw. As such, any wild fungus collected for the table should be cooked.

Humans are not the only animals to feed upon fungi. Numerous small insects and other invertebrates found in forest floor litter feed upon the hyphae and spores of the abundant microfungi that tend to grow there. Some small marsupials in Australia have diets largely composed of truffles, which have evolved to release pungent aromas at maturity in order to entice the animals to eat them. By digging for the belowground truffles, animals provide the essential ecosystem service of aerating the soil. Once the animal digests the truffle, the fungal spores will be dispersed throughout the forest in the animal's scats. Many fungal fruiting bodies are designed for this type of spore dispersal. However, if an animal is observed consuming a fungus, this does not mean that the fungus is edible to humans (Elliott and Stephenson 2018).

### 4.9.2 POISONS OR TOXINS

There are different kinds of toxic substances that can cause a fungus to be poisonous. These toxic substances (or toxins) can produce symptoms ranging from mild gastrointestinal discomfort to a painful death. Fortunately, the vast majority of fungal toxins are not fatal, but they can still cause nausea, vomiting, diarrhea, sweating, and hallucinations. Symptoms vary based on the quantity of the fungus consumed; on the victim's age, sensitivity to the compound, and any pre-existing health issues; and on the presence of other substances, such as alcohol, in the victim's body. Most nonfatal fungal toxins produce symptoms relatively quickly (often in less than four hours), but deadly toxins generally take six hours or longer to become evident. Perhaps the most dangerous toxins are the amatoxins, some of which are found in certain members of the genus *Amanita*. Amatoxins are one of the major causes of mushroom fatalities, and those found in some species of *Amanita* almost invariably result in death.

### 4.9.3 MEDICINAL COMPOUNDS

Various fungi have played a significant role in traditional folk medicine around the world, but the actual medicinal properties of even the most widespread examples are mostly unknown. New compounds isolated from fungi usually require a considerable period of time and several clinical trials to be approved for application in a modern medical setting. However, there are some well-known and valuable compounds that are extracted from fungi. Some prominent examples include cordycepin (a promising compound for leukemia and cancer treatment), cyclosporine (an immunosuppressant that makes organ transplants possible), lentinan (used in some cancer treatments), and penicillin (one of the key early antibiotics). There is little doubt that other potentially useful compounds remain to be discovered, and the "bioprospecting" processes carried out by some pharmaceutical companies to identify new medicinal compounds often include fungi.

## REFERENCES

Alexopoulos, C.J., C. W. Mims, and M. Blackwell. 1996. *Introductory Mycology*. 4th ed. New York: John Wiley & Sons, Inc.

Burnett, J. H. 2003. *Fungal Populations and Species*. New York: Oxford University Press.

Carmichael, J. W., W. B. Kendrick, I. L. Conners, and L. Sigler. 1980. *Genera of Hyphomycetes*. Edmonton, Canada: University of Alberta Press.

Elliott, T. F., and S. L. Stephenson. 2018. *Mushrooms of the Southeastern United States*. Portland, OR: Timber Press.

Jennings, D. H., and G. Lysek. 1999. *Fungal Biology: Understanding the Fungal Lifestyle*. 2nd ed. Oxford: BIOS Scientific Publishers Ltd.

Lado, C. 2005–2019. An on line nomenclatural information system of Eumycetozoa. Accessed July 15, 2019. http://www.nomen.eumycetozoa.com.

Landers, J. 2012. *Bambi Ate Thumper*. The Slate Group. November 16. Accessed April 2, 2020. https://slate.com/technology/2012/11/deer-eat-meat-herbivores-and-carnivores-are-not-so-clearly-divided.html.

Lang, G. E. 1974. Litter dynamics in a mixed oak forest on the New Jersey Piedmont. *Bulletin of the Torrey Botanical Club* (101): 277–286.

Martin, G. W., and C. J. Alexopoulos. 1969. *The Myxomycetes*. Iowa City: University of Iowa Press.

Raper, K. B. 1984. *The Dictyostelids*. Princeton, NJ: Princeton University Press.

Rollins, A. W., and S. L. Stephenson. 2013. Myxomycetes associated with grasslands of the western central United States. *Fungal Diversity* (59): 147–158.

Stephenson, S. L. 2010. *The Kingdom Fungi: The Biology of Mushrooms, Molds, and Lichens*. Portland, OR: Timber Press.

Stephenson, S. L., and H. Stempen. 1994. *Myxomycetes: A Handbook of Slime Molds*. Portland, OR: Timber Press.

Stephenson, S. L., and J. C. Landolt. 1992. Vertebrates as vectors of cellular slime molds in temperate forests. *Mycological Research* (96): 670–672.

Stephenson, S. L., and S. N. Rajguru. 2010. Dictyostelid cellular slime moulds in agricultural soils. *Mycosphere* (1): 333–346.

Wang, B., and Y. L. Qiu. 2006. Phylogenetic distribution and evolution of mycorrhizas in land plants. *Mycorrhiza* (16): 299–363.

# 5 Limiting Factors in Agriculture

*Noureddine Benkeblia*

## CONTENTS

## 5.1 INTRODUCTION

Agricultural history has paralleled humankind's technological and social development, and archaeological evidence indicates that food and animal domestication first occurred at least 10,000 years ago. However, according to the FAO (2017), agriculture is at present one of the riskiest and one of the most vulnerable activities of the modern day due to its negative impact on the environment (greenhouse gases, fertilizer pollution, etc.) and climate change threatening food production (Rosenzweig et al. 2013). Land use and crop production can be affected by many constraints and limitations, which fall into four categories: (1) meteorological factors, including global climate, $CO_2$ concentration, and the effects of local weather and precipitation; (2) land factors, including nutrient and water availability and soil-weather interactions; (3) biological factors, including pests and diseases; and (4) human factors, like social preferences or economics. Of these four types of constraints, this chapter will discuss in detail meteorological (temperature, $CO_2$ concentration, and water) and land factors (water and soil nutrient availability), since historically these are the most limiting factors affecting agriculture and have fewer or no mitigation potentials compared with biological (Chapters 6, 7, and 12–16) and human (Chapters 8–10) factors (Schlenker et al. 2007; Tanaka et al. 2006).

## 5.2 METEOROLOGICAL LIMITING FACTORS

### 5.2.1 EXTREME TEMPERATURE AS A LIMITING FACTOR

Extreme fluctuations of meteorological factors have always been considered major limiting factors in agriculture and food production. Temperature plays an important and complex role in crop yield, and all plant species have temperature thresholds above or below which they will not grow and develop properly. Unpredictable extreme climatic conditions such as high temperatures, low temperatures, drought, and excess periodical precipitation can significantly affect growth and yield of most crops (Assad et al. 2019; Zhao et al. 2015). Like any growth and development process, photosynthesis is strongly affected by temperature (Long 1991). The processes and variations of photosynthesis are described in more detail in Chapter 6.

DOI: 10.1201/9780429320415-5

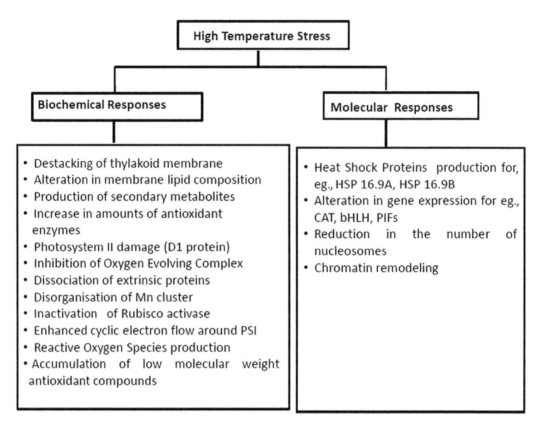

**FIGURE 5.1** A schematic diagram showing avoidance and tolerance mechanisms of a typical plant in response to high temperature stress. (Reprinted with permission from Elsevier: Mathur, S., and A. Jajoo. 2014. Photosynthesis: Limitations in response to high temperature stress. *Phytochemistry and Photobiology* 137: 116–26. Caption altered for clarity.)

### 5.2.1.1 High Temperatures

High temperature and heat stress have been reported to inhibit photosynthesis; notably, heat stress is often accompanied by drought stress (see Section 5.3) (Bita and Gerats 2013; Larcher 1995; Murchie et al. 2009; Niles et al. 2015; Salvucci and Crafts-Brandner 2004; Yamori et al. 2012). Although extensive literature reports on the effects of temperature on photosynthesis (see Moore et al. 2021; Song et al. 2014), the mechanisms of how plants sense high temperatures are still enigmatic (Ruelland and Zachowski 2010). However, inhibition of net photosynthesis and photosynthetic performance by high temperatures has been attributed to an inability of the enzyme Rubisco activase to maintain Rubisco in an active form (Salvucci and Crafts-Brandner 2004) and instability of the **thylakoid** membrane[1] under heat stress (see Figure 5.1) (Falk et al. 1996). Many reports have shown that high temperatures impair chlorophyll (pigment) function or accelerate its degradation (Dutta et al. 2009; Efeoglu and Terzioglu 2009; Mathur and Jajoo 2014). Furthermore, the replacement of chlorophyll is inhibited as its biosynthesis requires many enzymes (Rubisco, FBPase, SBPase, and PRK) which are **denatured** (deformed beyond function) in high temperatures. These enzymes in turn affect photosystem II (PSII) which is considered one of the most thermosensitive components of the photosynthetic machinery (Berry and Björkman 1980; Gujjar et al. 2020; Srivastava et al. 1997).

When the leaf concentrations of photosynthetic pigments (chlorophylls and carotenoids) are low, leaves show sustained decreases in PSII efficiency; low enough pigment concentrations result

[1] The internal membrane in plant cells in which photosynthetic proteins are embedded.

in **photoinhibition** (a decrease in photochemical efficiency in response to radiation damage from excess photon energy that cannot be used) or even damage to the PSII reaction center (Abadía et al. 1999; Ahammed et al. 2018). Impairments in photosynthetic rate in response to high or low temperatures are often reversible in a range varying from 10°C to 35°C (50°F–95°F) (Hikosaka et al. 2006; Yamori et al. 2014). Most plants possess considerable capacity in adjusting their photosynthetic characteristics to the environmental temperature under which they are growing in order to maximize the photosynthetic rate and efficiency (Yamori et al. 2014). However, when exposed to temperatures out of this range, plants might experience irreversible injury to the photosynthetic system (Bernacchi et al. 2013; Berry and Björkman 1980; Crafts-Brandner and Salvucci 2004; Makino and Sage 2007).

The photosynthetic reactions are variable in plants (see Chapter 6 in this volume), and the inherent ability for temperature acclimation of photosynthesis differs between C3, C4, and crassulacean acid metabolism (CAM) species, and among functional types within C3 plants as well (Yamori et al. 2014). In C4 plants, which are generally adapted to warm environments, photosynthesis is generally less plastic due to the rigid positioning of chloroplasts within bundle sheath cells,[2] while C3 plants show greater ability of temperature acclimation of photosynthesis across a broad temperature range (Sage and McKown 2006). CAM plants are also adapted to hot temperatures; however, the response of $CO_2$ fixation to temperature has not been elucidated in this group. CAM plants are known, however, to have lower but heat-stable photosynthetic rates at higher temperatures (Downton et al. 1984; Lin et al. 2006). In an interesting review, Lüttge (2006) presented a monographic survey of the neotropical genus Clusia, the only genus of real (**dicotyledonous** or woody) trees performing CAM, indicating that Clusia presents extraordinary flexibility in carbon acquisition under variable environmental conditions (see also Lüttge 1988, 1995, 2002, and 2004).

In an experiment carried out on heat-tolerant (HT) and heat-sensitive (HS) tuber-bearing *Solanum* species (potatoes *Solanum bulbocastaunum, S. chacoense, S. demissum,* and *S. stoloniferum*), a decrease in $CO_2$ fixation rates and loss of leaf chlorophyll were noted with increasing temperature. These effects were most pronounced in the HS tubers, confirming high temperature as a limiting factor of the photosynthesis in plants (Reynolds et al. 1990).

Numerous studies have investigated the impact of high temperature on crop growth, production, and yield. For example, high temperatures were shown to significantly decrease yields for maize (Gaile and Arhipova 2015), wheat (Kobza and Edwards 1987; Nicolas et al. 1984; Nuttall et al. 2018), grapevine (Hendrickson et al. 2004), cotton (Reddy et al. 1991), beans (Pastenes and Horton 1996), and rice (Nagai and Makino 2009), among other crops.

The temperature swing does not need to be major to have a large impact on yield volume and quality. For example, a 1°C increase in average temperature will lead to an 8%–10% decrease in corn yield and a 9% decrease in rice yield (Abrol and Ingram 1996). Warmer temperatures can lead to decreased rates of carbohydrate accumulation in corn crops and consequently lower yields (Wolfe et al. 2017). Increasing temperatures can also lead to lower marketable yields due to disruption of pollination and fruit development. For example, the pollen of corn (*Zea mays*) has decreased viability in temperatures above 35°C (95°F), and kernel growth can be delayed in temperatures above 30°C (86°F) (Hatfield and Prueger 2015; Hatfield et al. 2011). However, the response of plants to high temperatures depends on the developmental stage of the plant, and all stages of vegetative development from germination to initiation of floral structures are affected by high temperature, particularly earlier stages (Chen et al. 1982; Paulsen 1994).

Although extensively investigated, the mechanisms behind heat stress and the responses of plant species are complex and remain enigmatic for many reasons, among them the genetic diversity of plants and their varied responses to high temperatures. Plants seem to react more strongly to large variations in temperature, rather than to absolute temperatures (Falk et al. 1996; Howarth 1991; Nievola et al. 2017). In other words, we need to understand whether the temperature change rate

---

[2] The cells in which C4 plants isolate Rubisco; see Chapter 6 in this volume.

(variation) or the temperature by itself (absolute) is sensed by plants (Minorsky 1989). The temperature change rate might be the determinant factor when the exposure to temperature change is short, or when plant response is rapid (Nordin Henriksson and Trewavas 2003; Plieth 1999), while the absolute temperature becomes the determining factor if exposure is longer (Zarka et al. 2003). Nonetheless, plants can show remarkable responses to small changes in temperatures, even though how this temperature signal is perceived, including the early components of the temperature signal transduction pathway, remains a mystery (Penfield 2008; Ruelland and Zachowski 2010).

### 5.2.1.2 Low Temperatures

Low temperatures are also known to limit plant growth, development, and agricultural productivity (Xin and Browse 2001). It is important to distinguish between chilling temperatures, which range from 0°C to 15°C (32°F–59°F), and freezing temperatures, which are lower than 0°C (32°F) (Adam and Murthy 2014). Low temperatures are frequently noted in many regions to damage plant species; however, sub-tropical and tropical plants are more sensitive to chilling temperature (0–10°C) and have low or no ability to adapt to cold stress, while plants of temperate regions are more tolerant to chilling, might acclimate to cold, and may even show some tolerance to freezing (Levitt 1980; Zhu et al. 2007).

Under chilling temperatures, the fluidity of the cell is impaired, affecting the physiology of plants. The chloroplast membrane is one of the organelles most rapidly and drastically affected during cold stress (Banerjee and Roychoudhur 2019; Yadav 2010; Zhang et al. 2016). Under freezing temperatures, ice crystals form in the cytosol (internal fluid of the cell), damaging the membrane and inhibiting and reducing many cellular and physiological processes including the light-capturing reactions of photosynthesis, $CO_2$ fixation, and chlorophyll biosynthesis (Ensminger et al. 2006).

The tolerance of plants to cold stress is diverse depending on reprogramming gene expression to modify their physiology, metabolism, and growth (Chinnusamy et al. 2010; Janská et al. 2010). As cold stress has historically been more common than heat stress, many plant species have developed some mechanisms to enhance their tolerance to the cold.

Cold tolerance requires gene expression reprogramming resulting in adjusted metabolism and cell structure. These adjustments are dependent on how the cold signal is transduced, with first cold stress signal perception, then transcriptional cascades through both ABA (Abscisic Acid)-dependent and ABA-independent pathways to induce the expression of cold-regulated (COR) genes (Heidarvand and Maali Amiri 2010). The expressions of these COR genes trigger the biosynthesis of diverse metabolites known to be cold-protective, like fructans, which stabilize various cell membranes to reduce cold and freezing damage and might be involved in the regulation of osmosis (Hincha et al. 2003; Pontis 1989; Tarkowski and Van den Ende 2015). Many soluble sugars are also upregulated by COR genes (Kaplan and Guy 2004; Rekarte-Cowie et al. 2008; Tarkowski and Van den Ende 2015; Uemura et al. 2006), as they can stabilize membranes and have also been reported to have an antioxidant power (Cherkas et al. 2020) by quenching reactive oxygen species (ROS) (Bolouri-Moghaddam et al. 2010).

ROS, generated at low levels during normal metabolism, become much more prevalent and problematic in cold weather; the increase in their production due to stress may even act as a signal for the upregulation of COR genes (Suzuki and Mittler 2006). Many photosynthetic proteins are downregulated during low temperature stress, while carbohydrates active in energy transfer, ROS removal, and cell wall restructuring are upregulated in the cold. Many intermediate photosynthetic metabolites, which increase in concentration due to lower photosynthetic efficiency in the cold, are also thought to act as signaling molecules that help regulate special COR genes (Hu et al. 2016; Janmohammadi et al. 2015).

Figure 5.2 provides a proposed framework for understanding the physiological changes that winter wheat undergoes in cold weather and using this information to promote cold tolerance in other crops.

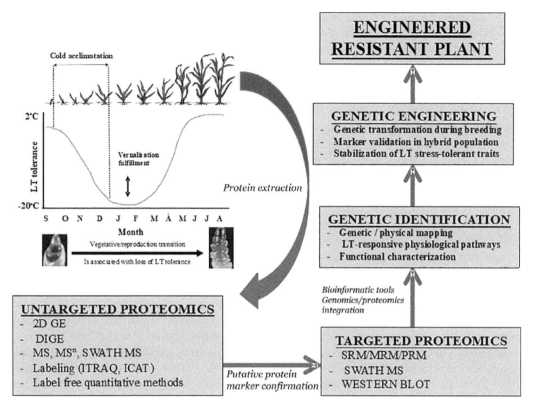

**FIGURE 5.2** A schematic framework for evaluating the proteome changes of winter wheat during the different growth stages and the next necessary steps for improving low temperature (LT) tolerance. The regulatory role of developmental stage progression on the expression of LT tolerance has been emphasized. Cool temperatures of autumn induce LT hardiness in wheat plants to protect them against freezing injury. Vernalization (exposure of seeds/plants to cold temperatures) gradually establishes capability for the vegetative/reproductive transition. Phenological development strictly controls level and duration of frost tolerance expression. Proteome investigations within a broad time frame and during the different phenological stages could provide important information on candidate proteins, biological processes, and mechanisms controlling LT tolerance. Consequently, conversions of targeted proteomics information to genomics data by bioinformatic tools facilitate the identification of LT-responsive quantitative trait loci (QTLs), genes, and alleles. This could provide the basis for breeding programs and for releasing cultivars with a well-timed switch from vegetative to reproductive growth. *Abbreviations*: 2D GE, two-dimensional gel electrophoresis; DIGE, difference gel electrophoresis; MSn, multi-stage mass spectrometry; SWATH MS, the definition of this method is given in Gillet et al. (2012); iTRAQ, isobaric tags for relative and absolute quantitation; ICAT, isotope-coded affinity tag; SRM, single reaction monitoring; MRM, multiple reaction monitoring; PRM, parallel reaction monitoring. (Modified from Elsevier: Janmohammadi, M., L. Zolla, and S. Rinalducci. 2015. Low temperature tolerance in plants: Changes at the protein level. *Phytochemistry* 117:76–89. Caption edited for length and clarity.)

### 5.2.1.3  Long-Term Effects of Extreme Temperatures

The responses of plants to past climate changes indicate that migration of plant communities was the most frequent reaction, rather than adaptation of species to the new local temperature paradigm (Huntley 1991). Overall, extensive literature is readily available on the effects of extreme temperatures on horticultural crops, cereals, and legume crops. Indeed, the impacts of hostile temperatures on plants, their adaptation, and their resilience are very important aspects of climate change risk assessment. However, further research is still required in order to fully understand how extreme temperatures affect yield and productivity of crops. Further information is also needed on the direction and degree of the impact of extreme temperature on yields and yield components. This information might contribute significantly to the improvement of crop models and will help with scoping appropriate adaptation options, including breeding and molecular engineering (see Chapter 6 in this volume; Luo et al. 2011).

### 5.2.2  $CO_2$ Concentration as a Limiting Factor

Carbon dioxide gas ($CO_2$) is necessary to the process of photosynthesis and hence to carbon fixation and energy storage in terrestrial ecosystems. When the limited amount of atmospheric $CO_2$ increases, plants experience increases in other resource use efficiencies, such as response to water and nutrient uptake. For example, immediate increases are seen in water use efficiency (WUE) and instantaneous transpiration efficiency (ITE) when atmospheric $CO_2$ can be enriched, ranging from no effect (Field et al. 1997) to +180% efficiency (De Luis et al. 1999). The observed increase of WUE with greater $CO_2$ is greater in plants under drought (Arp et al. 1998; Field et al. 1997). While the immediate change can be dramatic, these efficiencies may decline again over time as $CO_2$ levels remain high (Nogueira et al. 2004; Norby et al. 2001).

Prolonged exposure to $CO_2$ has been shown to elicit acclimatization mechanisms in plants, such as changes in key enzymes involved in the photosynthetic carbon reduction cycle, leading to an increase of nutrient use efficiency and decreased leaf conductance (net intake of $CO_2$) (Vico et al. 2013). The responses of leaf conductance and photosynthetic rate were found to be highly correlated in wheat, but the response varies over time at high temperatures (Del Pozo et al. 2005).

High levels of $CO_2$ have also been reported to enhance the carbon to nitrogen ratio (C/N) in plant tissues due to negative relations between WUE, the root mass ratio, and nitrogen use efficiency (NUE). That is, in high $CO_2$ concentrations, photosynthetically active volumes of $CO_2$ can be taken in more quickly, allowing the stomata to be closed more often, reducing water loss and the flow of water and nutrients from roots to leaves (McDonald et al. 2002). This results in large amounts of C fixation relative to N uptake. However, WUE was significantly enhanced with supplemental N application under high $CO_2$ concentration, with greater N application leading to further enhanced WUE, while greatest increase of NUE under high $CO_2$ was noted with lower N application (Hunsaker et al. 2000; Li et al. 2003). Thus, to maintain protein levels for important plant-based proteins, it might be necessary to increase the amount of nitrogen provided for fields, which could have serious negative effects on the global environment (see Chapter 9 in this volume).

Greater carbon fixation under global warming and rising $CO_2$ levels may increase vegetative growth of some plants but will not necessarily equate to higher agricultural yields. The responses of crop yields to high $CO_2$ have been actively researched; yields for some crops such as wheat and soybeans could potentially increase by 30% or more under a doubling of $CO_2$ concentrations, but yields for other major crops such as maize exhibit less response, and yield could potentially increase by only 10% (Drake et al. 1997). For some crops, global warming may substantially reduce yields; in warmer environments, for example, crops tend to grow faster and in cereals this extra growth may reduce the seed filling period, increase maturation speed, and thus reduce yields (Ferris et al. 1998; Stone and Nicolas 1995). Peng et al. (2004) analyzed weather data over 24 years (1979–2003) to assess the relationship between rice yield and temperatures, finding that grain yield decreased by 10% for each 1°C increase.

## 5.3  WHERE METEOROLOGICAL AND LAND FACTORS MEET: WATER AS A LIMITING FACTOR

From the turn of the 21st century, water shortage has been significantly exacerbated in many regions of the world, which have been affected by recurrent and consecutive droughts and rainfall records significantly below average. Besides extreme temperature, water unavailability is the second most important environmental factor that negatively affects plant growth and development, and thus crop production and yield (Bray 2007a). Recent studies are predicting that water deficit or drought will increase in severity alongside rising temperatures and that by the year 2100 the availability of water will have a substantial impact on our ability to produce crops. The effects of climate change will be reflected by either acute or chronic impacts associated with variable precipitation events and longer periods of drought. Africa will be among the most affected regions, with yields of major crops decreasing by more than 50% in 2050 and by up to 90% in 2100, relative to current yields (IPCC 2019; Li et al. 2009).

Agricultural drought is a disaster resulting from a deficiency in the available moisture in the soil, and even if of a short duration and low intensity, the reduction of moisture essential for plant growth and development results in low yields (Nagarajan 2009).

To address this challenge, improving WUE of crops is considered imperative and needs to be addressed urgently, as increased WUE is seen as one of the most important solutions in addressing water scarcity and drought resistance (Eslick and Hockett 1974; Hamdy et al. 2003; Tuberosa and Salvi 2006). There is a pressing need to improve WUE of both rain-fed and irrigated crops, and breeding new varieties with optimal WUE by using either conventional breeding or molecular engineering seems to be the most environmentally friendly and sustainable solution to face water shortage and drought in the future (Chaerle et al. 2005). However, prior to developing new crops or improving WUE of existing plants, we need to understand and decipher all the existing mechanisms developed by plants to face drought and their strategies to survive during short- and long-term dry periods. Indeed, plants encounter many unfavorable growth conditions, including drought, which limit their growth and development and have evolved many physiological responses to these (Krasensky and Jonak 2012).

Under water scarcity, both cellular (Mullet and Whitsitt 1996) and whole plant (Navari-Izzo and Rascio 1999) functions are disrupted, causing retardation of plant growth and reproduction, and yields can be reduced by c.a. 69% when plants are exposed to water deficit conditions in the field (Boyer 1982; Dodd and Ryan 2016).

From the physiological point of view, mechanisms to sense water deficit and the signal transduction following events are not fully understood. However, the reports suggest that water deficit causes a decrease in turgor pressure, and it is established that the plant hormone ABA is one intermediary in the signaling pathway (Nambara and Marion-Poll 2005), since numerous studies of plants in water deficit reported the increase of ABA content and its accumulation in all plant organs. This accumulation of ABA plays a major role in plant responses to water deficit, both physiological responses like closure of the stomata (Bauer et al. 2013; Bradford and Hsiao 1982; Bray 2007b) and molecular responses like the accumulation of osmoregulators such as fructans and proline (see Figure 5.3) (Bartels and Souer 2003; Bray 1993). As in temperature stress, excess ROS production is also one of the responses to drought, helping trigger the defense reaction in plants (Hasanuzzaman et al. 2014; Hussain et al. 2019).

Water deficit triggers stomata closure (Dodd 2013; Holbrook et al. 2002; Tardieu et al. 2010), hence, uptake of carbon dioxide is reduced, limiting the carbon assimilation rate of the plants (Chaves 1991). When water deficit persists for long periods, stomatal closure may reduce crop production and cause injury to the chloroplasts through the process of photoinhibition (Kaiser 1987; Navari-Izzo and Rascio 1999). Furthermore, water deficit may also interact with and be accentuated by other stresses, such as high temperatures and the reduction of transpiration (hence nutrient uptake) by closing or reducing stomata opening (Cohen et al. 2021; Haworth et al. 2019; Impa et al. 2019; Vile et al. 2012). ABA has been shown to play a major role in regulating the accumulation of essential proteins during a combination of water deficit and heat stress (Zandalinas et al. 2016).

Extensive literature has reported the limiting effects of drought and water deficit on growth and yield of crops. To give a few examples, water deficit was reported to be a limiting factor to the growth of coffee (Ribeiro et al. 2018), cotton (Zhao and Oosterhuis 1997), sunflower (Andrianasolo

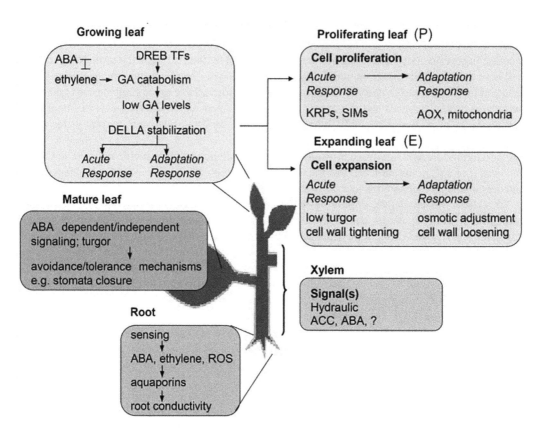

**FIGURE 5.3** Schematic representation of processes involved in growth regulation upon drought. Soil drying is sensed in roots activating a combination of hydraulic and chemical signals that are transported through the xylem to leaves where they initiate a number of tolerance mechanisms. Response of mature leaves can be described by the avoidance/tolerance model. In growing leaves stress leads to acute growth inhibition followed by growth adaptation both mediated by DELLA signaling. Ethylene promotes DELLA stabilization and growth inhibition while ABA is possibly involved in growth recovery. Negative ABA and ethylene cross-talk was demonstrated. While hormonal signaling is common between expanding (*E*) and proliferating (*P*) leaves, effector genes are distinct. In *P* leaves, inhibitors of CDKA might play a role in acute response while alternative respiration in growth adaptation. In expanding leaves, cell wall tightening and changes in cell turgor lead to growth cessation, while osmotic adjustment and cell wall loosening are important for growth adaptation. (Modified from Elsevier: Skirycz, A., and D. Inzé. 2010. More from less: plant growth under limited water. *Current Opinion in Biotechnology* 21:197–203. Caption edited for spelling.)

et al. 2016), corn (Claassen and Shaw 1970), and rice (Arbrecht and Carberry 1993; Lilley and Fukai 1994), among other crops. Higher temperatures and reduced water availability from effective precipitation increase water loss from transpiration (crops) and evapotranspiration (soils), increasing water requirements of plants (Mimi and Jamous 2010). Thus, increasing water deficits undoubtedly have adverse impacts on plant growth, productivity, and yields. Although water deficit impacts on plants and crops have been extensively investigated, it is of key importance to investigate further how water deficit and warm environments combine with $CO_2$ enrichment to affect plant growth and agricultural water use, especially irrigation requirements, given that irrigated land produces about 40% of the global harvest (Zhang and Cai 2013). Furthermore, it is also needed to understand further the complex mechanisms of response and resistance of crops to drought and their ability to compensate these effects, by investigating further the roles of osmotic regulators, photosynthetic metabolites, and endogenous hormones under water stress.

## 5.4  LIMITING LAND FACTORS: NUTRIENT SUPPLY

Soil is the source of macronutrients and micronutrients, essential elements for plant growth and development. In cultivated systems, nutrient availability can pose a real problem, since soils are continuously exploited, and even often over-exploited (used until the naturally occurring nutrients are depleted) (Kulcheski et al. 2015). Nitrogen (N), phosphorus (P), and potassium (K) are considered the three most important limiting minerals in the soil and should frequently be added in order to maintain fertility and a balanced soil chemistry (Jiang et al. 2006; Manna et al. 2007; Zhang et al. 2018). The lack of NPK, as they are abbreviated together, has been reported to impair yield of numerous crops such as potato (Yourtchi et al. 2013), maize (Gul et al. 2015), cassava (Adiele et al. 2020), sunflower (Nawaz et al. 2012), rice (Moe et al. 2019), mustard (Singh et al. 2009), tomato (Hebbar et al. 2004), carrot (Agbede et al. 2017), banana (Al-Harthi and Al-Yahyai 2009), Jerusalem artichoke (Matias et al. 2013), and pumpkin (Abayomi et al. 2012), among others. Although extensive literature has shown that NPK fertilizers are necessary for enhancing crop yields and sustaining soil fertility, inappropriate application does not guarantee increasing yields, and might cause low nutrient use efficiency and environmental problems in agroecosystems (see Chapter 9 in this volume). Therefore, balanced application of NPK is essential for producing top quality crops in high yields while maintaining environmental sustainability (Yadav et al. 2000; Yousaf et al. 2017).

### 5.4.1  NITROGEN (N)

Nitrogen is a major nutrient and essential mineral element for all living organisms and is considered the major limiting factor in terrestrial primary production, including agroecosystems. Acquisition and assimilation of nitrogen is necessary for plant growth and development (Newbould 1989) and is especially important for the construction of proteins. As a result, the production of high-quality and protein-rich plant products is extremely dependent upon the availability of nitrogen (Vance 1997). Indeed, many major crops, particularly cereals, respond favorably to high applications of nitrogen fertilizers. As a result, farmers have been ever increasing the use of chemical nitrogen fertilizer since it became available in the last century (see Chapter 9 in this volume).

Hence, one of the driving forces behind agricultural sustainability is effective management of nitrogen in the environment. Researchers are tackling the problems of how to improve soil fertility and N management for different types of climates, soils, and crop conditions around the world (Rütting et al. 2018). This includes the breeding of plants with increased yields and NUE under N application as well as investigating how to reduce N losses to the environment in the short and long term (Anas et al. 2020; Hirel et al. 2011; Sharma and Bali 2018). One of the most judicious methods to manage nitrogen is thought to be Biological Nitrogen Fixation (BNF) by soil microbes (Bohlool et al. 1992; Vance and Graham 1994). Many diverse associations contribute to symbiotic $N_2$ fixation (Sprent 1984), but in agricultural settings 80% of the biologically available nitrogen is fixed by soil bacteria, namely *Rhizobium, Bradyrhizobium, Sinorhazobium,* and the *Azorhizobium*-legume symbiosis (Vance 1996). This symbiotically fixed nitrogen may become available to crops through several methods, including the transduction of fixed N to plants that lack N-fixing bacterial symbioses via mycorrhizal connections (Peoples et al. 1995).

### 5.4.2  PHOSPHORUS (P)

Phosphorus is considered the second most important mineral for crops, and its role in soil fertility and crop nutrition has become obvious from long-term field experiments (Poulton and Johnston 2019). However, in contrast to nitrogen, P is present as mineral deposits in soil, often inaccessible to plants. Furthermore, P released from deposits quickly becomes immobile as it reacts with ions such as calcium and iron in soil. Soluble phosphorus fertilizers, once applied, largely revert to less soluble or insoluble forms, so that only 15%–20% of the applied quantity is available to crops (Prasad and Power 1997; Roberts and Johnston 2015). Therefore, to reach the requirements of continually cropped plants requires the addition of a large excess of P fertilizers (Loneragan 1997). To optimize plant P-use efficiency, the amount of P fertilizer applied should equal that amount taken up by the crop (Johnston et al. 2014; Poulton and Johnston 2019).

There is a strong relationship between P availability and crop yield. Cereal and forage crops especially require high levels of soil phosphorus, and because of the mineral's relative immobility, it is necessary to thoroughly mix P fertilizer with the soil to bring rooting zone P into the optimum level for crops (Brady 1990). However, phosphorus leaching occurs when the sorption capacity (P binding capacity) of soils is 17%–38% saturated (Holford 1997), so it can be frustrating to determine the correct levels of P fertilizer to apply without leading to negative environmental effects.

In many developing countries, farmers have limited access to P fertilizer, and the development of new varieties more efficient at low soil P is needed in order to minimize inefficient use of P-inputs and to reduce potential for loss of P to the environment (Richardson et al. 2011). Therefore, farmers should improve fertilizer P use by implementing fertilizer best management practices within the context of "4Rs": application of the *right nutrient source*, applied at the *right rate*, *right time*, and in the *right place* (Flis 2017, 2018; Johnston and Bruulsema 2014; Roberts and Johnston 2015). To achieve this goal, three strategies may improve P-use efficiency: (1) breeding or selecting plants with root-foraging strategies that improve P acquisition by lowering the critical P requirement of plant growth and allowing agriculture to operate at lower levels of soil P; (2) P-mining strategies to enhance the desorption, solubilization or mineralization of P from sparingly available sources in soil using root exudates (organic anions, phosphatases), and (3) improving internal P-utilization efficiency through the use of plants that yield greater harvest per unit of P uptake (Richardson et al. 2011). Manure application may also be an effective means of sustainably increasing levels of plant-available P. Organic compounds released during the decomposition of manures increase the availability of P from soil or fertilizers (Iyamuremye and Dick 1996), and this positive effect of manure was ascribed to mobilization of native soil phosphorus and improved physico-chemical properties of the soil by manure use (Reddy et al. 1999).

### 5.4.3  POTASSIUM (K)

Potassium, the third-most limiting nutrient for plant growth, is important for photosynthesis and the formation of amino acids and proteins. It is present in much larger quantities in soils than N or P and has the particularity to occur in inorganic (usable) form and be fairly well distributed throughout the soil profile (Prasad and Power 1997). Potassium is present in soils in several forms (Mulder 1950; Tisdale 1985; Sparks 1987):

- Solid mineral form (unavailable to plants)
- Non-exchangeable (slowly available to plants; acts as a storage form in soils)
- Exchangeable (attached to **colloids** [soil particles] and available to plants)
- In soil solution (available to plants)

In most soils, the exchangeable potassium adsorbed by clay or humus colloids is of much more direct importance for plant nutrition than that derived directly from soil minerals (Mulder 1950). Although little is known on potassium compared to the extensive literature existing on nitrogen and phosphorus, K is involved in numerous physiological functions related to plant health and tolerance to biotic and abiotic stress. Potassium deficiencies cause poor growth, lost yield, and reduced fiber quality of several crops such as tomato, bell pepper, and potato (Bidari and Hebsur 2011; Oosterhuis et al. 2014), although genotypic difference in crop species means that there is a range of potassium nutritional needs (Glass and Perley 1980).

# REFERENCES

Abadía, J., F. Morales, and A. Abadía. 1999. Photosystem II efficiency in low chlorophyll, iron-deficient leaves. *Plant and Soil* 215:183–92.

Abayomi, Y. A., M. O. Aduloju, M. A. Egbewunmi, and B. O. Suleiman. 2012. Effects of soil moisture contents and rates of NPK fertilizer application on growth and fruit yields of pepper (*Capsicum* spp.) genotypes. *International Journal of AgriScience* 2:651–63.

Abrol, Y. P. and K. T. Ingram. 1996. Effects of higher day and night temperatures on growth and yields of some crop plants. In *Global Climate Change and Agricultural Production: Direct and Indirect Effect of Changing Hydrological, Pedological and Plant Physiological Processes*, eds. F. A. Bazzaz, and W. G. Sombroek, 123–40. New York: Food and Agriculture Organization.

Adam, S. and S.D.S. Murphy. 2014. Effect of cold stress on photosynthesis of plants and possible protection mechanisms. In *Approaches to Plant Stress and Their Management*, eds. R.K. Gaur, and P. Sharma, 219–26. New Delhi: Springer.

Adiele, J. G., A. G. T. Schut, R.P.M. van den Beuken, et al. 2020. Towards closing cassava yield gap in West Africa: Agronomic efficiency and storage root yield responses to NPK fertilizers. *Field Crops Research* 253:107820. https://doi.org/10.1016/j.fcr.2020.107820.

Agbede, T. M., A. O. Adekiya, and E. K. Eifediyi. 2017. Impact of poultry manure and NPK fertilizer on soil physical properties and growth and yield of carrot. *Journal of Horticultural Research* 25:81–8.

Ahammed, G. J, W. Xu, A. Liu, and S. Chen. 2018. COMT1 Silencing aggravates heat stress-induced reduction in photosynthesis by decreasing Chlorophyll content, photosystem II activity, and electron transport efficiency in tomato. *Frontiers in Plant Science* 9:998. https://doi.org/10.3389/fpls.2018.00998.

Al-Harthi, K. and R. Al-Yahyai. 2009. Effect of NPK fertilizer on growth and yield of banana in Northern Oman. *Journal of Horticulture and Forestry* 8:160–7.

Anas, M., F. Liao, K. K. Verma, et al. 2020. Fate of nitrogen in agriculture and environment: Agronomic, eco-physiological and molecular approaches to improve nitrogen use efficiency. *Biological Research* 53:47. https://doi.org/10.1186/s40659-020-00312-4.

Andrianasolo, F. N., P. Casadebaig, N. Langlade, P. Debaeke, and P. Maury, P. 2016. Effects of plant growth stage and leaf aging on the response of transpiration and photosynthesis to water deficit in sunflower. *Functional Plant Biology* 43:797–805.

Arbrecht, D. G. and P. S. Carberry. 1993. The influence of water deficit prior to tassel initiation on maize growth, development and yield. *Field Crops Research* 31:55–69.

Arp, W. J., J. E. M. Van Mierlo, F. Berendse, and W. Snijders. 1998. Interactions between elevated $CO_2$ concentration, nitrogen and water: Effects on growth and water use of six perennial species. *Plant, Cell & Environment* 21:1–11.

Assad, E. D., R. R. R. Ribeiro, and A. M. Nakai. 2019. Assessments and how an increase in temperature may have an impact on agriculture in Brazil and mapping of the current and future situation. In *Climate Change Risks in Brazil*, eds. C. Nobre, J. Marengo, and W. Soares, 31–65. Tiergartenstrasse: Springer Nature.

Banerjee, A. and A. Roychoudhur. 2019. Cold stress and photosynthesis. In *Photosynthesis, Productivity and Environmental Stress*, eds. P. Ahmad, M. A. Ahanger, M. N. Alyemeni, and P. Alam, 27–37. London: Wiley & Sons.

Bartels, D. and E. Souer. 2003. Molecular responses of higher plants to dehydration. In *Plant Responses to Abiotic Stress. Topics in Current Genetics*, vol. 4, eds. H. Hirt, and K. Shinozaki, 9–38. Berlin: Springer.

Bauer, H., P. Ache, S. Lautner, et al. 2013. The stomatal response to reduced relative humidity requires guard cell-autonomous ABA synthesis. *Current Biology* 23:53–7.

Bernacchi, C. J., J. E. Bagley, S. P. Serbin, U. M. Ruiz-Vera, D. M. Rosenthal, and A. Vanloocke. 2013. Modelling C3 photosynthesis from the chloroplast to the ecosystem. *Plant, Cell and Environment* 36:1641–57.

Berry, J. A. and O. Björkman. 1980. Photosynthetic response and adaptation to temperature in higher plants. *Annual Reviews in Plant Physiology* 31:491–543.

Bidari, B. I. and N. S. Hebsur. 2011. Potassium in relation to yield and quality of selected vegetable crops. *Karnataka Journal of Agricultural Science* 24:55–9.

Bita, C. E. and T. Gerats. 2013. Plant tolerance to high temperature in a changing environment: Scientific fundamentals and production of heat stress-tolerant crops. *Frontiers in Plant Science* 4:273. https://doi.org/10.3389/fpls.2013.00273.

Bohlool, B. B., J. K. Ladha, D. P. Garrity, and T. George. 1992. Biological nitrogen fixation for sustainable agriculture: A perspective. *Plant and Soil* 141:1–11.

Bolouri-Moghaddam, M. R., K. Le Roy, L. Xiang, F. Rolland, and W. Van den Ende. 2010. Sugar signalling and antioxidant network connections in plant cells. *FEBS Journal* 277:2022–37.

Boyer, J. S. 1982. Plant productivity and environment. *Science* 218:443–8.

Bradford, K. J. and T. C. Hsiao. 1982. Physiological responses to moderate water stress. In *Physiological Plant Ecology II. Encyclopedia of Plant Physiology (New Series)*, vol. 12/B, eds. O. L. Lange, P. S. Nobel, C. B. Osmond, and H. Ziegler, 263–324. Berlin: Springer.

Brady, N. C. 1990. *The Nature and Properties of Soils*. 10th Ed. New York: John Wiley & Sons.

Bray, E. A. 2007a. *Plant Response to Water-Deficit Stress. Encyclopaedia of Life Sciences*, 1–7. Chichester: John Wiley & Sons.

Bray, E. A. 2007b. Molecular and physiological responses to water-deficit stress. In *Advances in Molecular Breeding Toward Drought and Salt Tolerant Crops*, eds. M.A. Jenks, P. M. Hasegawa, and S. M. Jain, 121–40. Dordrecht: Springer.

Bray, E. A. 1993. Molecular responses to water deficit. *Plant Physiology* 103:1035–40.

Chaerle, L., N. Saibo, and D. Van Der Straeten. 2005. Tuning the pores: Towards engineering plants for improved water use efficiency. *Trends in Biotechnology* 23:308–15.

Chaves, M. M. 1991. Effects of water deficits on carbon assimilation. *Journal of Experimental Botany* 42:1–16.

Chen, H. H., Z. Y. Shen, and P. H. Li. 1982. Adaptability of crop plants to high temperatures stress. *Crop Science* 22:719–25.

Cherkas, A., S. Holota, T. Mdzinarashvili, R. Gabbianelli, and N. Zarkovic. 2020. Glucose as a major antioxidant: When, what for and why it fails? *Antioxidants* 9:140. https://doi.org/10.3390/antiox9020140.

Chinnusamy, V., J. K. Zhu, and R. Sunka. 2010. Gene regulation during cold stress acclimation in plants. In *Plant Stress Tolerance. Methods in Molecular Biology*, ed. R. Sunkar, 39–55. Totowa: Humana Press.

Claassen, M. M. and R. H. Shaw. 1970. Water deficit effects on corn. II. Grain components. *Agronomy Journal* 62:652–5.

Cohen, I., S. I. Zandalinas, F. B. Fritschi, et al. 2020. The impact of water deficit and heat stress combination on the molecular response, physiology and seed production of soybean. *Physiologia Plantarum* 172:41–52.

Crafts-Brandner, S. J. and M. E. Salvucci. 2004. Analyzing the impact of high temperature and $CO_2$ on net photosynthesis: Biochemical mechanisms, models and genomics. *Field Crops Research* 90:75–85.

De Luis, I., J. J. Irigoyen, and M. Sanchez-Diaz. 1999. Elevated $CO_2$ enhances plant growth in droughted N2-fixing alfalfa without improving water status. *Physiologia Plantarum* 107:84–9.

Del Pozo, A., P. Pérez, R. Morcuende, A. Alonso, and R. Martínez-Carrasco. 2005. Acclimatory responses of stomatal conductance and photosynthesis to elevated $CO_2$ and temperature in wheat crops grown at varying levels of N supply in a Mediterranean environment. *Plant Science* 169:908–16.

Dodd, I. C. 2013. Abscisic acid and stomatal closure: A hydraulic conductance conundrum? *New Phytologist* 197:6–8.

Dodd, I. C and A. C. Ryan. 2016. *Whole-Plant Physiological Responses to Water-Deficit Stress*. Chichester: John Wiley & Sons.

Downton, W. J. S., J. A. Berry, and J. R. Seemann. 1984. Tolerance of photosynthesis to high temperature in desert plants. *Plant Physiology* 74:786–90.

Drake, B. G., M. A. Gonzàlez-Meler, and S. P. Long. 1997. More efficient plants: A Consequence of rising atmospheric $CO_2$? *Annual Reviews in Plant Physiology and Plant Molecular Biology* 48:609–39.

Dutta, S., S. Mohanty, and B. C. Tripathy. 2009. Role of temperature stress on chloroplast biogenesis and protein import in pea. *Plant Physiology* 150:1050–61.

Efeoglu, B. and S. Terzioglu. 2009. Photosynthetic responses of two wheat varieties to high temperature. *EurAsian Journal of BioScience* 3:97–106.

Ensminger, I., F. Bosch, and N. P. A. Huner. 2006. Photo stasis and cold acclimation: Sensing low temperature through photosynthesis. *Physiologia Plantarum* 126:28–44.

Eslick, R. F. and E. A. Hockett. 1974. Genetic engineering as a key to water-use efficiency. *Agricultural Meteorology* 14:13–23.

Falk, S., D. P. Maxwell, D. E. Laudenbach, and N. P. A. Huner. 1996. Photosynthetic adjustment to temperature. In *Photosynthesis and the Environment. Advances in Photosynthesis and Respiration*, ed. N. R. Baker, 367–85. Dordrecht: Springer.

FAO. 2017. *The Impact of Disasters and Crises on Agriculture and Food Security*. Rome: FAO. https://www.fao.org/3/I8656EN/i8656en.pdf (accessed June 17, 2021).

Ferris, R., R. H. Ellis, T. R. Wheeler, and P. Hadley. 1998. Effect of high temperature stress at anthesis on grain yield and biomass of field-grown crops of wheat. *Annals of Botany* 82:631–9.

Field, C. B., C. P. Lund, N. R. Chiariello, and B. E. Mortimer. 1997. $CO_2$ effects on the water budget of grassland microcosm communities. *Global Change Biology* 3:197–206.

Flis, B. S. 2017. Phosphorus management research and the 4Rs. *Crops & Soils* 55:28–67.

Flis, B. S. 2018. 4R history and recent phosphorus research. *Crops & Soils* 51:36–47.

Gaile, Z. and I. Arhipova. 2015. Influence of meteorological factors on maize performance in Latvia. *Proceedings of the Latvian Academy of Sciences B* 69:68–76.

Gillet, L. C., P. Navarro, S. Tate, et al. 2012. Targeted data extraction of the MS/MS spectra generated by data-independent acquisition: A new concept for consistent and accurate proteome analysis. *Molecular and Cellular Proteomics* 11:016717. https://doi.org/10.1074/mcp.O111.016717.

Glass, A. D. M. and J. E. Perley. 1980 Varietal differences in potassium uptake by barley. *Plant Physiology* 65:160–4.

Gujjar, R.S., P. Banyen, W. Chuekong, et al. 2020. Synthetic cytokinin improves photosynthesis in rice under drought stress by modulating the abundance of proteins related to stomatal conductance, chlorophyll contents, and rubisco activity. *Plants* 9:1106. https://doi.org/10.3390/plants9091106.

Gul, S., M. H. Khan, B. A. Khanday, and S. Nabi. 2015. Effect of sowing methods and NPK levels on growth and yield of rainfed maize (*Zea mays* L.). *Scientifica* 2015:198575. https://doi.org/10.1155/2015/198575.

Hamdy, A., R. Ragab, and E. Scarascia-Mugnozza. 2003. Coping with water scarcity: Water saving and increasing water productivity. *Irrigation and Drainage* 52:3–20.

Hasanuzzaman, M., K. Nahar, S. S. Gill, and M. Fujita. 2014. Drought stress responses in plants, oxidative stress, and antioxidant defense. In *Climate Change and Plant Abiotic Stress Tolerance*, eds. N. Tuteja, and S. S. Gill, 209–49. Berlin: Wiley-VCH Verlag GmbH.

Hatfield, J. L. and J. H. Prueger. 2015. Temperature extremes: Effect on plant growth and development. *Weather and Climate Extremes* 10:4–10.

Hatfield, J. L., K. J. Boote, B. A. Kimball, et al. 2011. Climate impacts on agriculture: Implications for crop production. *Agronomy Journal* 103:351–70.

Haworth, M., G. Marino, C. Brunetti, D. Killi, A. De Carlo, and M. Centritto. 2018. The impact of heat stress and water deficit on the photosynthetic and stomatal physiology of olive (*Olea europaea* L.) – A case study of the 2017 heat wave. *Plants* 7:76. https://doi.org/10.3390/plants7040076.

Hebbar, S., B. Ramachandrappa, H. Nanjappa, and M. Prabhakar. 2004. Studies on NPK drip fertigation in field grown tomato (*Lycopersicon esculentum* Mill.). *European Journal of Agronomy* 21:117–27.

Heidarvand, L. and R. Maali Amiri. 2010. What happens in plant molecular responses to cold stress? *Acta Physiologiae Plantarum* 32:419–31.

Hendrickson, L., M. C. Ball, J. T. Wood, W. S. Chow, and R. T. Furbank. 2004. Low temperature effects on photosynthesis and growth of grapevine. *Plant, Cell and Environment* 27:795–809.

Hikosaka K., K. Ishikawa, A. Borjigidai, O. Muller, and Y. Onoda. 2006. Temperature acclimation of photosynthesis: Mechanisms involved in the changes in temperature dependence of photosynthetic rate. *Journal of Experimental Botany* 57:291–302.

Hincha, D. K., E. Zuther, and A. G. Heyer. 2003. The preservation of liposomes by raffinose family oligosaccharides during drying is mediated by effects on fusion and lipid phase transitions. *Biochimica Biophysica Acta* 1612:172–7.

Hirel, B., T. Tétu, P. J. Lea, and F. Dubois. 2011. Improving nitrogen use efficiency in crops for sustainable agriculture. *Sustainability* 3:1452–85.

Holbrook, N. M., V. R. Shashidhar, R. A. James, and T. Munns. 2002. Stomatal control in tomato with ABA-deficient roots: Response of grafted plants to soil drying. *Journal of Experimental Botany* 53:1503–14.

Holford, I. C. R. 1997. Soil phosphorus, its measurements and its uptake by plants. *Australian Journal of Soil Research* 35:227–39.

Howarth, C. J. 1991. Molecular responses of plants to an increased incidence of heat shock. *Plant, Cell & Environment* 14:831–41.

Hu, Z., J. Fan, Y. Xie, et al. 2016. Comparative photosynthetic and metabolic analyses reveal mechanism of improved cold stress tolerance in bermudagrass by exogenous melatonin. *Plant Physiology and Biochemistry* 100:94–104.

Hunsaker, D. J., B. A. Kimball, P. J. J. Jr. Pinter, et al. 2000. $CO_2$ enrichment and soil nitrogen effects on wheat evapotranspiration and water use efficiency. *Agricultural and Forest Meteorology* 104:85–105.

Huntley, B. 1991. How plants respond to climate change: Migration rates, individualism and the consequences for plant communities. *Annals of Botany* 67:15–22.

Hussain, S., M. J. Rao, M. A. Anjum, et al. 2019. Oxidative stress and antioxidant defense in plants under drought conditions. In *Plant Abiotic Stress Tolerance*, eds. M. Hasanuzzaman, K. Hakeem, K. Nahar, and H. Alharby, 207–19. New Delhi: Springer.

Impa, S. M., R. Perumal, S. R. Bean, V. S. J. Sunoj, and S. V. L. Krishna. 2019. Water deficit and heat stress induced alterations in grain physico-chemical characteristics and micronutrient composition in field grown grain sorghum. *Journal of Cereal Science* 86:124–31.

IPCC. 2019. Climate Change and Land. An IPCC Special Report on climate change, desertification, land degradation, sustainable land management, food security, and greenhouse gas fluxes in terrestrial ecosystems. *Summary for Policymakers.* https://www.ipcc.ch/site/assets/uploads/2019/08/Edited-SPM_ Approved_Microsite_FINAL.pdf (accessed: September 14, 2020).

Iyamuremye, E. and R. P. Dick. 1996. Organic amendments and phosphorus sorption by soils. *Advances in Agronomy* 56:139–85.

Janmohammadi, M., L. Zolla, and S. Rinalducci. 2015. Low temperature tolerance in plants: Changes at the protein level. *Phytochemistry* 117:76–89.

Janská, A., P. Maršík, S. Zelenková, and J. Ovesná. 2010. Cold stress and acclimation – what is important for metabolic adjustment? *Plant Biology* 12:395–405.

Jiang, D., H. Hengsdijk, T. B. Dai, TW. de Boer, Q. Jing, and W. X. Cao. 2006. Long-term effects of manure and inorganic fertilizers on yield and soil fertility for a winter wheat-maize system in Jiangsu, China. *Pedosphere* 16:25–32.

Johnston, A. M., T. W. Bruulsema. 2014. 4R Nutrient stewardship for improved nutrient use efficiency. *Procedia Engineering* 83:365–70.

Johnston, A. E., P. R. Poulton, P. E. Fixen, and D. Curtin. 2014. Phosphorus. *Advances in Agronomy* 123: 177–228.

Kaiser, W. M. 1987. Effects of water deficit on photosynthetic capacity. *Physiologia Plantarum* 71:142–9.

Kaplan, F. and C. L. Guy. 2004. b-Amylase induction and the protective role of maltose during temperature shock. *Plant Physiology* 135:1674–84.

Kobza, J. and G. E. Edwards. 1987. Influences of leaf temperature on photosynthetic carbon metabolism in wheat. *Plant Physiology* 83:69–74.

Krasensky, J. and C. Jonak. 2012. Drought, salt, and temperature stress-induced metabolic rearrangements and regulatory networks. *Journal of Experimental Botany* 63:1593–608.

Kulcheski, F. R., R. Côrrea, I. A. Gomes, J. C. de Lima, and R. Margis. 2015. NPK macronutrients and microRNA homeostasis. *Frontiers in Plant Science* 6:451. https://doi.org/10.3389/fpls.2015.00451.

Larcher, W. 1995. *Physiological Plant Ecology: Ecophysiology and Stress Physiology of Functional Groups.* 3rd ed. Berlin: Springer-Verlag.

Levitt, J. 1980. Responses of plants to environmental stress. vol. 1: *Chilling, Freezing, and High Temperature Stress.* New York: Academic.

Li, F., S. Kang, J. Zhang, and S. Cohen. 2003. Effects of atmospheric $CO_2$ enrichment, water status and applied nitrogen on water- and nitrogen-use efficiencies of wheat. *Plant and Soil* 254:279–89.

Li, Y., W. Ye, M. Wang, and X. Yan. 2009. Climate change and drought: A risk assessment of crop-yield impacts. *Climate Research* 39:31–46.

Lilley, J. M. and S. Fukai. Effect of timing and severity of water deficit on four diverse rice cultivars I. Rooting pattern and soil water extraction. *Field Crops Research* 37:205–13.

Lin, Q., S. Abe, A. Nose, A. Sunami, and Y. Kawamitsu. 2006. Effects of high night temperature on crassulacean ccid metabolism (CAM) photosynthesis of *Kalanchoë pinnata* and *Ananas comosus*. *Plant Production Science* 9:10–9.

Loneragan, J. F. 1997. Plant Nutrition in the 20th and perspectives for the 21st century. *Plant and Soil* 196:163–74.

Long, S. P. 1991. Modification of the response of photosynthetic productivity to rising temperature by atmospheric $CO_2$ concentrations: Has its importance been underestimated? *Plant, Cell & Environment* 14:729–39.

Luo, Q. 2011. Temperature thresholds and crop production: A review. *Climatic Change* 109:583–98.

Lüttge, U. 1988. Day–night changes of citric-acid levels in crassulacean acid metabolism: Phenomenon and ecophysiological significance. *Plant, Cell & Environment* 11: 445–51.

Lüttge, U. 1995. Ecophysiological basis of the diversity of tropical plants: The example of the genus Clusia. *Scientia Guaianae* 5:23–36.

Lüttge, U. 2002. $CO_2$-concentrating: Consequences in crassulacean acid metabolism. *Journal of Experimental Botany* 3:2131–42.

Lüttge, U. 2004. Ecophysiology of crassulacean acid metabolism (CAM). *Annals of Botany* 93:629–52.

Lüttge, U. 2006. Photosynthetic flexibility and ecophysiological plasticity: Questions and lessons from Clusia, the only CAM tree, in the neotropics. *New Phytologist* 171:7–25.

Makino, A. and R. F. Sage. 2007. Temperature response of photosynthesis in transgenic Rice transformed with 'Sense' or 'Antisense' rbcS. *Plant and Cell Physiology* 48:1472–83.

Malingreau, J. P., H. Eva, and A. Maggio. 2012. *NPK: Will There Be Enough Plant Nutrients to Feed a World of 9 billion in 2050?* Brussels, Belgium: Joint Research Centre, European Commission. http://citeseerx.ist.psu.edu/viewdoc/download?doi=10.1.1.397.2079&rep=rep1&type=pdf (accessed January 11, 2022).

Manna, M. C., A. Swarup, R. H. Wanjari, and H. N. Ravankar. 2007. Long-term effects of NPK fertiliser and manure on soil fertility and a sorghum–wheat farming system. *Australian Journal of Experimental Agriculture* 47:700–11.

Mathur, S. and A. Jajoo. 2014. Photosynthesis: Limitations in response to high temperature stress. *Phytochemistry and Photobiology* 137:116–26.

Matías, J., J. González, J. Cabanillas, and L. Royano. 2013. Influence of NPK fertilisation and harvest date on agronomic performance of Jerusalem artichoke crop in the Guadiana Basin (Southwestern Spain). *Industrial Crops and Products* 48:191–7.

McDonald, E. P., J. E. Erickson, and E. L. Kruger. 2002. Can decreased transpiration limit plant nitrogen acquisition in elevated $CO_2$? *Functional Plant Biology* 29:1115–20.

Moe, K., A. Z. Htwe, T. T. P. Thu, Y. Kajihara, and T. Yamakawa. 2019. Effects on NPK status, growth, dry matter and yield of rice (*Oryza sativa*) by organic fertilizers applied in field condition. *Agriculture* 9:105. https://doi.org/10.3390/agriculture9050109.

Moore, C. E., K. Meacham-Hensold, P. Lemonnier, et al. 2021. The effect of increasing temperature on crop photosynthesis: From enzymes to ecosystems. *Journal of Experimental Botany* 72:2822–44.

Mulder, E. G. 1950. Mineral nutrition of plants. *Annual Review of Plant Physiology* 1:1–24.

Mullet, J. E. and M. S. Whitsitt. 1996. Plant cellular responses to water deficit. In *Drought Tolerance in Higher Plants: Genetical, Physiological and Molecular Biological Analysis*, ed. E. Belhassen, 41–6. Dordrecht: Springer.

Murchie, E. H., M. Pinto, and P. Horton. 2009. Agriculture and the new challenges for photosynthesis research. *New Phytologist* 181:532–52.

Nagai, T. and A. Makino. 2009. Differences between rice and wheat in temperature responses of photosynthesis and plant growth. *Plant and Cell Physiology* 50:744–55.

Nagarajan, R. 2009. Agriculture. In *Drought Assessment*, ed. R. Nagarajan, 121–59. Dordrecht: Springer.

Nambara, E. and A. Marion-Poll. 2005. Abscisic acid biosynthesis and catabolism. *Annual Review of Plant Biology* 56:165–85.

Navari-Izzo, F. and N. Rascio. 1999. Plant response to water-deficit conditions. In *Handbook of Plant and Crop Stress*, ed. M. Perassalki, 231–70. New York: Marcel Dekker.

Nawaz, N., G. Sarwar, M. Yousaf, T. Naseeb, A. Ahmad, and M. J. Shah. 2012. Yield and yield components of sunflower as affected by various NPK levels [2003]. *Asian Journal of Plant Sciences* 2:561–2.

Newbould, P. 1989. The use of nitrogen fertiliser in agriculture. Where do we go practically and ecologically? *Plant and Soil* 115:297–11.

Nicolas, M. E., R. M. Gleadow, and M. J. Dalling. 1984. Effects of drought and high temperature on grain growth in wheat. *Australian Journal of Plant Physiology* 11:553–66.

Nievola, C. C., C. P. Carvalho, V. Carvalho, and E. Rodrigues. 2017. Rapid responses of plants to temperature changes. *Temperature* 4:371–405.

Niles, M T., M. Lubell, and M. Brown. 2015. How limiting factors drive agricultural adaptation to climate change. *Agriculture, Ecosystems & Environment* 200:178–85.

Nogueira, A., C. A. Martinez, L. L. Ferreira, and C. H. B. A. Prado. 2004. Photosynthesis and water use efficiency in twenty tropical tree species of differing succession status in a Brazilian reforestation. *Photosynthetica* 42:351–6.

Norby, R. J., D. E. Todd, J. Fults, and D. W. Johnson. 2001. Allometric determination of tree growth in a $CO_2$-enriched sweetgum stand. *New Phytologist* 150:477–87.

Nordin Henriksson, K. and A. J. Trewavas. 2003. The effect of short-term low temperature treatments on gene expression in Arabidopsis correlates with changes in intracellular Ca2+ levels. *Plant, Cell & Environment* 26:485–96.

Nuttall, J. G., K. M. Barlow, A. J. Delahunty, B. P. Christy, and G. J. O'Leary. 2018. Acute high temperature response in wheat. *Agronomy Journal* 110:1296–308.

Oosterhuis, D. M., D. A. Loka, E. M. Kawakami, and W. T. Pittigrew. 2014. The physiology of potassium in crop production. *Advances in Agronomy* 126:203–33.

Pastenes, C. and P. Horton. 1996. Effect of high Temperature on photosynthesis in beans (II. $CO_2$ assimilation and metabolite contents). *Plant Physiology* 112:1253–60.

Paulsen, G. M. 1994. High temperature responses of crop plants. In *Physiology and Determination of Crop Yield*, eds. K. J. Boote, J. M. Bennett, T. R. Sinclair, and G. M. Paulsen, 365–89: Madison, WI: ASA-CSSA-SSSA Publisher.

Penfield, S. 2008. Temperature perception and signal transduction in plants. *New Phytologist* 179:615–28.

Peng, S., J. Huang, J. E. Sheehy, et al. 2004. Rice yields decline with higher night temperature from global warming. *Proceedings of the National Academy of Sciences USA* 101:9971–5.

Peoples, M. B., D. E. Herridge, and J. K. Ladha. 1995. Biological nitrogen fixation: An efficient source of nitrogen for sustainable agricultural production? *Plant and Soil* 174:3–28.

Plieth, C., 1999. Temperature sensing by plants: Calcium-permeable channels as primary sensors—a model. *Journal of Membrane Biology* 172:121–7.

Pontis, H. G. 1989. Fructans and cold stress. *Journal of Plant Physiology* 134:148–50.

Poulton, P. R. and A. E. Johnston. 2019. Phosphorus in agriculture: A review of results from 175 years research at Rothamsted, UK. *Journal of Environmental Quality* 48:1133–44.

Prasad, R. and J. F. Power. 1997. *Soil Fertility Management for Sustainable Agriculture*. Boca Raton, WI: CRC Press.

Reddy, D. D., A. S. Rao, K. S. Reddy, and P. N. Takkar. 1999. Yield sustainability and phosphorus utilisation in soybean-wheat system on vertisols in response to integrated use of manure and fertilizer phosphorus. *Field Crops Research* 62:181–90.

Reddy, V. R., D. N. Baker, and H. F. Hodges. 1991. Temperature effects on cotton canopy growth, photosynthesis, and respiration. *Agronomy Journal* 83:699–704.

Rekarte-Cowie, I., O. S. Ebshish, K. S. Mohamed, and R. S. Pearce. 2008. Sucrose helps regulate cold acclimation of *Arabidopsis thaliana*. *Journal of Experimental Botany* 59:4205–17.

Reynolds, M. P., E. E. Ewing, and T. G. Owens. 1990. Photosynthesis at high temperature in tuber-bearing Solanum species 1: A comparison between accessions of contrasting heat tolerance. *Plant Physiology* 93:791–7.

Ribeiro, W. R., A. A. Pinheiro, D. S. Ferreira, M. S. Gonçalves, C. A. daSilva Martins, and F. Fialho dos Reis. 2018. Water deficit as a limiting factor to the initial growth of coffee conilon variety diamante. *Journal of Experimental Agriculture International* 22:1–11.

Richardson, A. E., J. P. Lynch, P. R. Ryan, et al. 2011. Plant and microbial strategies to improve the phosphorus efficiency of agriculture. *Plant and Soil* 349:121–56.

Roberts, T. L. and A. E. Johnston. 2015. Phosphorus use efficiency and management in agriculture. *Resources, Conservation and Recycling* 105:275–81.

Rosenzweig, C., J. W. Jones, J. L. Hatfield, et al. 2013. The agricultural model intercomparison and improvement project (AgMIP): Protocols and pilot studies. *Agricultural and Forest Meteorology* 170:166–82.

Ruelland, E. and A. Zachowski. 2010. How plants sense temperature. *Environmental and Experimental Botany* 69:225–32.

Rütting, T., H. Aronsson, and S. Delin. 2018. Efficient use of nitrogen in agriculture. *Nutrient Cycling in Agroecosystems* 110:1–5.

Salvucci, M. E. and S. J. Crafts-Brandner. 2004. Inhibition of photosynthesis by heat stress: The activation state of Rubisco as a limiting factor in photosynthesis. *Physiologia Plantarum* 120:179–86.

Sage, R. F. and A. D. McKown. 2006. Is C4 photosynthesis less phenotypically plastic than C3 photosynthesis? *Journal of Experimental Botany* 57:303–17.

Schlenker, W., W. M. Hanemann, and A. C Fisher. 2007. Water availability, degree days, and the potential impact of climate change on irrigated agriculture in California. *Climate Change* 81:19–38.

Sharma, L. K. and S. K. Bali. 2018. A review of methods to improve nitrogen use efficiency in agriculture. *Sustainability* 10:51. https://doi.org/10.3390/su10010051.

Singh, P., M. Agrawal, and S. B. Agrawal. 2009. Evaluation of physiological, growth and yield responses of a tropical oil crop (*Brassica campestris* L. var. Kranti) under ambient ozone pollution at varying NPK levels. *Environmental Pollution* 157:871–80.

Skirycz, A. and D. Inzé. 2010. More from less: Plant growth under limited water. *Current Opinion in Biotechnology* 21:197–203.

Song, Y., Q. Chen, D. Ci, X. Shao, and D. Zhang. 2014. Effects of high temperature on photosynthesis and related gene expression in poplar. *BMC Plant Biology* 14:111. https://doi.org/10.1186/1471-2229-14-111.

Sparks, D. L. 1987. Potassium dynamics in soils. *Advances in Soils Science* 6:1–63.

Sprent, J. I. 1984. Evolution of nitrogen fixing symbioses. In *Advances in Plant Physiology*, ed. M. B. Wilkins, 249–76. London: Pitman.

Srivastava, A., B. Guissre, H. Greppin, and R. J. Strasser. 1997. Regulation of antenna structure and electron transport in photosystem II of *Pisum sativum* under elevated temperature probed by the fast polyphasic chlorophyll a fluorescence transient: OKJIP. *Biochimica Biophysica Acta* 1320:95–106.

Stone, P. J. and M. E. Nicolas. 1995. A survey of the effects of high temperature during grain filling on yield and quality of 75 wheat cultivars. *Australian Journal of Agricultural Research* 46:475–92.

Suzuki, N. and R. Mittler. 2006. Reactive oxygen species and temperature stresses: A delicate balance between signaling and destruction. *Physiologia Plantarum* 126:45–51.

Tanaka, S.K., T. Zhu, J. R. Lund, et al. 2006. Climate warming and water management adaptation for California. *Climatic Change* 76:361–87.

Tardieu, F., B. Parent, and T. Simonneau. 2010. Control of leaf growth by abscisic acid: Hydraulic or non-hydraulic processes. *Plant, Cell and Environment* 33:636–47.

Tarkowski, Ł. P. and W. Van den Ende. 2015. Cold tolerance triggered by soluble sugars: A multifaceted countermeasure. *Frontiers in Plant Sciences* 6:203.

Tisdale, S. L., W. L. Nelson, and J. D. Beaton. 1985. *Soil Fertility and Fertilizers*. New York: MacMillan.

Uemura, M., Y. Tominaga, C. Nakagawara, S. Shigematsu, A. Minami, and Y. Kawamura. 2006. Responses of the plasma membrane to low temperatures. *Physiologia Plantarum* 126:81–9.

Vance, C. P. 1996. Root-Bacteria interactions: Symbiotic $N_2$ fixation. In *Plant Roots: The Hidden Half*, eds. Y. Waisel, A. Eshel, and U. Kafkafi, 723–56. New York: Marcel Dekker.

Vance, C. P. 1997. Enhanced agricultural sustainability through biological nitrogen fixation. In *Biological Fixation of Nitrogen for Ecology and Sustainable Agriculture*, eds. A Legocki, H. Bothe, and A. Pühler, 179–86. Berlin: Springer-Verlag.

Vance, C. P. and P. H. Graham. 1994. Nitrogen fixation in agriculture: Application and perspectives. In *Nitrogen Fixation: Fundamentals and Applications*, eds. A. I. Tikhonovich, N. A. Provorov, V. I. Romanov, and W. E. Newton, 77–86. Dordrecht: Kluwer Academic.

Vico, G., S. Manzoni, S. Palmroth, M. Weih, and G. Katul. 2013. A perspective on optimal leaf stomatal conductance under $CO_2$ and light co-limitations. *Agricultural and Forest Meteorology* 182–183:191–9.

Vile, D., M. Pervent, M. Belluau, et al. 2012. Arabidopsis growth under prolonged high temperature and water deficit: Independent or interactive effects? *Plant, Cell & Environment* 35:702–18.

Wolfe, D. W., A. T. DeGaetano, G. M. Peck, et al. 2017. Unique challenges and opportunities for northeastern U.S. crop production in a changing climate. *Climatic Change* 146:231–45

Xin, Z. and J. Browse. 2001. Cold comfort farm: The acclimation of plants to freezing temperatures. *Plant, Cell & Environment* 23:893–902.

Yadav, R., B. S. Dwivedi, K. Prasad, O. K. Tomar, N. J. Shurpali, and P. S. Pandey. 2000. Yield trends, and changes in soil organic-C and available NPK in a long-term rice–wheat system under integrated use of manures and fertilisers. *Field Crops Research* 68:219–46.

Yadav, S. K. 2010. Cold stress tolerance mechanisms in plants. A review. *Agronomy* 30:515–27.

Yamori, W., C. Masumoto, H. Fukayama, and A. Makino. 2012. Rubisco activase is a key regulator of non-steady-state photosynthesis at any leaf temperature and, to a lesser extent, of steady-state photosynthesis at high temperature. *Plant Journal* 71:871–80.

Yamori, W., K. Hikosaka, and D. A. Way. 2014. Temperature response of photosynthesis in C3, C4, and CAM plants: Temperature acclimation and temperature adaptation. *Photosynthesis Research* 119:101–17.

Yourtchi, M. S., M. H. S. Hadi, and M. T. Darzi. 2013. Effect of nitrogen fertilizer and vermicompost on vegetative growth, yield and NPK uptake by tuber of potato (Agria CV.). *International Journal of Agriculture and Crop Sciences* 5:2033–40.

Yousaf, M., J. Li, J. Lu, et al. 2017. Effects of fertilization on crop production and nutrient-supplying capacity under rice-oilseed rape rotation system. *Scientific Reports* 7:1270. https://doi.org/10.1038/s41598-017-01412-0.

Zandalinas, S. I., D. Balfagón, V. Arbona, A. Gómez-Cadenas, M. A. Inupakutika, and R. Mittler. 2016. ABA is required for the accumulation of APX1 and MBF1c during a combination of water deficit and heat stress. *Journal of Experimental Botany* 67:5381–90.

Zarka, D. G., J. T. Vogel, D. Cook, and M. F. Thomashow. 2003. Cold induction of Arabidopsis CBF genes involves multiple ICE (inducer of CBF expression) promoter elements and a cold-regulatory circuit that is desensitized by low temperature. *Plant Physiology* 133:910–18.

Zhang, J., J. Balkovič, L. B. Azevedo, et al. 2018. Analyzing and modelling the effect of long-term fertilizer management on crop yield and soil organic carbon in China. *Science of the Total Environment* 627:361–72.

Zhang, J., H. Yuan, Y. Yang, et al. 2016. Plastid ribosomal protein S5 is involved in photosynthesis, plant development, and cold stress tolerance in Arabidopsis. *Journal of Experimental Botany* 67:2731–44.

Zhang, X. and X. Cai. 2013. Climate change impacts on global agricultural water deficit. *Geophysical Research Letters* 40:1111–7.

Zhao, D. and D. Oosterhuis. 1997. Physiological response of growth chamber-grown cotton plants to the plant growth regulator PGR-IV under water-deficit stress. *Environmental and Experimental Botany* 38:7–14.

Zhao, J., J. Guo, and J. Mu. 2015. Exploring the relationships between climatic variables and climate-induced yield of spring maize in Northeast China. *Agriculture, Ecosystems & Environment* 207:79–90.

Zhu, X. G., E. de Sturler, and S. P. Long. 2007. Optimizing the distribution of resources between enzymes of carbon metabolism can dramatically increase photosynthetic rate: A numerical simulation using an evolutionary algorithm. *Plant Physiology* 145:513–26.

# 6 Principles of Crop Breeding and Productivity

*Elizabeth "Izzie" Gall*

## CONTENTS

## 6.1 INTRODUCTION

Low availability of any one limiting factor (Chapter 5) can reduce plant growth and productivity. Therefore, increasing the supply of known limiting factors in the field helps plants thrive and produce more of the products we desire from them. When plants have access to resources beyond the needs of primary metabolism, the range of phenotypes related to flavor, color, and other desirable crop factors expands, increasing the number of traits that can be targeted by natural or artificial selection. However, each generation of an annual crop receives only one year's (or season's) worth of additions like fertilizer and irrigation. In order to drive long-term changes in quality and productivity, humans must also control crops' evolutionary environment.

Following crop sexual reproduction, farmers retain seeds from the plants with the most desirable (fittest) phenotypes to sow in the following season. Thus, unusual but favorable phenotypes become the average abnormally fast. Most crop plants are bred for specific traits which become very exaggerated compared with the same traits in their ancestors. However, there are practical limits to how far human selection pressure can push a phenotype.

## 6.2 ARTIFICIAL EVOLUTIONARY PRESSURE

Humans alter the evolutionary pressures acting on a crop plant by controlling certain aspects of the agricultural environment. Introducing an irrigation system lifts the growth limitations of a low water supply, but because all plants in the population have ample access to water, drought-tolerant and drought-intolerant individuals have equal fitness. Likewise, adding fertilizer to a field removes the fitness benefit of roots that efficiently take in nutrients. In general, standardizing input conditions

DOI: 10.1201/9780429320415-6

relaxes the pressures that differentiate plant fitness in the wild, so the crop population can accumulate mutations that might otherwise be deadly. Some of those mutations might encode phenotypes that are useful to humans, providing targets for human selection toward more flavorful, useful, and nutritious crops.

Allowing plants to finish their sexual life cycles is a vital part of agriculture. Humans select seeds from the plants with the traits most suited to human use, plant them, and cultivate the next generation. By encouraging the reproduction of individuals with certain phenotypes, humans give plants with certain genes more chances to reproduce (higher fitness). Meanwhile, because selection occurs over several sexual generations, new traits can be introduced and selected for (or against) within the artificial, human-controlled environment. In other words, crop domestication is achieved through **artificial evolutionary pressure**.

The genetic changes that accumulate in a species throughout the selective breeding process represent modifications that humans have intentionally performed on a species or variety, altering its genetic code so it will express certain desirable traits. Therefore, selective breeding is one type of **genetic modification** – a controlled change to the genes or gene expression of an organism. The terms "genetic engineering" or "molecular engineering" are currently used to describe a faster, more directed type of genetic manipulation that takes place in laboratories (see Section 6.6), but the process has been occurring naturally at a much slower rate for thousands of years. With humans driving artificial selection every season, population genotypes can shift in comparatively little evolutionary time, leading to the development of species ideally suited to human-controlled environments and very unlike their ancestors.

The domestication of maize (*Zea mays*) is a classic example of genetic modification through selective breeding. Ancient Mesoamerican humans realized that they needed to reserve some grains from each harvest to use as seed stock for the following season. Wild teosinte, the ancestor of maize, has wind-dispersed seeds that break easily off the stalk and blow away. Plants that did not release their seeds as easily were more likely to have their grains retained by humans for harvest and re-seeding. Over thousands of generations, the domesticated plants have developed heavy kernels that stay firmly attached to their cobs when ripe. In the wild, having seeds that stay firmly on the stalk instead of dispersing would be a low-fitness trait; in agricultural settings, that same trait is an important adaptation.

Crop varieties are often bred for product size (e.g., beans, fruits, or roots), vibrant coloration (squash, stone fruits, and lentils), strong flavor (herbs, berries, and roasting vegetables), nutrient concentration (dark leafy greens, garlic, and legumes), and pleasing texture (cabbage, melons, and apples). They can also be selected for their ability to efficiently use artificial inputs, to germinate in multiple seasons, or to devote more of their resources to yields. For example, the development of a shorter wheat variety in the 1950s increased the amount of fertilizer that plants could devote to harvestable products by allowing kernels to grow large without risk of snapping long, delicate stalks (see Chapter 9 in this volume).

In addition to sowing seeds from plants that have adapted favorably to agricultural systems, human farmers can perform directed crosses between individuals with important traits. For example, when two plants with particularly sturdy stems flower, a farmer might **cross-pollinate** them to try and achieve offspring with even stronger stems. In this process, farmers carefully exchange the pollen between the flowers of the plants with preferred traits. Farmers might also place bags over the flowers or remove additional blooms to make sure that uncontrolled crosses do not occur. The dwarf wheat developed in the 1950s was created via cross-pollination of dozens of varieties, requiring more than 10,000 hand pollinations per year for the five years leading up to its release. The results of a cross are not immediately known; the desired traits may not appear in new varieties until their fourth or fifth generations (Anderson 1975). Nonetheless, cross-pollination is so effective at directing artificial evolution that it is still the preferred method for developing new crop varieties. Recently, molecular tools have been used by international crop development organizations to reinforce, but not replace, manual crossing (Folger 2014).

**TABLE 6.1**

**Total Energy Use and Tissue Production in an Uncontrolled Hardwood Temperate Forest**

| Production Type | Production (kcal/m²/year) | Percentage of Total Plant Production | Percentage of Primary Production |
|---|---|---|---|
| Total plant production | 10,400[a] | 100.00 | N/A |
| Respiration/maintenance | 5,720[a] | 55.00[b] | N/A |
| Primary production | 4,680[a] | 45.00[b] | 100.00 |
| Foliage | 1,485[a] | 14.28[b] | 31.73[b] |
| Seeds | 80[a] | 0.77[b] | 1.71[b] |
| Other production | 3,115[b] | 29.95[b] | 66.56[b] |

[a] Gosz et al. (1978, p. 96).
[b] Calculations based on Gosz et al. (1978).

## 6.3 METABOLITE LIMITS OF SELECTIVE BREEDING

While many limits on plant growth can be overcome with artificial inputs, plants will always need to devote a certain amount of their energy to growth and to maintaining existing cells (**primary metabolism**). They also must defend against threats like herbivory and disease to keep existing cells intact and healthy. Even when humans minimize the number of weeds, diseases, and herbivores in a field, many plants still produce defensive compounds. In fact, some of the chemicals historically most desired in a crop, such as nicotine (from tobacco), capsaicin (chili peppers), and caffeine (coffee), began as defensive chemicals against herbivory. Plants that engage in mycorrhizal partnerships also need to devote a certain amount of sugar to their symbionts. As a result, about 55% of the energy fixed by native plants in a temperate ecosystem is used for maintenance (Gosz et al. 1978). The maintenance cost is even higher in warmer climates (Ricklefs and Miller 2000). In major agricultural crop species, maintenance costs between 30% and 60% of the carbon brought in by photosynthesis (Zhu et al. 2010).

After maintenance costs are paid, plants are able to generate new growth (**primary production**). Slightly less than a third of wild plants' primary production is devoted to the growth of foliage (calculations based on Gosz et al. 1978). Once photosynthesizing tissues are in place, plants are finally able to devote their remaining energy to the creation of flowers, seeds, storage tissues, fruits, or extra flavor and defensive compounds. Human artificial selection therefore acts on the flexibility of this remaining energy, which represents approximately 30% of a wild plant's total energy budget (see Table 6.1).

The mustard family is a marvelous showcase of artificial selection; purely through selective breeding, the species *Brassica oleracea* has given rise to the wildly diverse vegetables broccoli, Brussels sprouts, cabbage, cauliflower, kale, and kohlrabi. Human farmers selected individuals of *B. oleracea* with the thickest stems to reproduce over thousands of generations, giving plants with genes coding for stunted, thick growth a fitness advantage. Because the plants were selected for the amount of energy devoted to their stems, that energy could be taken from other tissues without negative fitness effects. The modern results of that process are known as kohlrabi. Independently, certain other *B. oleracea* individuals were selected for having larger leaves than their peers; due to the fitness advantage conferred by human preferences, and the ability to reduce energy investment in seeds, flowers, and stalks, those generations have ultimately generated the complex, characteristic leaves of kale. Brussels sprouts resulted from selection for large lateral buds, broccoli and cauliflower from selection for flower clusters, and cabbage from selection for large terminal buds (Figure 6.1).

**FIGURE 6.1** The differentiation of wild mustard, *Brassica oleracea*, demonstrates the extent of genetic variation within a single species and the extreme traits made possible through extensive artificial selection. Two examples out of six are shown. (a) The original variant of *B. oleracea*, wild mustard. (b) Selection for large lateral buds led to the creation of Brussels sprouts. (c) Selection for a thick stem led to the creation of kohlrabi. [Photos reprinted with permission from Shutterstock. (a) Martin Fowler; (b) Andre Muller; (c) audaxl.]

However, as with so many things in biology, the selective breeding process is not always clear-cut. Partially because of the way that phenotypes and genes are cross-linked, and partially because the genes available to artificial evolutionary pressure rely on the energy left after maintenance is taken care of, breeding for a particular trait will often require sacrifices in other areas. There is no descendant of *B. oleracea* which has large, ruffled kale leaves emerging from juicy Brussels sprouts buds along a short, squat kohlrabi stem and topped with a cabbage head. Each of the exaggerated traits is so expensive that the other physical aspects of the plant are sacrificed, even in human-controlled environments where normally limiting factors are freely available.

For agricultural plant species, the selective advantages conferred by strong foraging mechanisms or drought tolerance have disappeared. Instead, plants have access to a single major advantage if they pour as much energy as possible into the trait most desired by the humans cultivating them. Yields can be measured in terms of total product weight, wet or dried; the mass of the living (or recently living) plant is called its **biomass**. However, a plant with a high biomass does not necessarily have a high yield, as any gardener with a big, leafy tomato vine and no tomatoes can attest. When evaluating how efficient a plant is at generating products for humans, it is more useful to consider yields in terms of **harvest index** (**HI**; plural, *indices*). The HI is measured as the dry mass of harvest product a plant has produced (such as soybeans) relative to its overall dry biomass.[1] HI can reveal which crop varieties are most efficient at creating the products humans desire. Because

---

[1] Dry mass is used because it tells us how much carbon and solar energy went into making the product; fresh mass includes water weight, which can be less informative.

it is an index, the HI value is written as a decimal and does not have a unit (like percent or grams). The highest possible HI is 1.0, but because of the energy costs of maintenance and primary growth, very few plants have harvest indices close to the maximum.

The HI measurement is roughly equivalent to the primary production percentage divided by 100. Consider foliage to be the "harvest product" of the plants in the temperate forest of Table 6.1. Foliage represented 31.73% of primary production (gray box in Table 6.1). The HI for the foliage is 31.73/100, or 0.31. In either expression, we know that slightly less than a third of the plant's new growth energy was devoted to making foliage.

The harvest indices for several grain crops are shown in Table 6.2. The data in this table come from several different studies spanning many years and farm types in Australia. Whenever possible, the minimum and maximum recorded HI are shown in the table to illustrate how various factors like weather patterns, irrigation methods, farm management methods, and disease patterns might affect HI. Note that despite the broad range for some of the observed grains, the mean yield for these grain crops tends to be at least one-third of the plant's total dry mass – a considerable increase over the wild seed yields in Table 6.1.

Measuring yield efficiency with HI is usually much easier with annual crop plants, which can be uprooted, weighed, and processed intact at harvest (see Table 6.2). Estimates of primary production for perennial crops like trees or shrubs are less accurate, since the woody structures and roots represent multiple seasons' energy and carbon investments. Yield efficiency is usually measured differently for perennial plants, such as by comparing apple yields to the thickness of the parent tree's trunk (Autio 2019).

## 6.4 PHOTOSYNTHETIC LIMITS OF SELECTIVE BREEDING

After supplying plenty of essential nutrients and water and selecting for plants that efficiently use those inputs, there is another hurdle to maximizing harvest. Recall that solar radiation is the source

**TABLE 6.2**
**Harvest Indices of Several Australian Grain Crops**

| Crop | Species | Harvest Index | | | Mean Percentage of Primary Production |
| --- | --- | --- | --- | --- | --- |
| | | Mean | Minimum | Maximum | |
| Barley | *Hordeum vulgare* | 0.38 | 0.09 | 0.57 | 38 |
| Canola | *Brassica napus* | 0.28 | 0.04 | 0.41 | 28 |
| Chickpea | *Cicer arietinum* | 0.37 | 0.06 | 0.55 | 37 |
| Faba bean | *Vicia faba* | 0.44 | 0.11 | 0.58 | 44 |
| Field pea | *Pisum sativum* | 0.36 | 0.06 | 0.58 | 36 |
| Lentil | *Lens culinaris* | 0.33 | 0.06 | 0.51 | 33 |
| Lupin | *Lupinus angustifolius* | 0.3 | 0.04 | 0.5 | 30 |
| Maize | *Zea mays* | 0.52 | 0.41 | 0.62 | 52 |
| Mungbean | (Not given) | | | 0.55 | |
| Oat | *Avena sativa* | 0.21 | 0.11 | 0.48 | 21 |
| Peanut | *Arachis hypogea* | 0.33 | 0.02 | 0.57 | 33 |
| Rice | (Not given) | | | 0.7 | |
| Sorghum | *Sorghum bicolor* | 0.46 | 0.14 | 0.7 | 46 |
| Soybean | (Not given) | | | 0.35 | |
| Sunflower | (Not given) | | | 0.48 | |
| Triticale | *Triticum durum* x *Secale cereale* | 0.34 | 0.28 | 0.46 | 34 |
| Vetch | *Vicia sativa* | 0.38 | 0.16 | 0.47 | 38 |
| Wheat | *Triticum aestivum* | 0.37 | 0.08 | 0.56 | 37 |

*Source:* Data compiled from Unkovich et al. (2010).

of effectively all the sugar energy in a terrestrial ecosystem. Photosynthetic organisms capture the energy from photons in microscopic processes that generate sugar and other high-energy molecules that can be used in self-maintenance, growth, and reproduction. **Photosynthetic efficiency**, the rate at which plants can convert the energy of solar radiation into primary production, has not increased during human agricultural history. In fact, the photosynthetic efficiency of green plants has been fairly stable for more than 20 million years.

Very briefly, photosynthesis begins when a photon carrying a certain **wavelength** (energy level) of light strikes a pigment molecule, usually the green pigment **chlorophyll**, in a leaf. The energy is harnessed by a protein, **Photosystem II** (PSII),[2] and used to split a water molecule, then the electrons and hydrogen atoms are used to transport the energy to the site of sugar creation. (The oxygens from the water molecules combine to form gaseous oxygen, $O_2$, which is released.) An enzyme called **Rubisco** uses some of that energy to affix gaseous carbon dioxide, $CO_2$, onto a solid carbon skeleton, storing a considerable amount of energy in the six-carbon sugar molecule **glucose**. The first stable product generated when Rubisco fixes $CO_2$ has three carbons; therefore, plants that use this type of photosynthesis, including most agricultural crops, are known as **C3 plants**.

One molecule of glucose is generated for every six $CO_2$ molecules **fixed** (put into solid form) by Rubisco. These sugars are used to transport energy around the plant and also serve as the backbones for many biological molecules, allowing the construction of new biomass. This extremely simplistic overview of photosynthesis will suit our needs for this chapter.

Plants can only use a staggeringly low amount of the abundant solar energy reaching Earth. First, not all of the light from the sun can be used in photosynthesis. The photosynthetically active wavelengths are mostly visible to humans, but only 44% of the sunlight reaching the Earth's surface is in the visible range. Because of the changes in light level due to shade, weather changes, and seasonal light variation, producing lots of expensive photosynthetic molecules does not carry a fitness benefit. Therefore, plants create just enough photosynthetic machinery to function best at low light, achieving their highest photosynthetic efficiency at light levels between 10% and 25% of full sunlight. Stronger light can actually decrease the rate of photosynthesis, as the molecular machinery shuts itself off to protect from radiation damage (Ricklefs and Miller 2000). Furthermore, in order to use high- and low-energy photons with equal efficiency, plants actually harness the low-energy amount from photons of every wavelength. Therefore, plants store the same amount of energy whether they receive the highest-energy, violet photons or the lowest-energy, red photons (Zhu et al. 2010). Due to these limits, only 1%–2% of the photosynthetically active energy that reaches plants during the growing season is available for storage and use (Ricklefs and Miller 2000; Gosz et al. 1978) (Figure 6.2).

Unfortunately, plants cannot fully utilize even this comparatively small amount of solar energy. When Rubisco is working perfectly, it uses the energy from one photosynthetically active photon to fix one molecule of $CO_2$. This job is so important that Rubisco makes up 30%–50% of the protein in a leaf![3] Unfortunately, Rubisco has a difficult time differentiating between carbon dioxide and oxygen, and frequently fixes oxygen onto a carbon backbone, resulting in a "dead-end" molecule (Eisenhut et al. 2017) that cannot be used in any metabolic process. The fixation of oxygen to the planned sugar backbone must be reversed in a nine-step process known as **photorespiration**[4] before the sugar can be completed properly. Because molecular oxygen is one of the products of photosynthesis, it can easily become concentrated in the leaf, making photorespiration extremely common. In fact, almost one out of four Rubisco reactions uses oxygen instead of carbon dioxide (Peterhansel et al. 2010). While photorespiration has an important role in stress signaling, high light tolerance, and removal of reactive oxygen species (ROS; see Chapter 5 in this volume), the process can also

---

[2] Photosystem II was the second discovered photosynthetic protein system, but it is the first molecular complex used in photosynthesis.

[3] Rubisco constitutes 30% of the soluble proteins in a C4 leaf and 50% in a C3 leaf (Feller et al. 2007).

[4] Normal metabolic "respiration" is the digestion of sugars for energy.

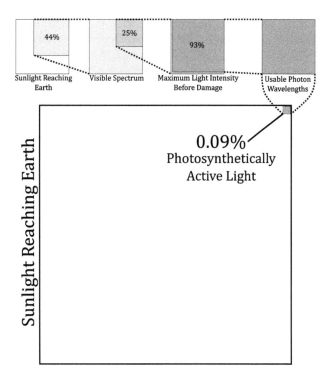

**FIGURE 6.2**    Very little of the sunlight reaching plants is photosynthetically active.

consume up to 30% of the energy transfer molecules available (Walker et al. 2016) and release up to 30% of photosynthesized carbon dioxide (Zhu et al. 2010) in a leaf!

Oxygen fixation is more common at high temperatures, when Rubisco is less able to distinguish between oxygen and carbon dioxide. It also occurs more often in hot and dry conditions because plants close the pores in their leaves (**stomata**) to prevent water loss. Unfortunately, this action also increases the concentration of oxygen gas inside leaves, leading to greater oxygen fixation by Rubisco and hence to increased photorespiration and decreased photosynthetic efficiency.

Therefore, the amount of solar energy stored by plants is not only limited by the amount of light a leaf can physically capture and use but also by Rubisco's frequent oxygen uptake.

## 6.5    EVOLVED ALTERNATIVES TO C3 PHOTOSYNTHESIS

As evidenced by the success of photosynthetic plants and the ecosystems that rely on them, the low photosynthetic efficiency of C3 plants still translates a massive amount of solar energy into biological molecules. However, there is clearly an evolutionary advantage to increasing photosynthetic efficiency, and several plant families have found ways to do so by reducing photorespiration.

### 6.5.1    C4 Photosynthesis

In contrast with C3 plants, **C4** plants fix $CO_2$ temporarily into a four-carbon form. These plants also reduce the likelihood of photorespiration by physically separating Rubisco from oxygen gas. The **mesophyll** (outer) cells of C4 plant leaves do not contain Rubisco; instead, they contain the specialized enzyme **PEP carboxylase**, which has a high affinity for $CO_2$. PEP carboxylase fixes $CO_2$ into a four-carbon compound that is shuttled to **bundle sheath** cells deeper inside the leaf. These innermost cells do contain Rubisco and are the sites of sugar generation. When the four-carbon compound arrives at these inner cells, the carbon dioxide is released. In this low-oxygen environment, the $CO_2$ is rapidly picked up by Rubisco, which then proceeds with the same glucose generation process used in C3 photosynthesis. By moving Rubisco away from an oxygen-rich environment, C4 plants achieve lower photorespiration rates than C3 plants. With all that saved energy and carbon, C4 plants can turn about 12.3% of photosynthetically active light into new biomass during a growing season, compared with 9.4% for C3 plants (Zhu et al. 2010). That means that C4 plants can harness almost 30% more solar energy than C3 plants!

In hot and dry conditions, C4 plants have an additional advantage. In high heat, plants close their stomata to reduce water loss. When C3 plants close their stomata, oxygen builds up in the leaves and leads to additional photorespiration. In contrast, because they already separate oxygen from Rubisco, plants that use C4 photosynthesis can close their stomata without increasing photorespiration rates. This gives C4 plants much higher water efficiency than C3 plants. However, C4 species are very sensitive to shade; anywhere that taller C3 plants can establish and shade out C4 plants, the environment will rapidly become C3-dominated (Sage 2016). Hence, C4 plants are most common in areas where C3 plants have trouble establishing, especially in dry habitats like scrubland and plains. Nineteen different plant families include plants that use C4 photosynthesis, with some families containing multiple species that developed C4 photosynthesis independently. In fact, the development of C4 photosynthesis, a process which involves hundreds of gene mutations, may have occurred more than 65 separate times (Sage 2016)!

Despite making up only 3% of the world's known flowering plant species, C4 plants conduct 23% of the world's carbon fixation (Kellogg 2013). The most agriculturally important C4 plants are maize, sugarcane, sorghum, millet, amaranth, fonio, and teff. More than 60% of C4 plants are grasses (Sage 2016), many of which have deep roots and important soil-stabilizing effects on their ecosystems. Many important pasture grasses also use C4 photosynthesis, contributing to human health and nutrition through animal consumption.

The fact that C4 plants have not totally outcompeted C3 plants and taken over the world demonstrates that there are also evolutionary disadvantages to C4 photosynthesis. Creating and maintaining a stock of PEP carboxylase requires plenty of energy, and C4 plants still need just as much Rubisco as C3 plants do to fix carbon dioxide into sugar. Creating an intermediate four-carbon molecule and shuttling it through cells also requires extra energy not expended by C3 plants. In conditions when Rubisco is operating at high efficiency, such as in lower temperatures or high environmental $CO_2$ concentrations, C4 plants have little to no fitness advantage over C3 plants (Lattanzi 2010). As humans continue to increase the global concentration of $CO_2$, that fitness gap is closing. Experimentally increasing the $CO_2$ concentration in a soybean field 15%, to match future projected $CO_2$ concentrations, led to a 22.6% increase in soybean C3 photosynthesis (Zhu et al. 2010) – largely closing the 30% energy advantage of the C4 method. In fact, due to the rate at which humans are increasing the atmospheric concentration of $CO_2$, C3 and C4 photosynthesis could be equally efficient by the year 2100 (Zhu et al. 2010).

However, elevated temperatures under global climate change create an opposite pressure, favoring C4 plants (which are more water efficient and saturate Rubisco with $CO_2$) over C3 plants (whose Rubisco will fix more oxygen at higher temperatures due to lower specificity). Photosynthesis is also only one process among many that contribute to the success or "failure" of a plant (Lattanzi 2010). Though changes in photosynthetic rate have been heavily investigated in different temperatures and $CO_2$ concentrations in many plant families, it is still unclear how different photosynthetic techniques will rank in the forthcoming high-$CO_2$, high-temperature paradigm (Taub 2010).

## 6.5.2 CAM Photosynthesis

In extremely hot and dry conditions, even C4 plants risk becoming dehydrated as they conduct photosynthesis. In these areas, some plants have developed a protective mechanism to guard against dehydration while also reducing photorespiration. The Crassulacean Acid Metabolism

(**CAM**) pathway is named for the family of succulent plants in which it was first discovered, the Crassulaceae. CAM plants isolate Rubisco and oxygen temporally (in time), opening their stomata and harvesting carbon dioxide at times when photosynthesis cannot occur, so Rubisco is not active. CAM plant stomata are usually closed during the day and open at night, when it is cool and wet and $CO_2$ can be gathered without risk of water loss. Once inside, $CO_2$ molecules are fixed onto four-carbon intermediates by PEP carboxylase, as in C4 photosynthesis. However, CAM plants exercise more control over this process with **PEP carboxylase kinase** (**PEP-CK**), a molecule that allows PEP carboxylase to be active in low temperatures but shuts it down in hot conditions. Rather than going directly to Rubisco, the four-carbon intermediate compounds build up inside leaves during the cool period (e.g., at night). When temperatures increase and humidity drops, the stomata are closed to reduce dehydration, the four-carbon compounds release their $CO_2$, and Rubisco is able to conduct its sugar creation in a high-$CO_2$ environment. Unfortunately, because the stomata are closed when photosynthesis is being conducted, the concentration of oxygen can increase within the leaf. Therefore, CAM plants endure a similar rate of photorespiration to C3 plants. The major advantage of CAM photosynthesis over the C3 method is water efficiency, which is a whopping 600% higher in CAM plants (Silvera et al. 2010)!

Roughly 6% of flowering plant species, in 35 plant families, use CAM photosynthesis. Unlike the C4 system, which relies on permanent, specialized structures in the leaf, the CAM method can be used on a sliding scale. A "typical" CAM plant is one which fixes 70%–77% of its carbon at night, but there are also plants that use CAM for only 30% or less of their carbon fixation (Silvera et al. 2010). Some plants express CAM proteins **constitutively** (all the time), while some only engage in **facultative** CAM when the environment becomes dry enough (from the Latin *facultas*="ability"). CAM plants dominate in areas with extreme rain variation such as deserts, tree canopies, and the mountains of tropical regions (Silvera et al. 2010). Agriculturally important CAM plants include pineapples, agaves, and vanilla orchids.

### 6.5.3 EVOLUTIONARY LIMITS

Plants that can capture and use more of the solar energy that reaches them have a fitness advantage over the plants that acquire less biomass for the same amount of energy. Meanwhile, plants that can conduct photosynthesis in arid environments have a fitness advantage over those that shut off biomass production when conditions are hot and dry. The degree of any fitness advantage depends on the severity of the problem being overcome, which is why C3 plants still dominate C4 and CAM plants in temperate climates and account for about 75% of carbon fixation worldwide.

**TABLE 6.3**

**Comparison of the Three Types of Photosynthesis**

| | C3 Plants | C4 Plants | CAM Plants |
|---|---|---|---|
| Carbon atoms in first intermediate molecule | Three | Four | Three or four |
| Enzyme creating first intermediate molecule | Rubisco | PEP carboxylase | PEP carboxylase, regulated by PEP carboxylase kinase |
| Separation of oxygen and Rubisco | None | Physical | Temporal |
| Stomata closed | Hot, dry conditions | Hot, dry conditions | Daylight hours |
| Solar energy lost to photorespiration | 6.1%[a] | 0%[a] | 0%–6.1%[b] |
| Portion of global carbon fixation | 75% | 23% | 2% |
| Flexibility of use | Constitutive | Constitutive | Facultative |

[a] Data from Zhu et al. (2010).
[b] This rate may be higher for facultative CAM plants.

The development of PEP carboxylase and PEP-CK is due to billions of years of evolutionary pressure to achieve higher photosynthetic efficiency with the resources available in nature: to lose less photosynthetic product to photorespiration and to conduct more photosynthesis in dry environments, respectively. All the same, the observed conversion of solar energy to chemical energy in the field remains just a third of the theoretical limit (Zhu et al. 2010). Even though humans have been supplying extra nutrients, water, and care to crops for thousands of generations, crop production is still limited by the inherent inefficiencies of photosynthesis.

## 6.6   OTHER METHODS TO INCREASE YIELDS AND EFFICIENCY

Beyond the selection of varieties, production of hybrids, and ample supply of inputs, there are other techniques farmers can use to increase harvest volume or quality.

### 6.6.1   INTERCROPPING

**Intercropping** is the process of planting multiple crop species near each other; in other words, it is the opposite of monocropping. Subsistence farms and home gardens often use intercropping. The most efficient intercropping techniques involve planting crops next to other species that interact with them in a positive way, or at the very least do not compete for the same resources.

For nearly 5,000 years, native American peoples have been using an intercrop known as "the three sisters". Each of the "sisters" – squash, beans, and maize – provides a service for the other two. Maize grows quickly, providing a scaffold for the beans to climb and shade to protect the light-sensitive squash plant. The squash grows low to the ground, laying out large leaves that reduce evapotranspiration, helping maintain soil moisture, and out-compete weeds nearby. Finally, the beans host nitrogen-fixing bacteria (in root nodules) that help fertilize the squash and maize as they grow. When grown together, the "three sisters" are more photosynthetically efficient, and produce more protein, than any one of them in a monocrop. If harvested together, the "sisters" also provide a complete panel of amino acids, several vital minerals, and the vitamins A, B2, and B6 (Hirst 2020). This partnership is still highly productive today and has inspired research into other favorable intercropping systems.

### 6.6.2   CROP ROTATION

Crop rotation is another traditional method of increasing yields. In the most basic form of crop rotation, nitrogen-fixing crops like beans are planted in a field for one year (or season) and followed by a grain crop the next. As in the "three sisters" system, crop rotation increases the nitrogen content of the grain crop without requiring (as much) nitrogen fertilizer. Crops can also use different proportions of minerals or scavenge resources from different zones of the soil, helping maintain soil fertility and reduce soil mineral depletion. For example, alfalfa roots reach up to 10 feet downwards, bringing minerals from deep in the soil toward the surface and storing them in plant tissues. Following harvest, the decomposition of alfalfa residues will release those deep-soil minerals into shallower soil, where crops with shallow root systems but higher mineral requirements, such as lettuce, can access them (Mohler and Johnson 2009). While each field might be planted with a single species, crop rotation can be viewed as a temporal intercropping: crops of different species still benefit one another by being present in the same space, just at different times.

Another major benefit of crop rotation is to prevent the establishment of crop diseases. Many plant diseases can become established in a monoculture during the growing season, reproduce, and lie dormant through the winter. In a continual monocrop, the disease organisms emerge from hibernation to find their preferred hosts ready and waiting. However, since biotrophic pests are extremely specific, a disease of maize is unlikely to cause problems for a bean plant. In a rotating crop system, diseases tailored to last season's crops emerge to find their preferred host gone, considerably reducing disease-related harvest losses.

Crop rotation can also involve reversion to pasture, allowing farm animals to wander freely on the temporarily unmanaged field and deposit manure to fertilize the soil. The long roots of native grasses will also help stabilize the soil in the following seasons, and the pasture reduces the need to import animal feed to the farm.

### 6.6.3 Directed Genetic Modification

Evolution can only act on traits that arise spontaneously in a population. Multiple mutations can lead to the same phenotype, as exhibited by the independent development of C4 and CAM pathways in multiple plant families. This is known as **convergent evolution** because different evolutionary routes have moved toward, or converged on, the same adaptation. However, a C3 plant cannot suddenly develop C4 photosynthesis, or another alternative pathway, if the underlying genes are not present. This is true for human-influenced evolution as well as for wild processes. Unless the right genes are present, selective breeding will come up against the hard limit of photosynthetic efficiency.

However, there is another way to encourage development of new photosynthetic pathways. **Genomic**, or gene-altering, techniques make it possible for scientists to manage the expression of individual genes or gene clusters, changing the number or type of molecules a cell creates, and observe the effects rapidly. For example, a team of scientists has achieved greater water efficiency in a plant simply by "instructing" the plant cells to increase the amount of a photosynthesis protein they already make[5] (Głowacka et al. 2018). In laboratory and greenhouse settings, the effects of these changes can be observed and evaluated before plants reach sexual maturity, potentially allowing for selection to occur more than once per generation. Already, altered genomes that reduce photorespiration have resulted in significant biomass increases in tobacco, a C3 plant commonly used as a model for many crops (South et al. 2019).

Currently, the acceptance of genetically modified organisms (**GMOs**) is hotly contested. It is not within the scope of this text to explore this important issue in the depth it deserves. Interested readers should refer to the textbook *Genetically Modified Foods: Basics, Applications, and Controversy* by Salah E.O. Mahgoub. Briefly, GM organisms represent a valuable source of biological variation that could be used to significantly increase the nutrition value and yield of important global food sources (Agre et al. 2016). Though the matter is still under investigation, so far, the process of molecular engineering has not been shown to make plants or their products any more dangerous or allergenic than traditionally bred foods (Norris 2015; Xu 2015). Some genetic modifications have allowed farmers to reduce the amount of synthetic inputs used on their fields, for example, by allowing plants to produce their own anti-pest compounds and reducing the need for pesticides (Folger 2014). However, as in any evolutionary arms race, the changes induced by genetic modification are never one-sided. Pests can develop resistance to artificially placed defensive genes just as pests can adapt to naturally evolved plant defenses or synthetic pesticides. Furthermore, some genetic modifications in crops encourage greater use of synthetic inputs, such as with the creation of herbicide-resistant GMOs. Because the crops are resistant to herbicides, farmers may be encouraged to apply excessive herbicides to kill off weeds in fields. Many of these herbicides are known to be dangerous to human health (Cohen 2020). It is also important to consider how genetic modifications may affect closely related plant species, or weeds, if cross-breeding occurs between fields and wild species. Planting GM organisms could also lead to sociocultural issues in certain areas. Finally, genetically modified crops could represent a possible threat to farmers' livelihoods if large

---

[5] Unfortunately, when the plants have ample access to water, the increase in water efficiency is accompanied by a decrease in biomass. Biology is messy!

corporations are able to patent genetically altered crop species and charge unreasonable rates for seeds in high demand (see Chapter 8 in this volume).

## 6.7  DEVELOPING COOPERATIVE VARIETIES OF PLANTS AND FUNGI

This chapter has explored the possibilities and limitations of selective breeding from an evolutionary perspective. The metabolite limitations of crop growth prevent yield from increasing perfectly in step with fertilizer input or even with overall plant mass. Even when plants have sufficient water, nutrients, and sunlight, the amount of solar energy harnessed by photosynthetic pigments and the rate of photorespiration limit how many biological molecules a plant can craft. Maintenance requirements further limit how much carbon fixation can go directly to added biomass, and from there to yields. Photosynthetic efficiency can be increased to some extent through selective breeding, but that process relies on the existing genetic diversity in a population. It is currently possible to genetically modify crops to gain certain traits more quickly and directly than traditional breeding systems allow, though there is some controversy surrounding this type of genetic modification. However, the best way to increase yields and photosynthetic efficiency will likely combine multiple techniques – including the use of fungal partners.

Many of the principles for developing new plant varieties do not apply to fungal crops, which are typically grown clonally. Mutations can be induced by irradiation or by mutagenic chemicals or can arise purely by accident. For example, in 1925 a strain of common brown button mushrooms, *Agaricus bisporus*, generated a white button mushroom instead. The pigment-free mycelium was easily cultured and, as with the rise of white bread around that time, rapidly became the more popular variety. Since then, many growers have sought mutations in other species in hopes of generating white strains (Gall 2016).

It is also possible to develop new varieties by selecting certain spores or hyphae from mating-compatible organisms and observing the resulting dikaryotic mycelium for favorable traits. This method parallels crossbreeding of plant varieties, though the results can be discerned much faster. However, the experimental success rate of basidiomycete crosses is just above 25%, with the chance of favorable mutations even lower. In 2011, one basidiomycete breeding experiment led by Dr. Jia Lee had just one out of 190 crosses result in a mushroom strain more productive than its parents – a 1% success rate (Lee et al. 2011). Still, because fungi reach maturity and express favorable phenotypes more quickly than plants, strains with favorable traits can quickly be put into production.

However new varieties are generated, human selection can change the fungal population of a growth house almost immediately by simply inoculating the next round of growth boxes with a different strain of mycelium. Wild mycelia are not nearly so contained, so removing an old strain from a growing house or field would be more difficult; however, allowing the old strain to exhaust its substrate and letting the soil dry out between crops allows a smoother transition to new varieties.

Many mushrooming fungi are excellent candidates for intercropping with agricultural plants. Among their many benefits, discussed extensively in Chapters 11 and 12, mycorrhizal fungi can increase the water efficiency and photosynthetic efficiency of their host plants. Because plant breeding is more established than fungal breeding, it makes most sense to develop a plant species, then expose a proposed fungal partner to it. Inducing the fungal partners to mutate within the partnership may lead to the creation of a more suitable mycorrhizal partner or one which develops larger fruiting size within the field environment, such as the larger fruiting mushroom mutants developed by Dr. Lee's team in 2011.

Fungi also do not have the same energy efficiency limits as plants; while plants harness only a fraction of the sun's energy, fungi can harness all of the digestible nutrients in their substrate. Because they lack photosynthetic processes including Rubisco, fungi can also keep and use every carbon atom they take in – none is lost to photorespiration. Hence, fungal composting represents a highly carbon-efficient technique for turning farm waste into usable soil (see Chapters 16 and 17 in this volume). Likewise, mycorrhizal inoculation is a highly carbon-efficient method to bolster plants' nutrient uptake, energy efficiency, and stress resistance.

## REFERENCES

Agre, P., Alferov, Z.I., Altman, S., et al. 2016. Nobel laureates to the leaders of Greenpeace, the United Nations and governments around the world, June 29, 2016. In *Support Precision Agriculture*, Managed by Sir Richard J. Roberts. https://www.supportprecisionagriculture.org/nobel-laureate-gmo-letter_rjr.html.

Anderson Jr., A. 1975. The green revolution lives. *New York Times*. 27 April 1975, p. 240. https://www.nytimes.com/1975/04/27/archives/the-green-revolution-lives-the-green-revolution.html (accessed 5 Dec 2019).

Autio, W. 2019. Apple yield efficiency. Extension apples. *USDA Cooperative Extension*, August 22, 2019. https://apples.extension.org/apple-yield-efficiency/.

Cohen, P. 2020. Roundup maker to pay $10 billion to settle cancer suits. *New York Times*. 25 June 2020, p. B1. https://www.nytimes.com/2020/06/24/business/roundup-settlement-lawsuits.html (accessed 15 Jan 2021).

Eisenhut, M., Bräutigam, A., Timm, S., et al. 2017. Photorespiration is crucial for dynamic response of photosynthetic metabolism and stomatal movement to altered $CO_2$ availability. *Molecular Plant* 10(January 2017):47–61. https://doi.org/10.1016/j.molp.2016.09.011.

Feller, U., Anders, I., and T. Mae. 2007. Rubiscolytics: fate of Rubisco after its enzymatic function in a cell is terminated. *Journal of Experimental Botany* 59(7)(May 2008):1615–24. https://doi.org/10.1093/jxb/erm242.

Gall, I. 2016. *Morel Dilemma Episode 4: The Taming of the Shroom*. Directed by Izzie Gall. Podcast.

Głowacka, K., Kromdijk, J., Kucera, et al. 2018. Photosystem II Subunit S overexpression increases the efficiency of water use in a field-grown crop. *Nature Communications* 9:868. DOI:10.1038/s41467-018-03231-x.

Gosz, J.R., Holmes, R.T., Likens, G.E., and F.H. Bormann. 1978. The flow of energy in a forest ecosystem. *Scientific American* 238(3):92–102. DOI:10.1038/scientificamerican0378-92.

Hirst, K. 2020. The three sisters: The traditional intercropping agricultural method. *ThoughtCo.com*. https://www.thoughtco.com/three-sisters-american-farming-173034.

Kellogg, E.A. 2013. C4 photosynthesis. *Current Biology* 23(14): R594–599. DOI: 10.1016/j.cub.2013/04/066.

Lattanzi, F.A. 2010. C3/C4 grasslands and climate change. In *Grassland in a Changing World*, eds. H. Schnyder, J. Isselstein, F. Taube, et al. vol. 15: Grassland Science in Europe. ISBN 978-3-86944-021-7.

Lee, J., Kang, H., Kim, S., Lee, C., and H. Ro. 2011. Breeding of new strains of mushroom by basidiospore chemical mutagenesis. *Mycobiology* 39(4):272–77. DOI:10.5941/MYCO.2011.39.4.272.

Mohler, C.L., and S.E. Johnson. 2009. Crop rotation effects on soil fertility and plant nutrition. In *Crop Rotation on Organic Farms: A Planning Manual*, eds. C.L. Mohler and S.E. Johnson. Sustainable Agriculture Research and Education (SARE) Outreach. https://www.sare.org/publications/crop-rotation-on-organic-farms/physical-and-biological-processes-in-crop-production/crop-rotation-effects-on-soil-fertility-and-plant-nutrition/.

Norris, M. L. 2015. Will GMOs hurt my body? The public's concerns and how scientists have addressed them. *Science in the News*. Harvard University. http://sitn.hms.harvard.edu/flash/2015/will-gmos-hurt-my-body/

Peterhansel, C., Horst, I., Niessen, M., et al. 2010. Photorespiration. *The Arabidopsis Book*, 2010(8). https://doi.org/10.1199/tab.0130.

Ricklefs, R. E., and G. L. Miller. 2000. Energy flow in ecosystems. Chapter 10. In *Ecology*, eds. R.E. Ricklefs and G.L. Miller, 192–97. New York: W. H. Freeman and Comp.

South, P. F., Cavanagh, A. P., Liu, H. W., and D. R. Ort. 2019. Synthetic glycolate metabolism pathways stimulate crop growth and productivity in the field. *Science* 363(6422):45–55. DOI: 10.1126/science.aat9077.

Taub, D.R. 2010. Effects of rising atmospheric concentrations of carbon dioxide on plants. *Nature Education Knowledge* 3(10):21. https://www.nature.com/scitable/knowledge/library/effects-of-rising-atmospheric-concentrations-of-carbon-13254108/ (accessed March 8, 2022).

Unkovich, M., Baldock, J., and M. Forbes. 2010. Variability in harvest index of grain crops and potential significance for carbon accounting: examples from Australian agriculture. Essay. In *Advances in Agronomy*, ed. D.L. Sparks, 105:173–219 Amsterdam: Elsevier, Academic Press.

Walker, B. J., VanLoocke, A., Bernacchi, C.J., and D.R. Ort. 2016. The cost of photorespiration to food production now and in the future. *Annual Review of Plant Biology* 67:107–29. DOI:10.1146/annurev-arplant-043015-111709.

Xu, C. 2015. Nothing to sneeze at: the allergenicity of GMOs. *Science in the News*. Harvard University, August 15, 2015. http://sitn.hms.harvard.edu/flash/2015/allergies-and-gmos/.

Zhu, X., Long, S.P., and D.R. Ort. 2010. Improving Photosynthetic Efficiency for Greater Yield. *Annual Review of Plant Biology* 61:235–61. DOI:10.1146/annurev-arplant-042809-112206.

# 7 Pests and Pest Management Methods

*Juan F. Barrera*

## CONTENTS

## 7.1 INTRODUCTION

Pests have been a headache for humanity since prehistoric times; most of the time, however, humans have won confrontations with pests thanks to their ingenuity. The aim of this chapter is to highlight the importance of pests of agricultural interest and to explore the strategies and control methods humanity has used to avoid collapsing under their siege. We are still far from claiming victory in the fight against pests; for this reason, at the end of this chapter, I reflect on the current pest management paradigm, which is largely dominated by pesticides. In the interest of providing a solution that helps reduce the environmental deterioration that Earth suffers from agricultural intensification, deforestation, and population growth, I summarize some recent alternative pest management strategies.

## 7.2 HUMANS VERSUS PESTS

Competition with pests for resources such as food, fiber, and wood (Headley 1975; Norris et al. 2003) has caused huge disasters that devastated entire human populations. There have been numerous instances of human groups that left their communities or countries due to attacks by agriculturally important pests. One of the most representative historical cases was the forced migration of

DOI: 10.1201/9780429320415-7

the Irish to the United States and other countries in the 19th century (Edwards and Williams 1994). The migration – and widespread starvation within Ireland itself – was caused by a phytopathogenic oomycete, *Phytophthora infestans*. This "late blight" devastated farmers' fields, depriving the entire population of its most important food: potato (*Solanum tuberosum*). Even centuries later, *P. infestans* continues to cause serious damage to various cultivated plants of the Solanaceae family around the world (Cooke et al. 2012; Chowdappa et al. 2015; Fry 2016).

The biblical desert locust (*Schistocerca gregaria*) is yet another devastating pest. According to the World Food Program of the United Nations, the locust is the most dangerous migratory pest in the world. In 2019, there was a huge population outbreak of this pest, the most serious outbreak in the last 25 years for Ethiopia and Somalia and the worst in the last 70 years for Kenya (UN 2020). By destroying crops and fodder, this pest threatened the food security, livelihoods, and economies of eight countries in the Greater Horn of Africa, a region whose people rely heavily on agriculture and pastoralism (FAO 2020).

"Coffee malaria" or "coffee rust", caused by the phytopathogenic fungus *Hemileia vastatrix*, has jeopardized entire regions whose main source of livelihood is the coffee crop (*Coffea* spp.). This disease was first discovered in southern India and Ceylon (now Sri Lanka) in 1869; since then, *H. vastatrix* spread to the Pacific and Indian Ocean Basin (1870–1920), West Africa (1950–1960), and finally the Americas in the late 1960s, causing economic losses ranging from slight disturbances to the destruction of more than 90% of crops (McCook 2006). However, coffee rust reached one of its largest and most serious infestations in Colombia, Mexico, Central America, Peru, and Ecuador in 2008–2013. The impacts of this rust epidemic on crop production seriously affected the income and livelihoods of farmers and workers, causing a crisis in regional food security (Avelino et al. 2015). Most of the people in these countries have managed coffee plantations quite successfully to reduce the losses caused by the "Big rust", but fears persist of another epidemic of similar magnitude in the future.

## 7.3   PEST MANAGEMENT STRATEGIES AND TACTICS

Throughout the millennia, humans have resorted to all forms of control to eliminate, or at least reduce, the impacts of pests. Everything has been used in this all-out fight, from netting and swatting insects to chemical pesticide bombs and mRNA vaccination of humans and animals.[1] Due to the experience gained in this ongoing war between humans and pests, pest management has evolved into what is now understood as Integrated Pest Management (**IPM**). The IPM paradigm can be summarized as the strategy that seeks to reduce or eliminate the economic impact of a pest while preserving the integrity of the environment and the well-being of society (Norris et al. 2003).

The two components of IPM are strategy and tactics. According to Pedigo (1999), a pest management **strategy** is the plan or program to reduce – or eliminate if possible – a real or potential pest problem. The design of the strategy is based on the bioecological traits of pests, socioeconomic factors associated with their management, and the pests' interaction with the species that we are interested in protecting. Once the strategy is designed, the most appropriate methods, also called **tactics**, are chosen to meet the objectives and goals of the plan.

Five types of pest management strategies are recognized (Norris et al. 2003):

- **Prevention**. The goal of this strategy is to prevent the arrival or establishment of a pest in a non-infested area. Given the difficulty of eliminating an economically important pest once it is established in an area, its introduction should be avoided at all costs. This strategy benefits from pest risk analysis, which informs and guides decisions from domestic pest surveillance to imported product quarantine (Devorshak 2012).

---

[1] Such as the iPED RNA vaccine, used in pigs to prevent Porcine Endemic Diarrhea Virus (Aida et al. 2021).

- **Temporary alleviation**. This strategy is followed to "extinguish" outbreaks or hot spots of pest infestation. It is less expensive to reduce pest outbreaks before they reach high levels of infestation.
- **The farm-level management of populations**. This strategy is used in cases where the pest is well established on the farm and losses can be considerable if its population is not managed. This means that humans must live with the pest; in other words, it must be assumed that from time to time pest populations will increase and some management action will be necessary. Among the strategies mentioned here, this strategy is the most common in pest management.
- **Area-wide pest management**. This strategy is used for pests that need to be managed on a regional scale to lower their populations. Many of the most important pests must be controlled with an area-wide pest management approach, as pests do not respect boundaries between farms or countries. To achieve greater success with this strategy, the cooperation of all sectors involved in pest management is essential.
- **Eradication**. The purpose of this strategy is to eliminate the entire population of a pest in an area or region. Because organisms can have ecosystem roles that are not always obvious to human observers, this strategy is desirable only when dealing with an invasive alien species.

Depending on the strategy pursued, agricultural pest management tactics or methods can be grouped into the following approaches (Luckmann and Metcalf 1975; Norris et al. 2003):

- **Manipulation of the pest species**. These tactics have a direct effect on the pest organism or its behavior. Tactics can include legal norms or laws for prevention (quarantines and suppression or eradication programs); use of non-chemical methods such as biological control (introduction, conservation, and augmentation of natural enemies); ethological control (attractants and repellants); mechanical control (destruction by hand, e.g., destruction by slapping, crushing, or grinding); exclusion barriers or sticky traps; physical control (heat, cold, humidity, light, sound); and the use of prophylactic and curative pesticides (sterilants, growth inhibitors, insecticides, fungicides, herbicides, bactericides, etc.). Pesticides should be used as a last option to avoid their undesirable effects.
- **Manipulation of the cultivated plants**. These tactics are used to increase the tolerance or resistance of the crop to attack by pests, or to change the crop in use for one that is not susceptible to the pest. There are two groups of tactics: (1) cultural control that includes crop rotation, destruction of crop residues, tillage of soil, variation of sowing or harvest dates, pruning, fertilization, sanitation, water management, and planting of trap crops; and (2) generating or enhancing host-plant resistance through genetic improvement, either through crossing and selection techniques or by the insertion of genes (genetic engineering).
- **Manipulation of the environment**. The purpose of these tactics is to alter the environment so that pest populations do not cause economic losses. Physical methods are used to modify the micro-habitats of the crop (e.g., opening the crop canopy to improve airflow and reduce fungal infection) or the environment that surrounds the crop (area-wide pest management programs).

## 7.4   ECONOMICS AND ECOLOGY OF PESTS

A **pest** can be defined as a species that interferes with human activities and desires; this means that any species, from a tiny viral particle to the most complex multicellular being, can be a blacklisted pest. In the case of agricultural systems, pests can be defined as "all organisms within the crop environment that cause injury to the crop and are capable of reducing yield or quality" (Norris et al. 2003). The term "pest", according to this definition, has a clear economic consequence in terms of the monetary losses that a species can cause. In other words, "pest" is a purely anthropocentric term – without ecological validity – because it exclusively conforms to the interests of the human being. Furthermore, the label lacks temporal stability, since a species that is perceived as a pest in a certain place and time, under other circumstances could play an important role in the ecology of an ecosystem (Pedigo 1999) or in the economy and culture of a society (Olckers 2011). For example, the species of desert locust that led to major food security issues for an estimated 20 million people in 2019–2021, is also a traditional part of the diet and trade in more than 20 African and Asian countries (Egonyu et al. 2021).

In this section the concept of a pest will be addressed under two approaches: economic and ecological. Both approaches are useful because they guide the choice of the most appropriate strategy and tactics for pest management.

### 7.4.1 THE ECONOMIC APPROACH

From the viewpoint of pest management, especially in the field of agriculture, pests can be classified according to the taxonomic group to which they belong, their morphological and life cycle features, the range of hosts or goods they affect, the type of problem they cause, and their permanent or transient residence in the system, among other categories (Conway 1981; den Belder and Sediles 1985). However, here we are interested in classifying them according to their economic importance. In this sense, the term "pest" is often assigned based on the population density of a species: the greater the number of individuals, the greater the potential to cause economic losses.

Stern et al. (1959) were the first to conceptualize the economics of the term "pest" in a decision-making process, and their work has been fundamental to developing and implementing the IPM paradigm (further discussed in Chapter 13). Later, Higley and Pedigo (1996) deepened and broadened the scope of these concepts. The proposal of Stern et al. (1959) was based on defining two management action/inaction thresholds for IPM, the **economic-injury level** (EIL) and the **economic threshold** (ET). These thresholds refer to the density or intensity of the pest infestation. The EIL was defined as the smallest number of individuals capable of causing economic losses, while the ET was defined as the number of individuals that would trigger the management action. Therefore, the ET will always be lower than the EIL. Because the ET anticipates the imminent danger of inaction if the organism population exceeds the EIL, it can be compared to the yellow signal that precedes the red signal of a traffic light. In other words, the ET is the operational decision rule for pest management.

In the case of an agriculturally important insect pest, the EIL is calculated with the following equation (Pedigo 1999):

$$EIL = \frac{C}{V \times I \times D \times K} \tag{7.1}$$

where $C$=cost of management per area (e.g., \$/acre); $V$=market value per unit of produce (e.g., \$/bushel); $I$=injury units per insect per production unit (e.g., percent defoliation/insect/acre, expressed as a proportion); $D$=damage per unit injury (e.g., bushels lost/acre/percent defoliation); and $K$=proportionate reduction in potential injury or damage in pest attack or the expected proportion of the population killed by the management tactic (e.g., 0.8 for 80%).

The ET is estimated with this equation:

$$ET = EIL \times C^{-x} \tag{7.2}$$

where $C$=factor of increase in the growth rate of the pest population per unit of time (e.g., 0.5/week, increasing 50% per week) and $x$=time (e.g., 4 weeks). ET is more difficult to estimate than the EIL since its value depends on being able to predict the direction that the pest population density will take at some point in the future. Practically, the ET is set as a percentage of the EIL to model time delays in management (*tdm*) (e.g., 80% EIL, expressed as a proportion):

$$ET = EIL \times tdm \tag{7.3}$$

Take the following case to illustrate the calculation of action thresholds. Higley and Wintersteen (1996) calculated the EIL and ET for control of the European corn borer, *Ostrinia nubilalis* (Hübner), with a soil-applied granular insecticide that used the bacterium *Bacillus thuringiensis* Berliner as the active ingredient. The component values of Equation 7.1 were: $C$ = \$16.90/acre (application cost = \$3.80/acre + insecticide cost = \$13.10/acre); $V$ = \$2.35/bushel; $I$ = 0.066 foliage lost/larva/plant; $D$ = 125 bushels/acre; and $K$ = 0.5 control (assuming only 50% of the pest population was killed by the insecticide). Applying Equations 7.1 and 7.3, the values of 1.74 and 1.39 larvae/plant were obtained for EIL and ET, respectively:

$$EIL = \frac{\$16.90 \, / \, acre}{\$2.35 \, / \, bushel \times (0.066 \, lost \, / \, larva \, / \, plant \times 125 \, bushels \, / \, acre) \times 0.5 \, control}$$

$$EIL = 1.74 \, larvae \, / \, plant$$

If 80% of the EIL is set for time delays in pest management (*tdm*), the ET can be calculated with Equation 7.3 as follows:

$$ET = 1.74 \, larvae \, / \, plant \times 0.80$$
$$ET = 1.39 \, larvae \, / \, plant$$

Therefore, for the system studied by Higley and Wintersteen (1996), management of the corn borers should be triggered when the average number of corn borer larvae is observed at 1.39 per plant (ET), before infestation reaches the economically damaging level of 1.74 larvae per plant (EIL).

The relationship between the EIL and the long-term mean density of a pest population (**general equilibrium position**, GEP) can be used to determine pest population status and identify the most appropriate management strategy. Based on the relationship between the EIL and the GEP, four types of pests can be recognized (Figure 7.1): non-economic, occasional, secondary, and primary (den Belder and Sediles 1985; Pedigo 1999).

- **Non-economic or subeconomic pests**. The GEP and the highest population fluctuations are well below the EIL (Figure 7.1a); "do nothing" is the best strategy, but these pests should be monitored to detect a change in status early or in case they appear as part of a pest complex.
- **Occasional or potential pests**. The GEP of these pests is below the EIL, but the highest fluctuations in population density can exceed it (Figure 7.1b). These pests are common and occur without causing economic loss, so a "wait and see" attitude is assumed. The best strategy is not to strive to reduce GEP, but rather to implement a therapeutic management approach with mainly chemical or biological pesticides[2] based on early detection to buffer the peaks of population outbreaks if they reach ET.
- **Secondary or perennial pests**. Most of the time these pests cause economic damage because the GEP is below but close to the EIL (Figure 7.1c); hence, some secondary pests eventually or intermittently become key pests. Population outbreaks of these pests result from changes in the management of the system that favor their population growth (e.g., elimination of their natural enemies). The management strategy is to reduce the GEP of the population.

---

[2] In some specific cases physical and mechanical tactics can be used such as direct destruction of pests (collection and destruction of pests), or behavioral control tactics of pests such as mass trapping or mating disruption with sex pheromones.

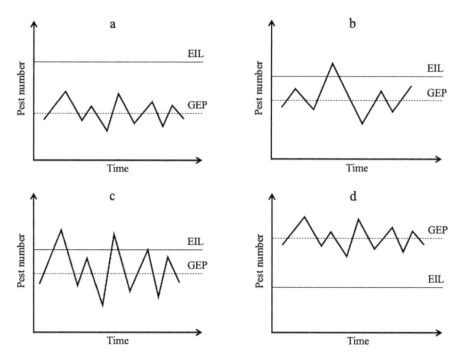

**FIGURE 7.1** Types of pests according to the relationship between the economic-injury level (EIL) and the general equilibrium position (GEP) of the population. (a) non-economic pest; (b) occasional pest; (c) secondary pest; (d) primary pest. (Own elaboration based on Pedigo 1999.)

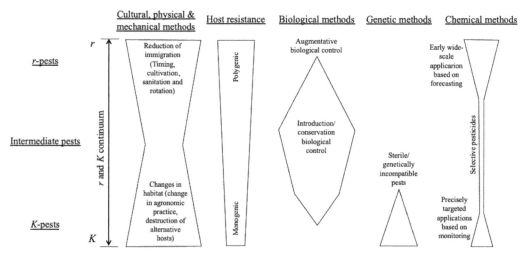

**FIGURE 7.2** Main control methods and their relationship with the continuum of pest strategies $r$ and $K$. (Own elaboration based on Conway 1981; den Belder and Sediles 1985.)

- **Primary or serious pests**. Because GEP is above EIL, these pests always represent a serious problem (Figure 7.1d); these pests are also called **key** pests and the management strategy is to implement tactics that lower the GEP of the population.

### 7.4.2 THE ECOLOGICAL APPROACH

Conway (1981) used the $r$ and $K$ strategies ecological species concept – first proposed by MacArthur and Wilson (1967) – to classify pests. He proposed the term "pest" as part of a continuum whose extremes are the strategies $r$ and $K$. As we will see, conceptualizing pests under this ecological approach has important implications for their management (Figure 7.2).

#### 7.4.2.1 *r*-Pests

*r*-Pests have high rates of population increase, as well as a high dispersal and host search capacity. The strategy of these species is to exploit plants with short life cycles that grow during an annual growing season. $r$-Pests are generally species that attack the foliage and roots of plants. These species are the most difficult to control. Because their invasions are frequent and the damage caused by their explosive population outbreaks is high, the use of chemical and biological pesticides (e.g., entomopathogens[3]) is the most popular control method; however, genetically resistant crop plants or animals (polygenic resistance) and cultural control (e.g., planting date and crop rotation) can also help reduce the size of $r$-pest populations. The temporality and management intensity of arable crops affects the stability of these systems and can limit the effectiveness of classical and conservation biological control against these pests. For this reason, periodic massive releases of biological control agents, in particular, the use of entomopathogens, have more chances of success in this type of agroecosystem. The high rates of population increase and the short biological cycles that produce several generations per year, along with the frequent use of chemical pesticides to control $r$-pests, favor the rapid development of pest resistance to control chemicals. The most practical strategy for dealing with this problem is either not to use pesticides at all or to use them judiciously to decrease the selection pressure they exert on pest populations (Figure 7.2).

#### 7.4.2.2 *K*-Pests

*K*-Pests have low growth rates, high ability to compete, and more specialized food preferences than $r$-pests. These species exploit plants with long life cycles (perennials) such as those in forests or orchards. Although the populations of these pests are low, they can cause serious crop damage, since many attack fruits or seeds. In theory, these species are more vulnerable to control, as their populations can be reduced to such low levels that the possibility of eradication is high. This is especially true in the case of insects that, among other characteristics, mate only once in their lives. In these cases, the sterile insect technique (SIT) is considered the most appropriate control. SIT consists of mass rearing of millions of insects in the laboratory, generally males, which are sterilized before being released into the field so that the wild females mate with them and fail to produce offspring. Precisely and timely applied pesticides can also work well for species with low populations that cause high damage. Because $K$-pests are specialists, cultural control methods that make the host less attractive or that destroy the pests' specialized niches are effective (e.g., changing stocking density, pruning, eliminating wild hosts). The monogenic resistance (inherent or developed through breeding) of plants or animals has a high potential to control this group of pests (Figure 7.2).

#### 7.4.2.3 Intermediate Pests

Most pests are grouped in this category. Normally, natural enemies regulate intermediate pests; therefore, these "intermediate" species become important pests when humans eliminate or reduce the efficiency of their natural enemies, such as secondary pests (Figure 7.1c). Because biological control is the most useful method against intermediate pests, the use of ecologically and physiologically selective pesticides is recommended to reduce the impact of pesticides on the natural predators or competitors of intermediate pests (Figure 7.2).

---

[3] Fungi that prey on arthropods.

## 7.5   THE FUTURE OF PEST MANAGEMENT

As the human population continues to grow, additional pressure is placed on agricultural systems. Agricultural intensification and deforestation contribute to landscape fragmentation and biodiversity loss, while globalization of agriculture creates more opportunities for introduction of invasive species, driving increases in pesticide use. These factors increase agricultural vulnerability to pests and make it more difficult to maintain the environmental services offered by natural control (Bianchi et al. 2006; Mainka and Howard 2010; Mishra et al. 2021; Morand and Lajaunie 2021; Sharma et al. 2019; Silva et al. 2021). Simultaneously, global warming and the globalization of the economy are threats that will accentuate the damage of some pests and favor the spread of others (Roques and Auger-Rozenberg 2019; Skendžić et al. 2021).

In this context, it is worth asking if in the fight against pests we are using the methods that will be most effective in the long term, or what should be done to tip the balance in our favor. Increasingly, agricultural scientists believe cultivation techniques must also prioritize land conservation (see Chapter 10 in this volume). Several alternatives to IPM have arisen that are based on agroecological pest management practices, so that crops do not depend on the use of pesticides or so that pesticides are used selectively and as a last weapon of defense. Therefore, actions must be taken to provide crops with mechanisms of "self-protection" or "autonomous control" (Vandermeer et al. 2010) – that is, "help crops help themselves" (Mann 2021). Many modern pest control proposals emphasize addressing pest problems under an integrated, total system approach, so that decision-making contributes to lower pest risk and vulnerability while increasing response capacity and the resilience of the socio-ecological system towards pests (Barrera 2020).

### 7.5.1   BIOLOGICALLY INTENSIVE IPM

Biologically intensive IPM follows a systems approach to pest management based on an understanding of the pest ecology. It begins with an accurate diagnosis of the nature and causes of the problems that result in the appearance of pests. It promotes plant health via resistant varieties, crop rotation, disruption of pest reproduction, and management of biological processes that keep pest populations within acceptable limits, such as biodiversification and encouragement of the pests' natural enemies (Benbrook et al. 1996; Frisbie and Smith 1991).

### 7.5.2   AGROECOLOGICALLY INTEGRATED PEST MANAGEMENT

This alternative proposes to design and build vegetation architectures that protect natural enemies of pests or that include plants with direct deterrent effects on pests. It emphasizes the restoration of natural control mechanisms to achieve optimal performance of natural enemies by adding a diversity of plants, rather than trying to force the establishment of biological control in monocultures where essential ecological elements are lacking. Management of diverse vegetation has the additional benefits of conserved energy, improved soil fertility, reduced erosion risks, and reduced dependence on external inputs (Altieri 1994; Andow 1991).

### 7.5.3   ECOLOGICALLY BASED PEST MANAGEMENT (EBPM)

Based on cultural and biological approaches, EBPM is built on the understanding of the managed system, including the natural processes of pest suppression. It takes advantage of technological advances to improve basic knowledge of the agroecosystem and, if necessary, uses chemical, biological, or physical inputs to achieve pest management strategies that are profitable, durable, and safe to farmers, workers, and consumers (NRC 1996).

### 7.5.4 TOTAL SYSTEM APPROACH

The Total System Approach of pest management considers ecosystem interactions and seeks solutions that have net benefits at the level of the entire ecosystem. Its management strategy begins by questioning how certain organisms have become identified as "pests" and trying to find the weak points of the agroecosystem that have led to the system imbalance. Unsustainable approaches that only treat symptoms of the problem, like pesticides, should only be used as a last resort. Therapeutic tactics (e.g., chemical and biological pesticides) are only required to reinforce natural regulators, such as the crop plants' inherent defenses, cover crops, intercropping, crop rotation, soil fertility conservation, natural enemies of the pests, and other inherent components of the system (Lewis et al. 1997).

### 7.5.5 INTEGRATED MANAGEMENT OF BIODIVERSITY (IMB)

IMB reconciles IPM and biodiversity conservation. As long as the densities of the pest species remain below the EIL, the species are to be conserved (e.g., managed to keep them above the extinction threshold). IMB requires sufficient habitats in the range of distribution of the species that are to be conserved, so that they complete their life cycles and their populations persist (Kiritani 2000).

### 7.5.6 INTEGRATED CROP MANAGEMENT (ICM)

ICM proposes to integrate the management of individual crops so the system can benefit from intercrop interactions, such as pest control or maintenance of soil fertility. Three components characterize the ICM: environmental aspects (a relatively more sustainable use of resources for producing food), economic viability (food production must be a profitable business), and modern technologies (e.g., judicious use of agrochemical and other inputs). It seeks the efficient and profitable use of inputs and considers crop rotation as an ideal long-term strategy to minimize weed and pest problems (Bradley et al. 2002).

### 7.5.7 INTEGRATED PRODUCTION

Integrated Production is a concept of sustainable agriculture based on agroecology and a system approach that aims at contributing to sustainable, resilient, profitable, and robust farming systems. It consists in producing high-quality food and other agricultural products using natural resources and regulatory mechanisms in place of polluting inputs to ensure the sustainability of the system. The components of Integrated Production are the central function of the agroecosystem, balanced nutrient cycles, production methods respecting environment and animal welfare, conservation and improvement of soil fertility, and a diversified environment; the observance of certain ethical and social criteria are also essential components of Integrated Production (Boller et al. 2004; Wijnands et al. 2018).

### 7.5.8 INTEGRATED AGROECOSYSTEM DESIGN AND MANAGEMENT (IADM)

IADM aims to proactively solve all pest problems through the Efficiency-Substitution-Redesign (ESR) concept, which emphasizes a "deep organics" ecosystem design (or redesign) approach in the ongoing development of food and agricultural systems. The "deep organics" concept applies to whole food and fiber systems, and considers all associated underlying aspects (philosophical, ethical, psychological, psychosocial, sociocultural, and spiritual). Such redesign might include the use of several cultivars with complementary functions (resistant, trap and insectary plantings), the design and management of vegetation structures, the careful balancing of plant nutrients, and the appropriate timing of all operations to favor the crops over pests and encourage natural pest controls. Additionally, ESR proposes to include personal (psychological) and cultural (psychosocial) development considerations and initiatives in all efforts to allow progressive change from conventional agroecosystems to the preventive stage of IADM. Disruptive curative interventions like chemical pesticides would be used only in emergencies (Hill 2014).

### 7.5.9 Crop Health and Salutogenesis

Under this paradigm, a healthy farming system is defined as "one that tends to achieve short-, medium-, and long-term goals of the farmer (or community of farmers), contributing to meeting the needs of society, enhancing ecological and social processes by minimizing reliance on external inputs and maximizing safety for humans and ecosystems" (Vega et al. 2019). Crop health is supported by **salutogenesis** (Antonovsky 1996), a science of health development that focuses on factors that promote health, rather than those that cause disease. Therefore, this approach adapts the postulates of salutogenesis, and proposes focus on the well-being of the cropping system instead of its limiting factors (pests, nutrients, or water); on viewing crop health as a continuum between health and disease; on understanding health as a process; on promoting crop health based on salutogenic factors rather than on risk factors; and on learning from organisms called pests rather than concentrating on destroying them.

### 7.5.10 Holistic Pest Management (HPM)

HPM is a participatory regional decision-making system in pest management aimed primarily at the welfare of the human population. It implements processes and products of low environmental impact, of high quality for self-consumption by farmers and workers, and which are strongly competitive in the market. These processes and products are generated from integral production systems. The producer manages these processes as a strategy to focus on the causes of pest outbreaks. Likewise, the producer implements agroecological tactics to minimize the economic, environmental, and social costs and risks derived from pest outbreaks and mismanagement of pest control techniques. Therefore, holistic pest management actions must place the grower at the center of the system (e.g., Farmer first approach, Matteson 1992) and must consider not only the pests but also other important components of the system (Barrera 2020).

## 7.6 CONCLUSION

If we want to avoid or reduce the problems caused by agricultural pests, it is not only necessary to know the biological, ecological, and behavioral aspects that govern the dynamics of their populations, it is also necessary for control strategies and methods to have the welfare of the human being as a guiding principle. To do this, science, technology, and innovation must make agribusiness profitable, but work towards that goal from a starting point of unrestricted respect for life and by conserving, strengthening, and improving the services provided by the environment. Such considerations and techniques will aid in developing agroecosystems that prevent population outbreaks of pests and rely less on external inputs and therapeutic tactics.

## ACKNOWLEDGMENTS

I am grateful to the editors for inviting me to contribute to this chapter and for the generosity with which they reviewed the manuscript. Special thanks to Elizabeth "Izzie" Gall for her suggestions and careful revision of the text.

*This chapter is dedicated to Dr. Jaime Gómez Ruiz (1961–2021), with whom I shared my passion for entomology, biological control, and pest management.*

# REFERENCES

Aida, V., Pliasas, V.C., Neasham, P.J., et al. 2021. Novel vaccine technologies in veterinary medicine: a herald to human medicine vaccines. *Frontiers in Veterinary Science*. April 15, 2021. https://doi.org/10.3389/fvets.2021.654289.

Altieri, M. A. 1994. *Biodiversity and Pest Management in Agroecosystems*. New York: Hayworth Press.

Andow, D. A. 1991. Vegetational diversity and arthropod population response. *Annual Reviews of Entomology* 35:561–8.

Antonovsky, A. 1996. The salutogenic model as a theory to guide health promotion. *Health Promotion International* 11(1):11–18. https://doi.org/10.1093/heapro/11.1.11.

Avelino, J., Cristancho, M., Georgiou, S. et al. 2015. The coffee rust crises in Colombia and Central America (2008–2013): Impacts, plausible causes and proposed solutions. *Food Security* 7:303–21.

Barrera, J. F. 2020. *Beyond IPM: Introduction to the Theory of Holistic Pest Management*. Cham: Springer.

Benbrook, C. M., Groth, E. III, J.M. Halloran, et al. 1996. *Pest Management at the Crossroads*. Yonkers, NY: Consumers Union.

Bianchi, F. J. J. A., Booij, C. J. H., and Tscharntke, T. 2006. Sustainable pest regulation in agricultural landscapes: A review on landscape composition, biodiversity and natural pest control. *Proceedings of the Royale Society B* 273:1715–27.

Boller, E.F., Avilla, J., Joerg, E., et al. 2004. Integrated production. Principles and technical guidelines. *IOBC WPRS Bulletin* 27(2):1–49.

Bradley, B. D., Christodoulou, M., Caspari, C., et al. 2002. Integrated crop management systems in the EU. *Amended Final Report for European Commission DG Environment. Agra CEAS Consulting*. https://ec.europa.eu/environment/agriculture/pdf/icm_finalreport.pdf (accessed November 2, 2021).

Chowdappa, P., Nirmal Kumar, B. J., Madhura, S., et al. 2015. Severe outbreaks of late blight on potato and tomato in South India caused by recent changes in the *Phytophthora infestans* population. *Plant Pathology* 64:191–199.

Conway, G. 1981. Man versus pests. In *Theoretical Ecology: Principles and Applications*, ed. R. M. May, 356–386. London: Blackwell Scientific.

Cooke, D. E. L., Cano, L. M., Raffaele, S., et al. 2012. Genome analyses of an aggressive and invasive lineage of the Irish potato famine pathogen. *PLoS Pathogens* 8:e1002940. https://doi.org/10.1371/journal.ppat.1002940.

den Belder, E., and Sediles, A. 1985. *Control integrado de plagas. Sanidad Vegetal. Entomología. Tomo I. Escuela de Sanidad Vegetal*. Managua, Nicaragua: Instituto Superior de Ciencias Agropecuarias.

Devorshak, C. 2012. Introduction. In *Plant pest risk analysis. Concepts and application*, ed. C. Devorshak, 3-6. London: CAB International.

Edwards, R. D., and Williams, T. D. 1994. *The Great Famine. Studies in Irish History 1845–52*. Dublin: The Lilliput Press.

Egonyu, J. P., Subramanian, S., Tanga, C. M., et al. 2021. Global overview of locusts as food, feed and other uses. *Global Food Security* 31:100574. https://doi.org/10.1016/j.gfs.2021.100574.

FAO. 2020. *Desert Locust Crisis: Appeal for Rapid Response and Anticipatory Action in the Greater Horn of Africa*. Roma: Food and Agriculture Organization of the United Nations. https://www.fao.org/emergencies/resources/documents/resources-detail/en/c/1263633/ (accessed February 22, 2022).

Frisbie, R. E., and Smith, Jr J. W. 1991. Biologically intensive integrated pest management: The future. In *Progress and Perspectives for the 21st Century, Centennial National Symposium*, eds. J. J. Menn, and A. L. Steinhauer, 151–164. Lanham, MD: Entomology Society of America.

Fry, W. E. 2016. *Phytophthora infestans*: New tools (and old ones) lead to new understanding and precision management. *Annual Reviews of Phytopathology* 54:529–47.

Headley, J. 1975. The economics of pest management. In *Introduction to Insect Pest Management*, eds. R. L. Metcalf, and W. H. Luckmann, 75–99. New York: John Wiley & Sons.

Higley, L. G., and Pedigo, L. P. 1996. *Economic Thresholds for Integrated Pest Management*. Lincoln: University of Nebraska Press.

Higley, L. G., and Wintersteen, W. K. 1996. Thresholds and environmental quality. In *Economic Thresholds for Integrated Pest Management*, eds. L. G. Higley, and L. P. Pedigo, 249–274. Lincoln: University of Nebraska Press.

Hill, S. B. 2014. Considerations for enabling the ecological redesign of organic and conventional agriculture: A social ecology and psychosocial perspective. In *Organic Farming, Prototype for Sustainable Agricultures*. eds. S. Bellon, and S. Penvern, 401–422. Dordrech: Springer.

Kiritani, K. 2000. Integrated biodiversity management in paddy fields: Shift of paradigm from IPM toward IBM. *Integrated Pest Management Reviews* 5(3):175–83.

Lewis, W. J., van Lenteren, J. C., Phatak, S. C., et al. 1997. A total system approach to sustainable pest management. *Proceedings of the National Academy of Sciences of the United States of America* 94:12243–8.

Luckmann, W. H., and Metcalf, R. L. 1975. The pest-management concept. In *Introduction to Insect Pest Management*, eds. R. L. Metcalf and W. H. Luckmann, 3–35. New York: John Willey & Sons.

MacArthur, R. H. and Wilson, E. O. 1967. *The Theory of Island Biogeography*. Princeton, NJ: Princeton University Press.

Mainka, S. A., and Howard, G. W. 2010. Climate change and invasive species: Double jeopardy. *Integrative Zoology* 5(2):102–111.

Mann, R. 2021. *Voluntary Best Management Practices to Control Pests without Pesticides*. Minnesota Department of Agriculture. https://www.mda.state.mn.us/voluntary-best-management-practices-control-pests-without-pesticides (accessed November 2, 2021).

Matteson, P.C. 1992. 'Farmer First' for establishing IPM. *Bulletin of Entomological Research* 82: 293-296.

McCook, S. 2006. Global rust belt: *Hemileia vastatrix* and the ecological integration of world coffee production since 1850. *Journal of Global History* 1:177–195.

Mishra, A. J., Arya, R., Tyagi, R., et al. 2021. Non-judicious use of pesticides indicating potential threat to sustainable agriculture. In *Sustainable Agriculture Reviews* 50, eds. V. Kumar Singh, R. Singh, and E. Lichtfouse, 383–400. Cham: Springer.

Morand, S., and Lajaunie, C. 2021. Outbreaks of vector-borne and zoonotic diseases are associated with changes in forest cover and oil palm expansion at global scale. *Frontiers in Veterinary Science* 8:661063. https://doi.org/10.3389/fvets.2021.661063.

Norris, R. F., Caswell-Chen, E. P., and Kogan, M. 2003. *Concepts in Integrated Pest Management*. Upper Saddle River: Prentice Hall.

NRC (National Research Council). 1996. *Ecologically Based Pest Management: New Solutions for a New Century*. Washington, DC: National Academy Press.

Olckers, T. 2011. Biological control of *Leucaena leucocephala* (Lam.) de Wit (Fabaceae) in South Africa: A tale of opportunism, seed feeders and unanswered questions. *African Entomology* 19(2):356–365.

Pedigo, L. P. 1999. *Entomology and Pest Management*. Hoboken, NJ: Prince Hall.

Roques, A, and Auger-Rozenberg, M.-A. 2019. Climate change and globalization, drivers of insect invasions. *Encyclopedia of the Environment*. https://www.encyclopedie-environnement.org/en/life/climate-change-globalization-drivers-of-insect-invasions/ (accessed November 2, 2021).

Sharma, A., Kumar, V., Shahzad, B., et al. 2019. Worldwide pesticide usage and its impacts on ecosystem. *SN Applied Sciences* 1:1446. https://doi.org/10.1007/s42452-019-1485-1.

Silva, G., Tomlinson, J., Onkokesung, N., et al. 2021. Plant pest surveillance: from satellites to molecules. *Emerging Topics in Life Sciences* 5(2):275–87.

Skendžić, S., Zovko, M., Živković, I.P., et al. 2021. The impact of climate change on agricultural insect pests. *Insects* 12(5):440. https://doi.org/10.3390/insects12050440.

Stern, V. M., Smith, R. F., van den Bosch, R., et al. 1959. The integrated control concept. *Hilgardia* 29(2):81–101.

UN (United Nations). 2020. La plaga de langostas es la epidemia migratoria más peligrosa del mundo. *Noticias ONU*. https://news.un.org/es/story/2020/03/1471322 (accessed November 2, 2020).

Vandermeer, J., Perfecto, I., and Philpott, S. M. 2010. Ecological complexity and pest control in organic coffee production: Uncovering an autonomous ecosystem service. *Bioscience* 60:527–37.

Vega, D., Gazzano Santos, M. I., Salas-Zapata, W., et al. 2019. Revising the concept of crop health from an agroecological perspective. *Agroecology and Sustainable Food Systems* 44(2):215–237.

Wijnands, F., Malavolta, C., Alaphilippe, A., et al. 2018. Integrated production. *IOBC-WPRS Objectives and Principles. 4th edition. IOBC–WPRS Commission IP Guidelines*. https://www.iobc-wprs.org/ip_integrated_production/IOBC-WPRS_IP_objectives_and_principles_4th_edition_2018_EN.pdf (accessed February 24, 2022).

# 8 Economic Factors in Agriculture

*Elizabeth "Izzie" Gall and Barbara Weil Laff*

## CONTENTS

## 8.1 INTRODUCTION

In economic systems, supply and demand are competing forces with an extremely complicated relationship. At its most basic, agricultural supply is governed by individual farmers' considerations of how to balance inputs and input costs with projected outputs and income; agricultural demand consists of consumers' preferences for certain agricultural products at certain volumes. In a free market, the supply and demand meet at an equilibrium where both the supplier and the consumer are relatively satisfied that they are exchanging the right volume and quality of product for the right amount of money. However, there are additional considerations that can shift the equilibrium and impact how suppliers and consumers interact with the market. Government interventions change how producers respond to market pressure; consumers and producers may be concerned about ecological impact or other factors not directly related to product consumption; and there may be legal considerations, like intellectual property protections, that impact the supply chain regardless of consumer or producer preferences.

## 8.2 SUPPLY AND DEMAND

Supply and demand are basic economic concepts that explain how markets work. **Supply** is the amount of any product that is being made available to buyers at the market; **demand** is the amount of the product that buyers are willing to purchase at a given price. In agricultural markets, the supply can be any food product, raw food material, or intermediate in a food production chain.

The very basic concepts of supply and demand are often shown as two curves on a graph (see Figure 8.1). The *x*-axis is defined as the quantity of a product, while the *y*-axis represents the

DOI: 10.1201/9780429320415-8

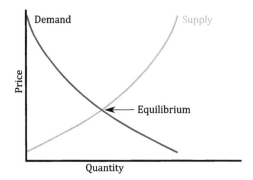

**FIGURE 8.1**   The equilibrium is the point at which suppliers and consumers agree on the quantity of a product that can be bought at a certain price.

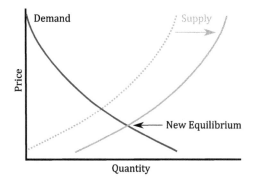

**FIGURE 8.2**   Graph of a supply glut. When supply exceeds demand, the equilibrium shifts right and downwards, representing that more of the product can be bought at a lower price relative to the status quo in Figure 8.1.

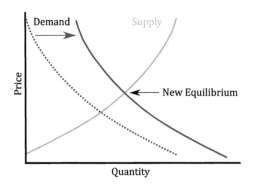

**FIGURE 8.3**   Graph of a supply deficit. When demand exceeds supply, the equilibrium shifts right and upwards, representing that consumers will pay a higher price for the product and demand more of it relative to the status quo in Figure 8.1.

price of the product. The curves of supply and demand are shown intersecting at the point where consumers demand exactly the quantity of product that is supplied. This point is known as the **equilibrium**.

Note that as price increases, demand decreases on the quantity axis. This represents that the amount of product consumers are willing to buy decreases as the price of the product goes up.

Meanwhile, the supply curve shows that suppliers are willing to supply more of the product if it will sell for a high price. This graph, and variations of it, can be useful for demonstrating why prices change when supply or demand are different than normal. We will explore some examples, with Figure 8.1 representing the typical market for corn in a hypothetical region.

Suppose that one year, rain is consistent, the weather is cool, and corn harvests are bountiful. Because the quantity of corn has increased, the supply curve shifts to the right on the graph as supply increases on the "quantity" axis. However, growing conditions have not changed the demand for corn, so the demand curve stays the same (Figure 8.2). Note that when the supply curve shifts to the right and demand does not change, the equilibrium is at a lower price and a higher quantity of product than it was in Figure 8.1. This represents that when the supply of a product is high relative to demand, prices go down. This is known as a supply **glut**. If the glut is extreme enough or lasts for a long time, producers will adjust by decreasing the amount of the product they are producing, reducing the supply to match demand at a new equilibrium.

However, suppose that rain has been inconsistent, the weather has been hot, and maybe there have been a lot of pests destroying corn plants across the region. The corn harvest suffers from these negative effects and the supply curve shifts left relative to the demand curve (or, put another way, the demand curve shifts right relative to the supply curve; Figure 8.3). Note that the equilibrium is now at a higher price and a higher quantity than it was in Figure 8.1. This represents that when the supply of a product is low relative to demand, consumers are willing to pay higher prices and desire more of the product because it is scarce. This is known as a supply shortage or **deficit**.

Note that the quantity at equilibrium will increase during either a glut or a deficit, but the price shifts in opposite directions during these two events.

There are many external factors affecting the interaction of supply and demand. For example, a shortage of popular health foods can occur because people who do not usually buy the food may be tempted to try it if they believe it will be beneficial to their health (an increase in demand). For example, quinoa is a staple grain crop that had been produced for thousands of years in the Andean region, with Bolivia and Peru leading world production and consumption. In the early 2000s, modern machine equipment made processing and exporting quinoa much easier. The health buzz around the protein-rich grain made it hugely popular in the United States, where increased demand led the annual price per tonne of quinoa to increase sevenfold within a decade, from about $400/tonne in 2006 to almost $3,000/tonne in 2014 (FAOSTAT 2020). Since then, other regions have begun producing quinoa, increasing the supply to better meet demand and consequently decreasing the price of quinoa down to about $1,000/tonne in 2018 (FAOSTAT 2020).

In contrast, a glut can occur when people who would normally buy a food avoid it instead, possibly because it has been determined to be "unhealthy" or unsafe (a decrease in demand). Occasionally an outbreak of a food-borne disease will cause a scare that results in a glut. For example, in 2018, an *E. coli* outbreak in the United States was traced back to Romaine lettuce, leading to a more than $70 million reduction in Romaine sales for that year (Taylor 2018). If the sale prices during a glut are low enough, farmers may find it more cost-effective to let produce rot in their fields rather than pay to transport it to a market where there is no demand. This results in considerable food waste and is an inefficient use of agricultural products, to say the least.

### 8.2.1  Agricultural Supply: Balancing Inputs and Outputs with Costs and Income

Farming is a type of investment. A farmer invests time, seeds, fertilizer, and other inputs hoping that the amount and value of the outputs will exceed the cost of the inputs. The supply of agricultural products is partially determined by photosynthetic rates, nutrient availability, and pest occurrence. Modern farmers have considerable control over nutrient availability and there are many techniques for reducing the incidence of pests. Agricultural supply also depends on farmers' management decisions, including which crops they will plant each season and the size of crop population they will attempt to grow.

To achieve high income, farmers need to minimize input costs and maximize yields and yield quality. The **marginal** (individual) **return** is the difference between the amount a farmer spends to cultivate each plant and the amount of money that each plant's yields earn for the farmer. Because each field has a carrying capacity, the marginal return is not equal for each seed. For example, if the carrying capacity of a field is 20,000 corn plants per acre, the marginal return of the 1,000th and 10,000th planted seed will be about equal: both should result in one adult plant and approximately the same amount of harvest and income. But the marginal return of the 20,200th planted kernel will be negative because the effort of sowing the seed will not result in an adult plant that can yield usable products or earn income for the farmer.

As discussed in Chapter 3, there are many factors impacting carrying capacity. One factor that has changed significantly for farmers in the modern era is the density of a population that fields can support. Corn yields in the United States have almost tripled in the last 80 years, from 63.5 bushels per acre in 1960 to 172 bushels per acre in 2020. Much of this increase has been due to the breeding of crowd-tolerant corn, which allowed surviving plant density to more than double from 13,600 plants per acre in 1963 to 31,800 plants per acre in 2020 (Elmore and Abendroth 2006; National Agricultural Statistics Service n.d.). Other factors allowing increased production are discussed in Chapter 9. As technology like new, crowd-tolerant hybrids is developed, marginal returns change and farm management plans must change as well. For example, crowd-tolerant varieties can be grown more densely, but the hybrid seeds are more expensive to buy, changes which have opposing effects on the marginal return of each additional seed.

Besides the cost and sowing density of seeds, farmers must contend with several unknown environmental factors when planning how to use a field. Soybean farmers in the United States are currently encouraged to try to achieve 100,000 adult plants per acre (Pedersen and De Bruin 2007), so farmers have the difficult task of achieving this number of adult plants after the growing season is interrupted by an unknown number of diseases, pests, and weather disruptions like drought. How many seeds should a farmer plant to achieve a certain ending population when the proportion of plants that will die over the season is unknown? Rather than matching carrying capacity for a field, it is more economically sound to try achieving a balance where the marginal cost is less than the marginal return (income) of each plant.

Seeds are expensive, so planting 200,000 seeds and hoping that around 100,000 will survive is not a cost-effective option. However, introducing a population far below the carrying capacity of the field wastes soil resources and space that could otherwise be turned into useful products. The timing of seed sowing can also impact yields and plant mortality, as early sowing can potentially lead to a longer crop growing season and higher yields, but a late frost can also kill off a considerable number of early-planted seedlings. To further complicate the issue, yields and plant population are not directly correlated; in a study that involved planting fields with 125,000 or 175,000 soy seeds per acre, the number of plants at harvest differed but the plots yielded equal soybean harvests (Pedersen and De Bruin 2007). In general, as the quality (and price) of seeds increases, it is possible to plant fewer seeds to achieve the same population, yield, and yield quality, but planting 1,000 fewer seeds can result in a lower yield than planting 1,000 extra seeds (Elmore and Abendroth 2006). Therefore, even the highest-quality seeds may be over-planted to try and achieve the perfect population density. After several years of research, Iowa University in the United States suggests that soybean farmers plant between 125,000 and 140,000 seeds per acre to achieve that target of 100,000 adult plants, but also notes that local factors and traits of individual fields have strong effects that must be evaluated on a field level.

## 8.2.2 AGRICULTURAL DEMAND: GLUT PRICE CYCLES

In addition to balancing the investment costs of seeds, equipment, labor, and management, farmers must respond to market demand in order to make income. One hundred thousand mature soybean plants are not useful to a farmer if no one is willing to buy the soybeans. Farmers must try to anticipate the demand for food products seasons or years in advance of their sale, a time delay that helps explain how food gluts and shortages can occur.

A **"free" market** is one in which producers and consumers interact based only on supply and demand. In a free market, if demand for a product increases and the product's price increases, producers will take note. To take advantage of the increased prices, more producers will create the desired product. Recall that the demand and price of quinoa skyrocketed between 2006 and 2014. While the price was still increasing, the amount of land devoted to quinoa worldwide increased from 101,000 hectares (ha) in 2011 to 172,000 ha in 2012; this reflects that many producers noticed the increased price of quinoa and began planting it to capitalize on the price increase. As a result, production of quinoa rose from 84,000 tonnes in 2011 to 186,000 tonnes in 2014, at the peak of quinoa's price hike. Since then, world production of quinoa has hovered around 186,000 tonnes but the price of quinoa has decreased (as discussed in Section 8.2). The price decrease from 2014 to 2018 illustrates that the supply of quinoa now matches demand much more closely.

If supply exceeds demand for long enough, farmers can get locked into a negative spiral of planting more and more of a product that is only decreasing in value. To break even on their planting investment, farmers must either sell their products or hold onto them and hope that the invested cost of storage is returned when prices rise again. However, if the price is stable or decreasing, or the products are perishable, farmers will sell at a low price rather than store their products.

There are different demand factors for **commodities**, products that are treated as though every individual item is identical. For example, when consumers buy corn cobs directly, they may care about kernel size and color, how straight the rows of kernels are, and how sweet the corn tastes. But when the corn goes to a grain elevator and is mixed with bushels of corn from lots of other farms, all that matters is general quality (like amount of dirt mixed in with the grains) and sheer quantity; the corn has become a commodity (Pollan 2016). About one-third of U.S. corn is sold at elevators, where it is combined and used for animal feed, ethanol production, and creation of food additives like high fructose corn syrup – so the appearance, texture, and taste of each raw kernel is unimportant (Pollan 2016).

Commodity markets often experience problems with gluts and exceptionally low prices. For example, in 2006, Iowa corn elevators were buying corn at $1.45 per bushel, but corn cost $2.50 per bushel to grow (Pollan 2016). In this case, demand was considerably lower than supply, so farmers had to accept this low price for their products. Accepting the low price was easier because of **subsidies** that made up some of the difference in cost, as discussed in Section 8.4.1. However, the elevator is often the only agricultural buyer in farming towns and the elevator only purchases commodities like corn or soybeans (Pollan 2016). Therefore, farmers are encouraged to grow the same crops again, but in order to break even, they attempt to plant more corn or soybeans than in previous years. When every farmer plants more corn and more soybeans, the supply of those products increases further as demand stays the same; this forces the purchase prices even lower. Farmers locked in this cycle for multiple seasons may not have enough savings to investigate different markets or invest in different types of seeds, forcing them to stay in corn or soy production and driving the negative spiral forward.

## 8.3   FARM SIZES AND TYPES

In 2016, data from the Food and Agriculture Organization of the United Nations (FAO) was used to estimate that there are more than 570 million farms in the world (Lowder et al. 2016). Even within one country or region, there is high variability in the traits of farms. To name just a few of these variables, farms can be many different sizes; may be owned and operated by an individual, family, corporation, cooperative, or government; may rent out lands to be operated by others; may be managed by the owners or by others; may be worked by hired labor or by family of the owner; and may include forests or fisheries in addition to land devoted to crops or livestock. When discussing production on farms, farm policy, and how different agricultural movements impact farms differently, it is often useful to put farms into economic categories.

One particularly useful category is that of the family farm. **Family farms** are farms owned (or mostly owned) by the farm operator and members of the operator's family (Whitt 2021). Some definitions also require most of the labor on a family farm to come from members of the operator's family as opposed to seasonal hired labor. Based on these criteria, more than 90% of farms worldwide may be considered family farms and family farms operate about 75% of the world's agricultural land (Lowder et al. 2016). Family-operated land has the advantage of generational knowledge: older family members can pass on understanding of local weather patterns, best practices for their fields, and what plants grow best on their lands, which gives them a competitive, long-term advantage over non-family-owned farms (Amadeo 2020).

There is an important distinction between family farms and small farms. Family farms *can* be small farms, but not all family farms are small. Farm size may be measured in terms of the area of land operated by each farm or in terms of the amount of income generated by the farm each year. A general definition of a **small farm** is one which operates on 2 ha of land or less (about five acres). Based on this definition, about 80% of global farms are small farms, operating 12% of global arable land (Lowder et al. 2016). Defining **large farms** is much more difficult, as some countries or regions are far more abundant in arable land than others. For example, in Australia and Latin America, the largest farms often exceed 1,000 ha (2,400 acres), while in parts of Sub-Saharan Africa and of South Asia, it is rare for farms to reach even 50 ha in size. Large farms are much less common than small farms, but they are significant holders of agricultural land; only 16% of the world's farms are larger than 2 ha, but these hold 88% of the world's farmland (Lowder et al. 2016).

Farm size can also be defined based on the farm's income. For example, the United States Department of Agriculture recognizes three sizes of farm; by their metric, "small" farms are those with an income less than $350,000 per year, "medium" farms are those with income between $350,000 and $1 million per year, and "large" farms have income of $1 million or greater per year (Whitt 2021). By this definition, 3% of all U.S. farms are "large" (Amadeo 2020) and 90% of large farms in the United States are family owned (MacDonald et al. 2018). Therefore, it is not the case that family farms must be small.

**Subsistence** farms are those which are used to feed a household but do not (usually) provide income. Because the products of a subsistence farm are not sold, usually the crop selection does not depend on market forces. However, subsistence farmers still need to consider the monetary cost of seeds and other inputs and may have to change their crop makeup in response to seed supply and other forces external to the household's preferences.

### 8.3.1 FARM CONSOLIDATION

Sometimes, a financially successful farmer will use their profits to buy more land to farm. When this happens, one farm becomes larger in a move known as farm **consolidation**. At the same time, there is another farm that has either become smaller (by selling some of its land) or has disappeared (by selling all of its land). Therefore, when farm consolidation occurs frequently, the number of total farms decreases and the relative number of small farms also decreases. Farm consolidation occurs most frequently in regions and countries with higher **per capita** (per person) income, so countries with lower incomes hold a greater proportion of (smaller) farms. As the average income in a region increases, so does the average farm size (Louder, Skoet, and Raney 2016). A 2016 FAO study demonstrated that the thirty highest-income countries sampled had experienced a "clear" increase in average farm size from 1960 to 2000. As a result, only 4% of global farms are in high-income countries (Lowder et al. 2016).

The United States presents a good case study for the process of farm consolidation. In the early 1900s, a typical farm in Iowa, USA, would have been family owned and would have reared several types of livestock as well as tree fruits and multiple types of grains (Pollan 2016). Manure from the livestock was used to fertilize the crops, while crop waste (like stalks, cobs, and leaves) could be used to feed livestock. This helped the farm exist as a closed "loop", where the product of one process drove the growth of another process. It was unnecessary to bring in animal feed from outside the farm or have waste hauled away and disposed of externally. However, in the mid-20th century, several factors led to a corn productivity boom (see Chapter 9 in this volume). With demand unchanged, the extra supply of corn drove the price down. Farmers were encouraged to use more of their land to grow corn in order to make the same profits. Livestock require grazing space, which was converted to corn growth, so livestock were slowly eliminated from farms that grew corn (Pollan 2016). Meanwhile, other farms became specialized for livestock and eliminated vegetable crops from their operations. An absence of livestock on crop-focused farms means that fertilizer must be purchased from outside the farm and crop waste must be hauled away for disposal. Meanwhile, farms that focus on livestock production devote their entire space to livestock and none to growing feed, so they must purchase and import feed from outside the farm. When farms consolidate, they specialize, and this leads to the formation of linear systems rather than closed loops.

In 1972, a Russian grain shortage brought American grain prices to a historic high. The U.S. government encouraged farmers to grow as much corn as possible in order to flood the market (increase the supply of corn) and push the prices back down. Farmers were encouraged to "get big or get out" – buy land to produce more corn or sell their land to farmers who would (Pollan 2016). From 1987 to 2007, consolidation in the United States was particularly rapid. The rate of consolidation has since slowed, but consolidation continues into the 21st century. In 1991, 31% of U.S. agricultural production was conducted on large farms; by 2015, it was 51% (MacDonald et al. 2018).

The United States has had multiple factors contributing to rapid farm consolidation, and mostly consolidation, over the last 100 years. Globally, farm consolidation is not always so consistent – or consistently successful. From 1950 to 1990, Europe and North America saw considerable farm consolidation, but during the same period, the number of farms increased (average size of the farms decreased) in Asia and Africa (Lowder et al. 2016). In general, countries with higher per capita income have fewer, larger farms because they are experiencing, or have experienced, farm consolidation. In the world's wealthiest countries, 70% of the agricultural land is operated by farms larger than 20 ha (about 50 acres). In the world's poorest countries, the same 70% of land is operated by farms smaller than 5 ha (about 12.4 acres; Lowder et al. 2016). This is because when wages are high and technology like farm machinery is highly available, hired labor is less cost-effective than investment in machines that can be operated by farm managers or family members (MacDonald et al. 2018). Large farms are also more common in land-abundant areas like Africa, Australia, Central Asia, Eastern Europe, and Latin America than in land-constrained areas (Lowder et al. 2016).

As farm consolidation proceeds in a country or region, smaller farms that previously grew 4–6 crops will tend to specialize, growing only 2–3 crops (MacDonald et al. 2018). Due to the lower diversity of crops, operators of small farms will usually focus on higher-value products. For example, in 2007, farms operating on 4 or fewer hectares (10 or fewer acres) operated 0.18% of U.S. farmland but accounted for 3% of total agricultural sale value due to their focus on high-value items like orchard fruits, nursery trees, vegetables, and flowers (Newton 2014). Therefore, while small farms still exist after an area experiences farm consolidation, most agricultural production will occur on a small number of large farms (MacDonald, Hoppe, and Newton 2018). In the United States, only 7.84% of U.S. farms are mid- to large-sized family farms, but those produce 65% of the United States' agricultural products (Whitt 2021).

## 8.4 MARKET DISTORTIONS

In a fully free market, producers will generate the products that the market demands, responding to changes in demand to keep supply and demand approximately equal. When there are distortions, producers react to **incentives** (pressures) other than consumer demand or consumer demand is affected by factors other than the supply and quality of products.

### 8.4.1 GOVERNMENT INTERVENTIONS

Because the agricultural sector is so important for regional food security, farms often receive intervention and income support from their governments. Farm support is often in the form of a **price stabilization**, where the government artificially maintains the price of certain products at a level that is affordable to the consumer, beneficial to the producer, or both. A **subsidy** is a price stabilization where the government pays farmers the difference (or part of the difference) between the projected and actual price of a commodity or product. There are different types of subsidies and other price supports, some of which allow farmers to respond to free market pressures (less distorting supports) and others of which separate the farmer from free market pressures (more distorting supports).

According to the World Trade Organization, the least disruptive supports to the market are subsidies for research and development and subsidies that help farmers access new infrastructure, like irrigation systems (Greenville 2020). These supports are less distorting because farmers can use them while also responding to free market demands. For example, subsidized research and development can be useful for finding new products to create with existing foods (increasing the demand for the raw materials) or for discovering the best methods to integrate a newly popular food into a farm (better meeting high demand for the food). Irrigation systems can help stabilize food production and allow farmers to adjust the amount they are watering based on climate conditions, such as the use of "smart" sprinklers that do not activate on rainy days. Less-distorting supports are those that encourage flexibility and allow adjustments that are important for the long-term viability of agricultural production (Abbott 2020).

The most market-disruptive supports are those that discourage farmers from responding to new market and climate conditions. For example, **output subsidies** are payments based on the sheer amount of a commodity farmers produce. If there is an output subsidy for corn, farmers are encouraged to produce the greatest volume of corn possible regardless of consumer demand for corn. **Input subsidies** are payments based on the amount of a certain input a farmer uses. A subsidy on nitrogen fertilizer makes nitrogen fertilizer less expensive to farmers, so they will be willing to use more fertilizer than they would purchase on their own, potentially leading to pollution from runoff of the excess. A **price minimum** artificially maintains the price of a product above the equilibrium, supporting farmer income regardless of consumer demand. When price minimums are in place, farmers do not have to respond to decreasing market demand for a crop and have no incentive to diversify their production or respond to demand for other products. Finally, **tariffs** are artificially increased prices for foreign-produced goods that indirectly also drive up the price of domestic goods. For example, if there is a tariff raising the price of imported grain, locally produced grain can also become more expensive because producers have no incentive to produce grain more efficiently; consumers will buy at high costs because alternative (foreign) grain is just as expensive or more expensive. Thus, subsidies decrease the amount that farmers rely on consumers and increases the amount they rely on the government, disconnecting them from natural market forces (Lincicome 2020).

High levels of disruptive supports can actually change the type and amount of food that is produced, as seen historically in the United States (Greenville 2020). In the United States, corn is subsidized with a price stabilization; the government pays farmers a certain amount per bushel when the market price dips below a certain threshold. This encourages farmers to produce and sell corn season after season, year after year, no matter what the market price or actual demand are. The increased supply of corn pushes the price down further, but government supports will make up the difference, so farmers are still incentivized to produce corn. As of 2006, the U.S. government spent $5 billion subsidizing corn every year, which represented almost half of net farm income annually (Pollan 2016). Decades of such subsidies have created a system where the supply of commodity corn is far above demand, a distortion of the free market. Farmers are effectively farming to earn subsidies rather than responding to consumer needs (Ross and Edwards 2012).

Another effect of strong subsidies is to reduce the diversity of crops on a farm. If corn or another commodity is guaranteed to deliver subsidy income, there is low incentive to devote space and resources to other crops. If inputs are also subsidized, there is low incentive to plant cover crops or rotation crops to let the land recover its natural fertility (Lincicome 2020). Farmers can simply keep using subsidized fertilizer to add nitrogen to the soil (Pollan 2016). Thus, highly disruptive supports encourage mono-culture, making the farm more susceptible to damage from pests, diseases, and droughts. On a diversi-fied farm, if one crop fails, profits from the other crops can help make up that loss of income. On a single-crop farm, if the crop fails, only input supports and insurance contribute to making up the loss.

High price supports in one country or region can also disrupt the international market; if the cost of a local commodity is artificially raised in a year, foreign buyers must spend more of their cur-rency to purchase the same amount of a product they received in previous years. This lowers global demand of the commodity, reducing global trade of the product (Amadeo 2020). For example, when the U.S. government subsidized cotton, U.S. farmers responded to the higher price by producing more cotton. The increased production in the United States increased global supply of cotton while the global demand for cotton remained the same, pushing the global price of cotton down (because supply was exceeding demand). Now the 2–3 million farmers in West Africa who rely on cotton sales for their income had to sell their cotton at a lower price than before, which drove many of them into debt (Lincicome 2020). A price support in a high-income country negatively affected farmers in low-income countries, demonstrating that local policies can have global consequences. Therefore, reducing price supports in high-income countries is beneficial to farmers in developing countries, especially because so much of the economy and labor force in developing countries is focused on the agricultural sector (Lincicome 2020).

Unfortunately, high-income countries are the ones with the greatest tendency to apply subsidies. A report from the Organization for Economic Cooperation and Development (OECD) found that the 54 most-developed countries in the world spend about $700 billion USD on farm subsidies annually, equal to $2 billion every day and making up about 12% of gross farm revenue worldwide. Seventy-five percent of the supports go to individual producers, with more than half of individual support coming from market-disruptive methods like price supports, output subsidies for a particu-lar crop or animal, or input subsidies for specific fertilizers. Less than 13% of the support from those countries went toward less disruptive measures like inspections, infrastructure, and research and development (Abbott 2020). Large farms tend to receive the most support because subsidies are usu-ally applied to commodities like corn, soybeans, wheat, cotton, and rice, which are mostly grown on large farms that can produce the sheer volume required to earn solid income from the subsidies (Amadeo 2020). For example, from 1995 to 2019, the most profitable 10% of U.S. farms received 78% of U.S. farm subsidies; the top earning 1% of farms received 26% of the overall payments. Meanwhile, 62% of U.S. farms received no subsidy earnings (Amadeo 2020).

High levels of support can also hurt local economies, as taxpayers bear the brunt of the subsidies and consumers pay higher prices for agricultural products. In fact, most of the negative impacts of high price supports are felt by the countries that put the supports in place. Over the long term, high subsidies decrease agricultural productivity, innovation, and sector growth (Greenville 2020). Out of all sectors of the economy, growth in the agricultural sector is the most effective at fighting pov-erty (Pingali 2012). Therefore, government interventions that stifle agricultural growth and develop-ment also stifle reductions in poverty for the regional or national population. Conversely, countries with lower relative subsidies tend to have greater involvement in the global agricultural trade and demonstrate effective growth in both farm income and the overall value of the agricultural sec-tor. Therefore, according to the OECD, the removal of market-distorting subsidies would increase global food production. As an example, Australian agriculture is the second-least subsidized in the world (second to New Zealand) and more than 70% of its agricultural production is exported. More than half the jobs in Australia are related to international trade (Greenville 2020). Because the producers respond to an undistorted market, they are very tuned in to the demands of consumers across the world.

Australia and New Zealand demonstrate that it is possible to support the agricultural industry with low governmental intervention and by using less-disruptive measures when subsidies are put into place. In 2018, the OECD found that on average, farms producing rice received 60% of their incomes from farm subsidies globally; the United States supplied about 10% of overall American farm income. During the same year, subsidies supplied only 2% of farm income in Australia – but the gross income from the agricultural sector increased by 14% over the previous year (Australian Bureau of Statistics 2019; Greenville 2020). Research and development subsidies represented 56% of Australian federal agricultural support in 2018, compared with the global average investment of 13% (Greenville 2020). Most of the subsidies the Australian government implements are input subsidies targeted toward drought resilience, such as irrigation infrastructure subsidies, which reduce farmer risk of loss and help stabilize the food supply (Greenville 2020).

New Zealand and Australia have not always had such low subsidies. In 1984, 40% of New Zealand sheep and lamb rancher income was directly paid by government subsidies, which encouraged production and breeding of the animals. However, because actual demand was much lower than supply, millions of lamb carcasses were being rendered down yearly without being used. From 1972 to 1984, New Zealand's agricultural sector was growing at a rate of about 1.8% per year. After the subsidies were removed in 1985, New Zealand farmers diversified their crops and developed new products to sell, responding to actual market demand (Ross and Edwards 2012). As a result, from 1985 to 1998 the average yearly growth rate for the agricultural sector exploded to 4% while the rest of the economy maintained a growth rate of 0.9% (St. Clair 2002). As of 2012, New Zealand exports about 90% of its agricultural output, demonstrating again that a lightly subsidized agricultural industry can be globally competitive (Ross and Edwards 2012).

## 8.5 OTHER FACTORS

### 8.5.1 ECOLOGICAL CONSIDERATIONS

As global climate change continues to cause more extreme weather events and pollution causes human health issues, many consumers are trying to turn away from "conventional" large-scale farming, which uses fertilizers, pesticides, and other synthetic inputs that can cause harmful ecological effects. For example, nitrogen fertilizer does not bind tightly to soil, so much of it washes away from fields and into rivers and other bodies of water. This results in algal blooms that can disrupt entire freshwater or marine ecosystems (see Chapter 9 in this volume). Nitrogen runoff can also make downstream drinking water dangerous, as the nitrate can poison children by blocking oxygen flow to their brains – an effect so common that several towns in Iowa, USA, issue "blue baby alerts" in the spring when fertilizer is applied (Pollan 2016). Mechanization of large farms can also have negative effects, such as reducing soil fertility (see Chapter 12 in this volume) and threatening marginal lands (see Chapter 9 in this volume). Additionally, the use of tractors and other large equipment requires considerable use of fossil fuels; as of 2016, it took upwards of 50 gallons of oil to fertilize, harvest, and transport an acre's worth of corn (Pollan 2016).

In the last several decades, many consumers have become more interested in ecologically responsible farming techniques, or at least those advertised as being more ecologically responsible, like non-GMO and organic farming. This impacts the market, increasing the demand for foods that are riskier and more expensive to produce. Unlike genetically modified hybrid varieties, non-GMO plant varieties cannot be planted as densely together, decreasing the number of adult plants a field can raise each season. Certified organic farms cannot use modern pesticides, so their crops may be more susceptible to pest damage or diseases. The increased cost on the production end increases the price of the products, so consumers must pay more for foods that are produced by more sustainable means.

Fortunately, some ecologically responsible methods can also offer a farm economic benefits. For example, long crop rotations involving four to six crop species reduce the establishment of pests and diseases and each crop provides insurance against the failure of another. However, farmers must then consider the market demand for every additional crop, as well as how each crop in the rotation will influence soil and microbial community structure (see Chapter 12 in this volume).

Many ecologically responsible methods are costly, so farmers avoid practices that might be beneficial in the long run due to high upfront costs. For example, in the United States during the early 20th century many farmers avoided soil conservation methods, ultimately leading to the ecological disaster known as the Dust Bowl (see Chapter 9 in this volume). Furthermore, when ecologically irresponsible practices like the overuse of fertilizer are subsidized by the government, farmers are incentivized to maximize productivity rather than to incorporate ecologically sound techniques (Pingali 2012). When the U.S. government began instituting sugar subsidies, sugar cane production in Florida increased substantially. The increased use of phosphorus fertilizer resulted in runoff that has polluted the Everglades (Lincicome 2020).

Fortunately, the loss of certain input or output subsidies demonstrably results in more sustainable farming practices. When Indonesia stopped subsidizing pesticides in the 1990s, pesticide usage dropped "dramatically" (Pingali 2012). Prior to New Zealand's subsidy reforms, farmers often applied far too much fertilizer and used marginal lands that were not sustainable sites of production. Following the reforms, fertilizer use and water pollution have decreased and marginal land has been taken out of production (Ross and Edwards 2012; St. Clair 2002).

## 8.5.2 Intellectual Property Law

Intellectual property is the expression of an idea. The United States Constitution gave Congress the power to protect intellectual property in order "to promote the Progress of Science and useful Arts, by securing for limited Times to Authors and Inventors the exclusive Right to their respective Writings and Discoveries" (U.S. Const. Art. I, Sec. 8, clause 8). There are four types of intellectual property covered by law: patents, trademarks, copyrights, and trade secrets. A **utility patent** provides the inventor the sole right to exclude anyone else from using, selling, or offering for sale the invention for 20 years from the date the patent is issued (35 U.S.C. §100 et seq). Essentially, a utility patent allows creators to hold a monopoly on their idea or product for 20 years. A utility patent offers more protections than any other kind of patent, so they can be difficult and expensive to obtain. A **trademark** provides ten years of renewable protection to a logo, word or phrase that signifies the origin of a product or service (15 U.S.C. §1051 et seq). For example, Monsanto® is not just a company name, it is a registered trademark representing the source of the company's products. (MONSANTO, U.S. Trademark Registration No. 4,470,392). A **copyright** protects the written, recorded, or artistic representation of an idea (17 U.S.C. §101 et seq). Finally, **trade secrets** consist of information that is valuable and of competitive use to a business; they are protected by law so long as the business takes steps to keep the information secret [Restatement (Third) of Unfair Competition § 39 (1995)].

### 8.5.2.1 Plants as Intellectual Property

In the United States, plants could first be patented under 35 U.S.C. §161 (U.S. Congress, The Plant Patent Act of 1930):

> Whoever invents or discovers and asexually reproduces any distinct and new variety of plant, including cultivated sports, mutants, hybrids, and newly found seedlings, other than a tuber propagated plant or a plant found in an uncultivated state, may obtain a patent therefor, subject to the conditions and requirements of this title.

Note that merely identifying a new variety is not enough; to qualify for a plant patent under the Plant Patent Act, a breeder had to be able to reproduce the plant asexually (for example, from a cutting or by grafting). Tubers may have been excluded because the same tissue is used for propagation and as food (Pottage and Sherman 2007). Plant patents were originally granted for 17 years, as opposed to the 20-year utility patents granted for "new and useful inventions" (Pottage and Sherman 2007).

In 1970, the Plant Variety Protection Act [PVPA, U.S. Congress 7 U.S.C. § 2402(a)] provided for 25 years of "patent-like" intellectual property rights to sexually reproduced plants, including tubers:

> The breeder of any novel variety of sexually reproduced plant (other than fungi,[1] bacteria, or first-generation hybrids) who has so[2] reproduced the variety […] shall be entitled to plant variety protection therefor.

In addition, the PVPA allowed farmers and researchers to save and use seeds from patented plants without violating the act (7 U.S.C. §2543). While this so-called "farmers' exemption" to the PVPA was limited in 1994, it did not go away entirely (Rives 2022). This meant that farmers could save and plant the seeds from the hybrid plants they purchased. The "farmers' exemption" reduced seed producers' incentive to invest in development of new crops because each farmer might only need to pay for the new hybrids once, failing to offset the cost of ongoing research.

Prior to 2001, the U.S. Patent and Trademark Office did not grant utility patents on sexually reproduced plants; therefore, seed producers had to try to protect their new plant lines as trade secrets. This type of protection depends on the information – in this case, the genetic line of the plants – staying secret and not becoming public at any time. Such protection can be difficult and expensive, especially when plants self-propagate (e.g. the seeds could wind disperse, thus becoming publicly available). Still, trade secret law provided protection for the developers of sexually reproduced plants, which were not otherwise protected until 2001, and courts could (and still can) grant damages to parties whose seeds have been taken or misused [*Pioneer Hi-Bred International v. Holden Foundation Seeds, Inc.,* 35 F.3rd 1226 (8th Cir. 1994)].

In 1980, the United States Supreme Court ruled for the first time that a living, human-made microorganism could be patented. In the case *Diamond v. Chakrabarty,* 447 U.S. 303 (1980), a scientist was able to patent a manufactured bacterium that could break down crude oil and help clean up spills. In 2001, the Supreme Court upheld the validity of a challenged plant utility patent [*J.E.M. Agricultural Supply v. Pioneer Hi-Bred*, 122 S. Ct. 593, 596 (2001)]. This decision further altered the legal landscape of agriculture, as a plant utility patent is infringed when the protected plant is produced either sexually or asexually (Stim 2022). As of this writing, seed producers can obtain protection for not only their seeds, but for specific "genes, traits, methods, plant parts, or varieties" under U.S. law (U.S.D.A. n.d.).

### 8.5.2.2   Patents and Seed Re-Use

A utility patent on a plant lasts for 20 years and includes the right to exclude others from asexually reproducing the plant as well as from using, offering for sale, or selling the plant covered by the patent (35 U.S.C. §162). Patent **infringement** (violation) is a **strict liability** offense – one does not need to intentionally violate a patent to be held responsible for the violation. Therefore, the ability of the patent holder to exclude others from *using* the patented plant has caused considerable distress among cultivators and farmers.

One seed that successfully matures will naturally produce more seeds. In the past, farmers saved seeds from one harvest to plant the next year's crops. However, in 1926, the Hi-Bred Corn Co. released the first commercial hybrid corn seed demonstrating **hybrid vigor** (Rives 2002). Sometimes, cross breeding plant varieties will result in a hybrid generation that is more vigorous (producing greater harvests) than either of its parent plants *and* more vigorous than the hybrids' own offspring. Now the seed producers had exclusive access to the seeds that would produce the most vigorous plants and farmers had incentive to buy new seeds every planting season. Seed producers also had new incentive to keep researching and testing new crosses that would produce vigorous hybrids.

Currently, ten multinational corporations hold approximately 67% of global commercial seeds for major crops (ETC Group 2008). Monsanto Corporation alone controls 60% of the global corn

---

[1] While fungi and bacteria are certainly not plants, they are often grouped with plants in public awareness.
[2] That is, sexually.

and soybean seed markets. In the United States, over 90% of soybean acreage and 80% of corn acreage is planted with Monsanto patented seeds (Fernandez-Cornejo et al. 2014) (see Figure 8.4). The control of such a large portion of agriculture by very few companies can lead to higher prices for seed, and eventually, for the food products grown (Hubbard 2009).

Companies such as Monsanto protect their seeds with "Stewardship" or "Technology Use" license agreements with growers, prohibiting them from using patented seeds from one year to the next. The agreements also govern weed and pest management methods as well as where harvest products can be distributed (Katiraee 2018).

There is a persistent myth that Monsanto (now part of Bayer Global) sues farmers who have inadvertently reared plants from Monsanto's patented seeds that were accidentally wind-dispersed into the farmers' fields. In fact, in 2012, the Organic Seed Growers & Trade Association (OSGATA) sued Monsanto, arguing that contamination of fields by Monsanto's patented seeds was inevitable and that the company's patents should therefore be declared invalid [*OSGATA v. Monsanto Co.*, 851 F. Supp. 2d 544 (S.D.N.Y 2012) *aff'd.* 718 F.3d 1350 (Fed. Cir. 2013), *cert. denied* (U.S. 2013)].

OSGATA asserted that it needed to protect its member farmers from possible lawsuits claiming patent infringement due to the presence of Monsanto seeds in members' organically grown crops. However, because none of OSGATA's members had actually been sued for, or even accused of, patent infringement, the case was considered premature and was eventually dismissed. This case is best known for the Court of Appeals' ruling that Monsanto was bound by its statement that it would not "take legal action

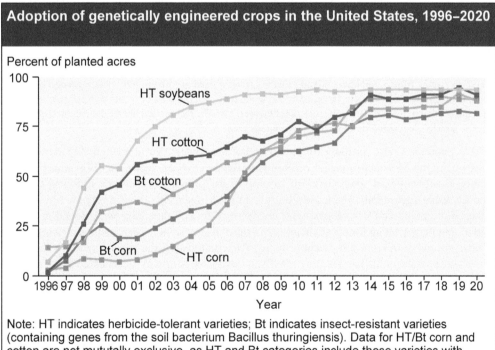

Note: HT indicates herbicide-tolerant varieties; Bt indicates insect-resistant varieties (containing genes from the soil bacterium Bacillus thuringiensis). Data for HT/Bt corn and cotton are not mututally exclusive, as HT and Bt categories include those varieties with overlapping (stacked) HT and Bt traits.
Source: USDA, Economic Research Service using data from the 2002 ERS report, *Adoption of Bioengineered Crops* (AER-810) for the years 1996–99 and National Agricultural Statistics Service, (annual) June Agricultural Survey for the years 2000–20.

**FIGURE 8.4**    The increase in agricultural genetically engineered crop usage in the United States from 1996 to 2020. Monsanto "Roundup Ready" (glyphosate-resistant) seeds are the dominant brand of herbicide-tolerant seeds used in the United States. (Fernandez-Cornejo et. al. 2014; chart reproduced from Dodson 2020, which is in the public domain.)

against growers whose crops might inadvertently contain traces of Monsanto biotech genes (because, for example, some transgenic seed or pollen inadvertently blew onto the grower's land)" (pp. 1358–59).

However, Monsanto has indeed sued farmers who use its seeds without license agreements. One well-known case involved a farmer, Percy Schmeiser, who discovered that his field was contaminated with Monsanto's Roundup Ready (glyphosate herbicide-resistant) canola seeds when land segments around his utility poles were sprayed with Roundup (Monsanto-brand glyphosate herbicide). The issue arose when Mr. Schmeiser planted the resulting seeds in the following years, which the Federal Court of Canada found to be patent infringement [*Monsanto Canada Inc. v. Schmeiser*, 2001 FCT 256 (Fed. Ct. 2001)].

Mr. Schmeiser argued in the case that the use was not deliberate, but as in the United States, lack of intent is not a defense to patent infringement in Canada. The farmer also argued that Monsanto had essentially lost the right to exercise a patent over the canola seed by releasing the gene into the environment with wind-dispersable canola seeds.

Mr. Schmeiser had not had to buy new canola seeds for six years, due to his practice of saving a portion of the canola he harvested to serve as seed for the next generation of crops. While the patented seeds may indeed have accidentally ended up in his field (through wind or a spilled bag of his neighbor's seed), he was only allowed to use those seeds (or rather, the patented genes within them) *once*.

Ultimately, the court found that both Mr. Schmeiser and his hired worker knew where the seeds had come from. The evidence showed that the hired worker had saved for replanting the seeds specifically from an area in which Mr. Schmeiser had sprayed plants with Roundup. The plants that survived the herbicide spray can only have grown from Monsanto's patented "Roundup Ready canola" seeds (para 102–105), so their offspring also contained the patented "Roundup Ready" gene. Mr. Schmeiser violated the patent by saving and reusing those patented seeds. The court ruled (para 92):

> Thus a farmer whose field contains seed or plants originating from seed spilled into them, or blown as seed, in swaths from a neighbour's land or even growing from germination by pollen carried into his field from elsewhere by insects, birds, or by the wind, may own the seed or plants on his land even if he did not set about to plant them. He does not, however, own the right to the use of the patented gene, or of the seed or plant containing the patented gene or cell.

The U.S. Supreme Court similarly ruled that an Indiana farmer, Vernon Hugh Bowman, was liable for infringing Monsanto's patent on Roundup Ready soybean seeds. Mr. Bowman purchased Roundup Ready soybean seeds for the first crop of his growing season and followed the terms of the licensing agreement, which allowed him to grow one crop. To save money, Mr. Bowman purchased his next seeds from a grain elevator, but still treated the plants with Roundup. This act killed all but the Roundup Ready soybeans from the elevator, so he – perhaps inadvertently – grew an entire second crop from the patented seeds. After harvesting the second crop, he then saved some of those seeds for his late-season crop. Even though his re-use of the patented seeds may have been unintentional, he was still liable for patent infringement [*Bowman v. Monsanto Co.*, 569 U.S. 278 (U.S. 2013)].

Most of patent law is based around the act of deliberate creation; the patent laws do not allow a purchaser of a patented item to keep making copies of that item. If someone buys a new, patented tractor, takes it apart to learn about the new design, and then creates hundreds of identical tractors to sell, they have clearly violated the original patent. While farmers do not conduct genetic research on the patented seeds and recreate them from scratch, replanting the patented seeds is still propagating the patented product and is legally controlled as such. The fundamental issue is that seeds, unlike tractors, are self-replicating. The "actor" truly at fault for patent infringement is the seed itself, which grows up and creates more seeds with the patented traits. This creates considerable friction between the seed producers and courts (which correctly recognize that patented objects have been illegally copied) and the agricultural community (which correctly recognizes that self-replicating systems cannot be controlled in the same way as deliberately created objects).

## 8.6   CONCLUSION

Farmers must invest time, energy, and other resources into food production, hoping that the value of their outputs exceeds the value of their inputs. One important way to ensure that the outputs are valuable is to observe the market and respond to demand. However, the signal of demand is often complicated by time delay, market distortions, government interventions, and ecological and legal factors, making it difficult for farmers to know which outputs are the most valuable to the end consumer and which inputs are most worth investing in. To protect against poor weather, low yields, and other random factors, it is best for farmers to diversify their crops, but natural economic forces like consolidation and government interventions like subsidies often discourage crop diversification. Additionally, patent and trade secret protections on new seed varieties can make inputs for even a single crop very expensive in both attention and monetary value. The patent laws grant inventors of genetically modified or hybrid seeds an exclusive right to use, sell, and control their invented seeds or plants for a period of 20 years, creating a legally granted monopoly. Because a small number of companies now own the rights to over two-thirds of the genetically modified seeds sold around the world, they have been able to control prices and access to those seeds. This lack of diversity on the market, coupled with enforcement of the companies' patent rights, can have the effect of raising the costs to grow and diversify food. Furthermore, since patented seeds can self-replicate, farmers must be careful of which seeds they sow regardless of whether they directly conduct business with seed producers.

## REFERENCES

Abbott, C. 2020. World farm subsidies hit $2 billion a day. *Successful Farming. Meredith Agrimedia*, June 30, 2020. https://www.agriculture.com/news/business/world-farm-subsidies-hit-2-billion-a-day (accessed July 22, 2021).

Amadeo, K. 2020. How farm subsidies affect you. ed. M. J. Boyle. *The Balance. com*, November 10, 2020. https://www.thebalance.com/farm-subsidies-4173885 (accessed July 7, 2021).

American Law Institute. 1995–2007. Restatement (Third) of Unfair Competition § 39 (1995).

Australian Bureau of Statistics. 2019. *Gross Value of Irrigated Agricultural Production, 2017–18 Financial Year.* https://www.abs.gov.au/statistics/industry/agriculture/gross-value-irrigated-agricultural-production/2017-18 (accessed July 7, 2021).

Dodson, L. 2020. *Recent Trends in GE Adoption.* U.S. Department of Agriculture, Economic Research Service. https://www.ers.usda.gov/data-products/adoption-of-genetically-engineered-crops-in-the-us/recent-trends-in-ge-adoption/ (accessed February 5, 2022).

Elmore, R., and L. Abendroth. 2006. Seeding rate relative to seed cost. Iowa State University Extension and Outreach. *Integrated Crop Management* IC-496(6):82–83. https://crops.extension.iastate.edu/encyclopedia/seeding-rate-relative-seed-cost.

ETC Group. 2008. *Who Owns Nature? Corporate Power and the Final Frontier in the Commodification of Life.* ETC Group Communique No. 100 (2008). http://www.etcgroup.org/content/who-owns-nature (accessed February 7, 2022).

FAOSTAT. 2020. *UN Food and Agriculture Organization, Corporate Statistical Database* https://www.fao.org/faostat/en/#data (accessed January 14, 2021).

Federal Court of Appeal. 2001. *Monsanto Canada Inc. v. Schmeiser,* 2001 FCT 256 (Fed. Ct. 2001).

Fernandez-Cornejo, J., Wechsler, S., Livingston, M., and L. Mitchell. 2014. *Genetically Engineered Crops in the United States, ERR-162.* U.S. Department of Agriculture, Economic Research Service. https://www.ers.usda.gov/webdocs/publications/45179/43668_err162.pdf (accessed December 23, 2021).

Greenville, J. 2020. *Analysis of Government Support for Australian Agricultural Producers.* Department of Agriculture, Water and the Environment. Australian Government. https://www.agriculture.gov.au/abares/research-topics/trade/analysis-of-government-support-agricultural-producers (accessed July 7, 2021).

Hubbard, K. 2009. Out of hand. Farmers face the consequences of a consolidated seed industry. *Farmer to Farmer Campaign on Genetic Eng'g, Nat'l Family Farm Coal,* p. 4. http://farmertofarmercampaign.com/Out%20of%20Hand.FullReport.pdf (accessed February 7, 2022).

Katiraee, L. 2018. Dissecting claims about Monsanto suing farmers for accidentally planting patented seeds. *Genetic Literacy Project.* https://geneticliteracyproject.org/2018/06/01/dissecting-claims-about-monsanto-suing-farmers-for-accidentally-planting-patented-seeds/ (accessed February 1, 2022).

Lincicome, S. 2020. *Examining America's Farm Subsidy Problem*. Cato.org. Cato Institute, December 18, 2020. https://www.cato.org/commentary/examining-americas-farm-subsidy-problem (accessed July 8, 2021).

Lowder, S.K., Skoet, J., and T. Raney. 2016. The number, size, and distribution of farms, smallholder farms, and family farms worldwide. Food and Agriculture Organization of the United Nations. *World Development* 87:16–29.

Lyons, J., and S. Sarkis. 2021. *Larger-Than-Average Gulf of Mexico 'Dead Zone' Measured*. National Oceanic and Atmospheric Administration. U.S. Department of Commerce. https://www.noaa.gov/news-release/larger-than-average-gulf-of-mexico-dead-zone-measured#:~:text=Today%2C%20NOAA%2Dsupported %20scientists%20announcedto%20fish%20and%20bottom%20species (accessed March 11, 2022).

MacDonald, J.M., Hoppe, R.A., and D. Newton. 2018. *Three Decades of Consolidation in U.S. Agriculture*. U.S. Department of Agriculture, Economic Research Service, March 2018. https://www.ers.usda.gov/webdocs/publications/88057/eib189_summary.pdf?v=2257.8 (accessed July 6, 2021).

MONSANTO. 1934. U.S. Trademark Registration No. 4,470,392, filed January 19, 1934, and registered July 24, 1934.

National Agricultural Statistics Service. n.d. *Quick Stats*. United States Department of Agriculture. https://quickstats.nass.usda.gov/ (accessed July 6, 2021).

Newton, D.J. 2014. *Working the Land with 10 Acres Small Acreage Farming in the United States*. U.S. Department of Agriculture, Economic Research Service, April 2014.

Pedersen, P. and J. De Bruin. 2007. How many seeds does it really take to get 100,000 plants per acre at harvest? *Iowa State University Extension and Outreach*. Iowa State University. IC-498(5):107–108. https://crops.extension.iastate.edu/encyclopedia/how-many-seeds-does-it-really-take-get-100000-plants-acre-harvest (accessed July 5, 2021).

Pingali, P.L. 2012. Green revolution impacts, limits, and the path ahead. *Proceedings of the National Academy of Sciences USA* 109(31):12302–8.

Pollan, M. 2016. *The Omnivore's Dilemma*. New York, NY: Penguin Books.

Pottage, A., and B. Sherman. 2007. Organisms and manufactures on the history of plant inventions. *MelbULawRw* 22(2007):31(2). Melbourne University Law Review 539.

Rives, E. 2002. Mother nature and the courts are sexually reproducing plants and their progeny patentable under the utility patent act of 1952? *Cumberland Law Review* 32:187.

Ross, M. and C. Edwards. 2012. *In New Zealand, Farmers Don't Want Subsidies. CATO*. CATO Institute, July 17, 2012. https://www.cato.org/commentary/new-zealand-farmers-dont-want-subsidies (accessed July 7, 2021).

St. Clair, T. 2002. *Viewpoint – Farming without Subsidies – A Better Way. Why New Zealand Agriculture Is a World Leader*. Politico.eu. Politico, July 17, 2002. https://www.politico.eu/article/viewpoint-farming-without-subsidies-a-better-way-why-new-zealand-agriculture-is-a-world-leader/ (accessed July 7, 2021).

Stim, R. 2022. *Can I Patent a Plant, Fruit, Seed, or Other Growing Thing?* Nolo.com. https://www.nolo.com/legal-encyclopedia/plant-patents.html (accessed February 22, 2022).

Supreme Court of United States. 1980. *Diamond v. Chakrabarty*, 447 U.S. 303 (1980).

Supreme Court of United States. 2001. *J.E.M. Agricultural Supply v. Pioneer Hi-Bred*, 122 S. Ct. 593, 596 (2001).

Supreme Court of United States. 2013. *Bowman v. Monsanto Co.*, 569 U.S. 278 (U.S. 2013).

Taylor, K. 2018. Romaine lettuce sales are down more than $71 million so far this year as the industry has been pummeled with food-poisoning outbreaks — and things are about to get worse. *Business Insider*, Nov 21, 2018. https://www.businessinsider.com/e-coli-outbreaks-drag-romaine-lettuce-sales-down-2018-11 (accessed July 7, 2021).

United States Court of Appeals, Eighth Circuit. 1994. *Pioneer Hi-Bred International v. Holden Foundation Seeds, Inc.*, 35 F.3rd 1226 (8th Cir. 1994).

United States Court of Appeals for the Federal Circuit. 2013. *OSGATA v. Monsanto Co.*, 851 F. Supp. 2d 544 (S.D.N.Y 2012) aff'd. 718 F.3d 1350 (Fed. Cir. 2013), cert. denied (U.S. 2013).

U.S. Congress. 1787. Copyrights and Patents. U.S.C. Art. I, § 8, cl. 8.

U.S. Congress. 1930. Plant Patent Act of 1930. 35 U.S.C. §§161–164.

U.S. Congress. 1946. Application for Registration; Verification. 15 U.S.C. §1051 et seq.

U.S. Congress. 1952. Inventions Patentable. 35 U.S.C. §100 et seq.

U.S. Congress. 1970. 1970 Plant Variety Protection Act, 7 U.S.C. § 2402(a).

U.S.D.A. n.d. *Plant Variety Protection*. U.S. Department of Agriculture. Agricultural Marketing Service. https://www.ams.usda.gov/services/plant-variety-protection (accessed February 22, 2022).

Whitt, C. 2021. *Farm Structure*. U.S. Department of Agriculture, Economic Research Service. January 6, 2021. https://www.ers.usda.gov/topics/farm-economy/farm-structure-and-organization/farm-structure/ (accessed July 7, 2021).

# 9 The Green Revolution
## *Agricultural Shifts of the 20th Century*

*Elizabeth "Izzie" Gall*

## CONTENTS

## 9.1 INTRODUCTION

**Agroecology** is the study of how human agriculture interacts with its surrounding ecology, what is often considered the "natural" world. It often incorporates elements of economics and social science, acknowledging different factors that may encourage farmers to use certain crop varieties or farming techniques. "Agroecology" can also refer to an agricultural management style that considers the ecosystem around a farm to be important and worth paying attention to, maintaining, and – sometimes – protecting.

During the 20th century, agroecology was generally used to make the natural world productive for human use. Repeated food crises and an expanding human population made increased crop production essential, so the ecological concept of sustainability was sacrificed in favor of field expansion and yield improvements. Technologies like hybrid seed varieties, irrigation, mechanization, and artificial fertilizer were adopted around the world to increase production of staple food crops. The agricultural breakthroughs of this century, which significantly increased global food availability and reduced the amount of land required to produce food (Figure 9.1), are jointly referred to as the Green Revolution.

Unfortunately, many of these techniques and the policies governing their use led to long-term losses in soil fertility, water cleanliness, and wild habitat quality. Today there are calls for a "Second Green Revolution", one in which nutritious yields and ecological sustainability will be given equal importance.

This chapter outlines some of the historical, social, and economic paths that led to the Green Revolutions of the 20th century. Like many social revolutions, these were multinational and

DOI: 10.1201/9780429320415-9

Arable land needed to produce a fixed quantity of crops (1961 = 1), 1961 to 2014

Arable land needed to produce a fixed quantity of crops is calculated as arable land divided by the crop production index (PIN). The crop production index (PIN) here is the sum of crop commodities (minus crops used for animal feed), weighted by commodity prices. This is measured as an index relative to 1961 (where 1961 = 1).

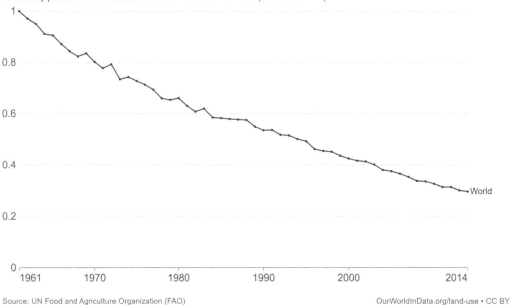

Source: UN Food and Agriculture Organization (FAO)                                                    OurWorldInData.org/land-use • CC BY

**FIGURE 9.1** Thanks to many technological advancements and increased input availability, the amount of land required to produce commodity crops (see Chapter 8 in this volume) dropped significantly during the Green Revolutions of the 20th century. (Chart reproduced under the CCBY license from Ritchie, H., and M. Roser. 2019. Land Use. *Our World in Data*. Global Change Data Lab. https://ourworldindata.org/land-use [accessed March 18, 2022].)

sometimes global movements with no clear beginning or end dates. Some authors consider the process as one long, nebulous Green Revolution. However, in our view, there are two clearly definable movements that produced dramatic increases in yields for the crops considered most vital at the time. In this text we refer to these as the Initial Green Revolution and the Extended Green Revolution.

The Initial Green Revolution began in Mexico in the 1940s, where a Mexican American partnership gave rise to the Mexican Agricultural Program. This first wave of new breeding and cultivation technologies introduced the first high-yielding hybrid varieties of corn (maize) and wheat to the global market. The major yield increase rates of the Initial Green Revolution began declining in the early 1970s.

The Extended Green Revolution focused on Asia and Southeast Asia, using similar technology to produce high-yielding hybrid varieties of rice and other grains before yield increase rates started to decline in the 1980s.

Examining the paths, policies, and technology changes of these Green Revolutions informs us about the current state of global agroecology. We hope that by understanding these factors, current and future agroecologists can avoid the mistakes of the 20th century's agricultural transitions while working toward a "Second Green Revolution".

## 9.2 THE UNITED STATES: FROM DUST BOWL TO GREEN REVOLUTION

As the world recovered from the Great War of 1918 (World War I), inexpensive grain probably seemed like a godsend for consumers. However, the low prices made life difficult for farmers, many of whom tried to compensate for their low incomes by increasing the volume of grain they produced. With affordable, industrial-scale fertilizer still decades away,[1] farmers on the American Great Plains increased their yields by planting on low-fertility, marginal or **submarginal** land that usually was not worth the trouble (National Drought Mitigation Center 2018). Many famers also decided to invest in time-saving machines like gas tractors and ploughs, often taking out large loans to support the purchases. After such purchases, indebted farmers felt even more financial strain, which increased their **incentive** (motivation) to produce grain on even their low-fertility lands.

In the Great Plains, submarginal lands were usually grasslands. Grasses native to the Great Plains region are adapted to dry conditions and have developed very deep roots to sustain their water needs. For example, the roots of the native grass *Sporobolus heterolepis* reach almost twice as far as the roots of cultivated wheat (Scherer and Steele 2019; Mid-America Regional Council 2019). The deep roots help to stabilize dry soil, reducing **erosion** (loss of soil) with the movement of water or wind; the deep roots also create vertical tunnels which allow rainwater to **infiltrate** (penetrate) deep into soil, increasing the amount of rainwater that can be absorbed and reducing flooding (Marapara et al. 2020). When American farmers uprooted the native grasses in favor of crop plants with smaller root systems, those submarginal lands were made highly susceptible to damage, especially wind erosion. If enough soil is picked up by the wind, it can move quickly across considerable distances in a weather event known as a **dust storm**. In addition to removing soil from fields, dust storms cause crop damage as soil particles whip along on the wind and tear through leaves. Dust storms also leave a thin layer of dry soil in their wake, which can disrupt water and fertilizer absorption along the storm's path. To make matters worse, fertilizers are already less effective on marginal land. Dry soils are less able to hold the compounds plants need, so nutrients are more likely to **leach** away, washing out of a field or below plants' reach. Additional dry soil deposited by a dust storm will only cause more fertilizer to leach past plants.

Erosion and leaching can be fought with soil conservation methods like **terracing**, where sloping fields are cut into steps, and **listing**, where farmers dig deep trenches at regular intervals to improve field drainage. While terracing raised an acre's land value by almost five times the terracing cost (Hurt 1984), the implementation costs of both soil conservation methods were too high for many indebted farmers in the early 20th century. Farmers often abandoned conservation in favor of aggressive machines that made planting easier but increased the risk of soil loss (National Drought Mitigation Center 2018). As a result, the American prairieland of the late 1920s was highly susceptible to damage from drought, high winds, and dust storms, which resulted in a high rate of crop failure. It should also be noted that with the loss of natural grassland, the wildlife of the prairie suffered untold damage.

The cycle of high production demand, overuse of submarginal land, and crop failure triggered an agricultural depression that contributed to the Great Depression of 1929. With the American breadbasket at its historically most frail, a series of extreme droughts hit the country. This marked the beginning of the Dust Bowl, the worst agricultural disaster in U.S. history. More than 10% of the nation's farms changed ownership during this decade as farmers defaulted on loans or abandoned drought-stricken lands. Millions of people abandoned their farms and moved to urban areas seeking work (National Drought Mitigation Center 2018).

From 1932 to 1938, the U.S. government devoted almost one billion dollars to programs that rolled back the damaging practices of the 1920s (National Drought Mitigation Center 2018). This funding, equivalent to more than 15 billion modern U.S. dollars, paid for the emergency listing of more than 11.8 million acres, installation of "shelter belts" of young trees to slow winds and reduce wind erosion, and the reversal of 1.28 million acres of marginal land from agricultural production back to grassland (Hurt 1981; National Drought Mitigation Center 2018; Official Data Foundation 2019). The American government also created the Soil Conservation Service to research

---

[1] The first commercially available granular fertilizer was produced in 1930 (Russel & Williams 1977. History of Chemical Fertilizer Development).

soil conservation techniques and educate farmers about the most effective conservation options. Experimental stations and demonstrations clearly illustrated the benefits of terracing and listing, techniques that nearly quadrupled the penetration depth of rain and resulted in 35% higher wheat yield over untreated soil (Hurt 1981).

As soil conservation techniques were re-adopted nationwide and grasslands were reestablished, U.S. food production began to stabilize. In 1941, rainfall in the United States returned to average levels; over the next few years, the disastrous Dust Bowl ended (National Drought Mitigation Center 2018). However, in hopes of averting such a catastrophe in the future, some organizations began investigating new ways to ensure sustainable food security for the region.

In 1943, the U.S.-based Rockefeller Foundation partnered with the Mexican government to research hybrid crop varieties that could perform well throughout North America. While the Rockefeller work would continue fruitfully in Mexico (see Section 9.3), involvement in World War II limited the United States focus on agricultural and conservation research. Meanwhile, development in chemical production had made commercial-scale fertilizer widely accessible. With plenty of fertilizer, a booming war economy, and the return of normal rainfall, many American farmers were able to maintain stable yields while slipping back into less sustainable agricultural practices (National Drought Mitigation Center 2018).

Dozens of ammunitions factories popped up in the United States during World War II; by 1945, the United States was generating more than 700,000 tons of ammonia per day (Hergert, Nielsen and Margheim 2015). When the war ended, the U.S. government advocated for the continued production of ammonium, now for use as nitrogen fertilizer. Government-sponsored educational campaigns advertised the utility of nitrogen fertilizer, increasing its demand and the frequency of its use (Russel and Williams 1977). In 1940, American farmers applied approximately 300,000 metric tons of nitrogen fertilizers; by 1980, they were applying more than 30 times that amount (9.8 million metric tons) each year (Cao et al. 2018). Prior to World War II, a variety of nitrogen fertilizers were in common use, but by 1930 chemical nitrogen fertilizer had become more than 83% of agricultural nitrogen input. By the 1970s, more than 80% of all the nitrogen used agriculturally was artificial ammonia fertilizer, $NH_4^+$ (Cao et al. 2018), which is the form of nitrogen most prone to releasing nitrous oxide (a greenhouse gas) if it is not taken up quickly enough by plants. The second most common form of synthetic nitrogen fertilizer is nitrate, $NO_3^-$, which has a high soil mobility that helps plants use the nitrogen more effectively but also allows the nitrogen to leach away quickly during heavy rains or with intensive irrigation, a technology which was also rising in popularity.

As the U.S. government encouraged more settlement in the arid West, millions of acres of land were brought under irrigation. Government-sponsored dam projects from the 1930s to 1950s made more surface water available for irrigation while also supplying hydropower and controlling floods. Between the start of U.S. irrigation practices in 1847 and the end of World War II, 7.5 million hectares of land were irrigated. Irrigation projects continued in the United States and elsewhere after the war, with new technologies like steam and electric pumps making it easier than ever to access **groundwater** below the surface and bring it up to fields. By 1970, 1.98 billion hectares of land were under irrigation worldwide. Unfortunately, while systems were put in place to bring surface water to fields, ecological impacts and drainage were not considered. As a result, many fields became **waterlogged**, unable to drain properly (Sojka et al. 2002). With poor drainage, the salts from irrigation waters could not leach away, leading to high soil salinity (**salinization**) in addition to the low oxygen, low nitrogen state caused by consistently soaked soil. Many fields with better drainage suffered from water erosion as unprecedented volumes of water were flushed through the soil. As a result of intensive irrigation, 45 states in the United States have issues with **ground subsidence**, where the ground level sinks due to the exhaustion of groundwater systems. More than 13,500 square miles of ground have subsided in the United States as a direct cause of groundwater exploitation (United States Geological Survey n.d.).

By the 1950s, American agricultural production had fully recovered from both the Great Depression and the Dust Bowl. With this victory, the U.S. government turned its attention to helping restore food production to the European continent. Once again, considerable federal funding was directed to agricultural research, with a new focus on guidelines for crop nutrition (fertilizer) requirements, the most effective uses of fertilizer, and development of high-yielding hybrid crop varieties (Hergert, Nielsen and Margheim 2015). Later, U.S.-based research groups would also turn their attention to Southeast Asia and Latin America (see Section 9.4).

## 9.3 MEXICO AND THE INITIAL GREEN REVOLUTION

At the start of the 20th century, only 10% of the Mexican rural population owned land; as a result, land reform was a major aspect of the new constitution following the Mexican Revolution. Rather than seizing land from the wealthy and redistributing it to the peasantry, the new Mexican government achieved land reform by bringing new lands, including forested areas, under agricultural production (Sonnenfeld 1992). As in the American Great Plains to the north, native species and trees with deep root systems were removed in favor of crop varieties with shallower roots, resulting in zones of destabilized soil across Mexico. The deforested areas and other marginal lands in Mexico underwent **desertification**, becoming sandy and infertile. Sadly, the newly distributed lands were quickly depleted of their fertility and the peasantry lacked productive agricultural land once again.

When World War II began, the U.S.-based Rockefeller Foundation had to cease its agricultural work in most of the world. In 1940, U.S. Vice President Henry Wallace was invited to Mexico for the inauguration of Mexican President Manuel Camacho. Camacho hoped to industrialize the Mexican economy, but unfortunately the country depended on food imports to feed its population. Wallace was startled to learn that Mexican farmers were laboring 20 times as long as U.S. farmers per bushel of maize (Ganzel 2007) but only achieving about one-tenth of U.S. production (Anderson 1975). After Wallace's visit, the Rockefeller Foundation began working with the Camacho administration to help increase Mexican food production. Camacho also believed that larger-scale, more intense agricultural production would free up workers to labor in industrial urban factories (Ganzel 2007).

The first phase of the plan involved once again expanding the amount of land under agricultural use. Unfortunately, methods for that process had not changed much since the Mexican Revolution's land reforms, but the agricultural market now had access to the type of industrial machinery used in the United States. With rural areas under high pressure to produce, the pattern of the American Great Plains began to play out again: Mexican smallholders abandoned traditional, sustainable methods in favor of high yield volumes and often went into debt to invest in modern equipment (Sonnenfeld 1992).

Fortunately, new methods and technologies were also being developed. In 1943, the Rockefeller Foundation and the Mexican government co-founded the Mexican Agricultural Program (MAP), intended to create and distribute new hybrid varieties that would have higher yields and better pest resistance than the current crops used in both countries. To achieve high yields, MAP focused on **fertilizer-responsive** crop lines, which could efficiently use the extra nutrients applied to fields (see Chapter 6 in this volume for measures of crop efficiency). We regard the crop hybridization research from this period as the start of the Initial Green Revolution.

The most famous single aspect of the MAP's operation was the hybrid wheat pioneered by Norman Borlaug, whose work would help avert multiple famines worldwide and later earn him a Nobel Peace Prize. Borlaug arrived in Mexico in 1944 and developed a unique method of shuffling experimental hybrid varieties between northern and southern Mexico, which allowed his team to use multiple growing seasons (and multiple plant generations) per year. This method also exposed the crops to a wide range of environments, creating varieties that could perform well in a variety of settings. The resulting high-yield varieties (HYVs) were highly productive on large, irrigated fields where modern farm machinery could distribute water and fertilizer to many plants at a time (Sonnenfeld 1992).

Thanks to the mobile breeding method and other new techniques, MAP released its first high-yielding hybrid varieties of wheat and corn[2] in just five years (Ganzel 2007; Rockefeller Archive Center n.d.). The preferred wheat variety expressed dwarfism, meaning the plant devoted

---

[2] The head of the corn breeding program during this period was Edwin Wellhausen (Ganzel 2007).

less energy to creating a tall, inedible stalk and more energy into packing the grains with nutrients. The shorter stalks were also less likely to break under the weight of the heavy heads of wheat that resulted from high fertilizer inputs. With the new varieties and input techniques, Mexico achieved self-sufficiency in corn production in 1948 and in wheat in 1956 (Rockefeller Archive Center n.d.). By 1958, Mexico was exporting surplus grain (Ganzel 2007).

Following the success of HYVs on industrially managed land, and hoping to continue industrial development of the economy, the Mexican government shifted focus firmly to supporting large agricultural corporations. The Mexican government invested in the River Basin Development Program, intended to continue agricultural intensification by nearly doubling the amount of irrigated land around the country. From 1947 to 1964, the equivalent of more than 33 trillion modern pesos (Official Data Foundation 2019) was devoted to building dams, generators, roads, and other rural infrastructure (Sonnenfeld 1992).

In contrast to this considerable investment in irrigation, agricultural research efforts by the government waned (Anderson 1975). Although Green Revolution varieties responded very well to irrigation, most of the yield increases of this era actually came from the rain-fed, un-irrigated, marginal lands granted to smallholders during the 1920s (Sonnenfeld 1992). While the soils there were generally poor and prone to erosion, fertilizer use facilitated huge increases in production. The heavy use of fertilizer and modern machinery on marginal lands led to topsoil loss, so maintaining yields required the additional use of fertilizer. Mexican farmers began to experience the same debt cycles seen in the 1920s Great Plains of the United States. By 1979, Mexican farmers were applying 350 times the amount of synthetic nitrogen fertilizer they used in 1940 (Sonnenfeld 1992). The expense of maintaining production was so great that by 1960, 15 million ha of reformed land had been abandoned and tens of thousands of rural inhabitants displaced to the cities (Sonnenfeld 1992). Erosion of high-altitude abandoned plots brought low-fertility, sandy topsoil to low-lying areas, slashing their productivity almost in half (Sonnenfeld 1992). Nonetheless, agricultural output in Mexico more than quadrupled between 1940 and 1970 (Anderson 1975; Sonnenfeld 1992). Within this period, Mexican corn yields increased by almost 20% per year, with even higher yield increases in the production of beans, wheat, and sorghum (Sonnenfeld 1992).

As yields were increasing, so was the Mexican population. The Mexican government continued to focus more on the urban-industrial sector and export markets than on agricultural development, especially as inexpensive labor flooded from abandoned farms to cities. To support the urban workforce, the government introduced controls that kept the price of staple food grains low. With no such controls on the price of cattle feed grains, and high sale prices on exportable luxury crops like asparagus and strawberries, Mexican farmers had little incentive to keep growing staple crops (Sonnenfeld 1992). As a result, Mexico's agricultural self-sufficiency only lasted from 1956 to 1971 (Ganzel 2007). Nonetheless, the incredible yield increases from the MAP hybrid breeding program were used by international agencies to promote the Green Revolution model worldwide (see Section 9.4).

In the early 1980s, the Mexican government began financing a program intended to bring the country back to agricultural self-sufficiency. Unfortunately, the oil crash of 1983 cut this project short and forced the end of state-sponsored seed production. Since then, Mexican farmers have depended on purchasing commercially available crop varieties (FAO 2015). Even though 90% of Mexico's agricultural production is still achieved on rainfed plots, heavy use of irrigation following the River Basin Development Program has gravely depleted the groundwater supply – and quality – in several areas. The overexploitation of aquifers near Mexico City has led the city to sink 30 feet during the 20th century; in 2001, the rate of ground subsidence reached three feet per year (Rudolph 2001).

## 9.4 THE PHILIPPINES, CHINA, AND SOUTHEAST ASIA: GLOBAL ATTENTION AND WILD SUCCESS

Based on Mexico's success, the Rockefeller Foundation and national governments worldwide encouraged farmers to follow the Green Revolution model: increase production by using high-yielding varieties in high-input farm settings. But as famines loomed in India, Pakistan, and surrounding areas, it became clear that the Green Revolution would need to expand to include other grains more prevalent in the Southeast Asian diet.

In 1960, two U.S.-based organizations, the Rockefeller and Ford Foundations, jointly founded the International Rice Research Institute (IRRI) in the Philippines. IRRI plant pathologist Peter Jennings ultimately developed a revolutionary strain of rice known as India Rice 8 (Folger 2014). Prior to the use of Green Revolution (GR) strains, India was producing 1 metric ton of wheat and 1.5 tons of rice per hectare of farmland each year. With the adoption of India Rice 8 and other HYVs, yields more than doubled, reaching 5 tons of wheat and 7 tons of rice per hectare per year (Anderson 1975). Some Indian farmers who converted to GR "miracle" seeds and GR techniques saw their yields per acre triple or even quadruple (Zwerdling 2009).

Less dramatic but substantial yield increases were also achieved in mainland China, the Philippines, and Thailand. The high-yielding strains also achieved success in the nations south and east of the Mediterranean, from Tunisia to Iraq and Jordan to Turkey (Anderson 1975).

In 1966, IRRI was joined by the International Maize and Wheat Improvement Center; additional international and public organizations would follow in the next few decades, creating a worldwide network of agricultural researchers (Pingali 2012). Together these organizations generated dozens of high-yielding, pest-resistant crop varieties every year. Ultimately, Asian production of rice and wheat doubled from the 1960s to the 1990s (Folger 2014).

Thus, the 1960s brought the success of the Green Revolution to the Eastern Hemisphere. Unfortunately, the high yields came at a financial and environmental cost that should by now be familiar. To achieve those massive increases in the 1960s, farmers did not just plant GR seeds; they also followed the advice of governmental and international organizations to use "modern, American" farming techniques (Zwerdling 2009). Farmers were encouraged to replace mixed-crop fields with **monocultures** of GR corn, wheat, rice, cotton, or other high-yielding crops. Such large fields sown with single crop varieties are more susceptible to pests and diseases. To overcome this issue, farmers were encouraged to apply pesticides to their fields. The high-yield crops also began stripping the soil of nitrogen, phosphorus, iron, and manganese, so that traditional fertilizers like cow dung had to be replaced with synthetic fertilizers with the right balance of nutrients to resupply the soil (Zwerdling 2009). As with the Initial Green Revolution, the high cost of inputs put financial strain on farmers; however, the Indian government subsidized chemical fertilizers, pesticides, and HYV seeds to help guarantee the success of the Green Revolution in their country (Zwerdling 2009).

While food crises had been averted worldwide, other shocks hit the agricultural market. In 1972, fertilizer consumption exceeded production for the first time, leading to a worldwide fertilizer shortage and an eightfold increase in the cost of fertilizer within two years (Anderson 1975). By 1974, worldwide inflation combined with poor weather conditions had quadrupled the cost of rice and more than tripled the price of wheat (Anderson 1975). In 1975, many governments had begun regulating the price of grains to keep them affordable, but fertilizer prices remained unregulated (Anderson 1975), increasing the pressure on farmers to produce high yields at low cost. A familiar pattern emerged once again, as indebted farmers were incentivized to invest in time-saving but soil-damaging machinery and put low-fertility land into production.

The Green Revolution seeds, developed in irrigated areas, also required more water than naturally falls in India. As a result, the adoption of high-yielding varieties in India during the 1960s and 1970s strongly followed the availability of groundwater (Pingali 2012). As in both the United States and Mexico, decades of groundwater exploitation have resulted in a drastically dropping Indian water table. In Punjab, the breadbasket of India, the water table has fallen an incredible 200 feet since 1960. Whenever the water table drops below the reach of existing equipment, farmers must pay for deeper access wells and more powerful pumps, often taking out large loans to do so. By 2009, this cycle of debt meant that Indian farmers taking out loans to reach the dropping water table might face interest rates of up to 24% (Zwerdling 2009).

In the 1980s, annual yield gains began to decrease, especially for rice, corn, and wheat (Folger 2014; Pingali 2012). Between 1966 and 1985, annual yield gains were estimated at 1% for wheat and slightly less for rice and maize (Pingali 2012), but by 2014, GR crops were achieving below 1% annual yield increases across the board (Folger 2014). This slowed yield increase does not necessarily indicate any decline in the quality of new hybrid HYVs. Instead, the declining growth rate reveals that the massive increases during the initial phases of the Green Revolution were partially due to the adoption of high-efficiency synthetic fertilizers, of modern pesticides, and of widespread irrigation (Pingali 2012). Now that these inputs are established in agricultural systems, their contribution is not measured in the yearly yield increases. However, while yields may be stabilizing in Green Revolution areas, the use of those artificial inputs is only increasing. For example, Indian farmers in 2009 had to apply three times as much fertilizer to their fields as they did in the 1980s to achieve the same yields (Zwerdling 2009). The Green Revolution might have averted several famines in the 20th century, but its methods are not sustainable. New techniques must come into widespread use to meet future food production demands (see Chapter 10 in this volume).

## 9.5   PATTERNS OF THE GREEN REVOLUTION

Green Revolution technologies were implemented in many parts of the world during different decades of the 20th century and several patterns seem to have characterized the movement as it spread. For example, there is a clear pattern of increasing farmer debt as the GR proceeds, beginning with expensive inputs and low crop sale prices and ending with the abandonment of smallholder land. This pattern is obviously a negative aspect of the Green Revolutions. However, the creation of international organizations devoted to developing high-yielding varieties of staple crops is a positive aspect of both historical Green Revolutions. Those international partnerships will be important in the further development of non-staple crops and in researching more sustainable methods of achieving high yields worldwide. As with so many historical movements, the Green Revolution of the 20th century achieved its successes at a social, environmental, and political cost. The proposed Second Green Revolution must avoid or, ideally, reverse these costs.

## 9.6   ENVIRONMENTAL IMPACTS OF HIGH-INPUT SYSTEMS

Perhaps the most criticized aspect of the Green Revolution is that hybrid varieties were developed on industrially managed, irrigated, and fertilized lands, so those are the conditions to which they are adapted; therefore, the supposedly high-yielding varieties have less of an advantage when planted on smallholder lands or lands managed in a more ecologically responsible manner (Anderson 1975). For example, switching to HYVs of wheat in irrigated areas with "modest" use of fertilizer increases yields by 40%, but in non-irrigated areas with similar fertilizer use, using HYVs only increases yields by 10%. In addition, poor farmers have smaller plots on more marginal land and less access to modern industrial technology like tractors, so the areas arguably most in need of crop improvements receive the least benefit from GR seed varieties (Pingali 2012).

Countering these criticisms, Dr. Keith Finlay, former deputy director of the International Maize and Wheat Improvement Center (CIMMYT),[3] emphasized that HYVs represent yield gains even in non-irrigated and non-fertilized areas relative to native varieties of the same grains (Anderson 1975) – for example, the 10% increase referenced above. In addition, the HYVs developed through Borlaug's shuttling method are not adapted to one specific location but are tolerant of a wide variety of environmental conditions, light intensities, soil types, and weather patterns. Thanks to the varieties developed during the Initial and Expanded Green Revolutions, global grain yields have

---

[3] The acronym is based on the Spanish-language name for the organization.

considerably increased; for example, wheat yields in developing countries tripled from 1960 to 2000 (Pingali 2012).

### 9.6.1 FERTILIZERS

During the GR, governments commonly encouraged the use of inputs like fertilizer but did not subsidize those inputs. Farmers were thus unprotected from the costs of the inputs their seeds responded best to, leading to additional farmer debt. At the same time, the high use of inputs like synthetic fertilizers can deplete natural soil fertility, pollute aquatic habitats downstream of agricultural systems, and even increase erosion rates; synthetic fertilizers contain significantly fewer **water stable aggregates** (which do not move or break in water) than natural fertilizers like manure (Ozlu and Kumar 2018), so soil covered with synthetic fertilizer is more prone to water erosion than organically fertilized soil (Hillemann 2018).

Each of the two most used synthetic nitrogen fertilizers represents an important source of pollution. Ammonia fertilizers readily release nitrous oxide to the air (Cao et al. 2018). The agricultural release of nitrous oxide from soils is one of the major sources of greenhouse gas emissions worldwide. More than half of nitrogen applied as ammonium fertilizer can be lost to the air within days of application (Cao et al. 2018). Meanwhile, nitrate fertilizers are responsible for considerable water pollution. Nitrate is highly mobile in soil, which makes it very available to plants but also makes it very quick to leach during high water flow events, such as heavy rain or intensive irrigation use.

High nitrogen levels in water systems lead to **algal blooms**, the growth of enough algae on the water's surface to starve subaquatic plants like seagrass of sunlight. As subaquatic plants stop releasing oxygen through photosynthesis, the oxygen level in the water drops. Upon death, the algae, plants, and animals in the water sink to the bottom and decompose by the action of saprotrophic microorganisms, which use additional oxygen and release carbon dioxide. This results in a severely oxygen-depleted, or **hypoxic**, environment in which aquatic organisms can suffocate in massive die-offs (World Resources Institute n.d.). These eutrophication events have become much more common since the advent of chemical fertilizers; in 1960, only 10 eutrophication events were reported globally, but in 2007 there were 169 documented cases (World Resources Institute n.d.). Since 1985, scientists have surveyed an annual hypoxic zone in the Gulf of Mexico which appears due to runoff from farms along the Mississippi River watershed. Its size fluctuates, but in 2021 it measured 6,334 square miles (Lyons and Sarkis 2021) – an area larger than that of Connecticut, USA. Some "blooming" algae also produce harmful toxins that poison animals all along the food web. Application of synthetic fertilizer often takes place in the spring or fall, just before planting or harvest, when rain is the most intense. This fertilizer timing, which does not always contribute to the growth of desired plant products, only worsens the soil leaching effects (Cao et al. 2018).

Nonetheless, the benefits of fertilizer use should also be at the forefront of the debate. In response to objections over the amount of fertilizer HYVs require, Norman Borlaug compared a hypothetical crop plant that does not require fertilizer to a human that does not require food (Anderson 1975). In order to produce products that humans need or desire, a plant must first have its basic nutritional needs met (see Chapters 1, 5, and 6 in this volume). Increased use of fertilizer was responsible for 26% of the yield increases seen in the 20th century for U.S. production of top crops including corn, wheat, and rice (Cao et al. 2018). In addition, nitrogen use efficiency has increased 30% in the past 40 years largely due to the release of additional HYVs that are more fertilizer responsive (Cao et al. 2018). The continuing development of hybrid varieties responds to the technologies in use, creating a positive feedback loop[4] of food production and efficiency. However, the issue of fertilizer pollution can only be resolved by studying the movement of fertilizers through different soil types and educating farmers on best practices to apply only the amount of fertilizer needed by their crops and only at times when the crops can divert the nutrients toward harvestable mass.

---

[4] A *positive feedback loop* is a process that results in an effect or product that allows the process to continue.

### 9.6.2 IRRIGATION

The amount of water needed for irrigated lands is staggering. During 2013, American farmers applied 74 million acre-feet of water to cropland – the equivalent of more than 100 million Olympic - sized swimming pools (Hrozencik 2019). This would be enough water for the entire population of the United States[5] to shower for 25 days straight! More than 85% of the world's freshwater is currently devoted to agricultural use (D'Odorico et al. 2020; Głowacka et al. 2018). In addition to overall water use, criticisms of irrigation use include the issues of salinization, waterlogging, and the leaching of fertilizers into groundwater and surface water systems. Overexploitation of groundwater for irrigation also increases the risk of land subsidence and can exhaust important natural water sources.

However, the benefits of the irrigation systems already in place have been enormous. Irrigation increases the yield and value of crops by as much as three times the baseline (Sojka et al. 2002). Measuring global irrigated cropland is a difficult task due to several issues, including (Thenkabail et al. 2009):

- Spotty data sets (e.g., not all countries report their irrigated acreage every year, or reports may be inaccurate)
- Varied intensity and temporality of use (e.g., some irrigated croplands produce multiple annual harvests, some produce only one, and some are left fallow, producing no annual harvest)
- Constant fluctuations in irrigated spaces (e.g., new land becoming irrigated or irrigated land being abandoned due to waterlogging or salinization)

Nonetheless, current estimates indicate that in the year 2000, 20.8%–29.7% of global cropland was irrigated (own calculations based on Ritchie and Roser 2019 and Thenkabail et al. 2009). That 20%–30% of cropland produces a whopping 60% of globally produced grain and 30%–40% of global food production overall (Falkenmark and Rockström 2004; Sojka et al. 2002).

Furthermore, instead of irrigating more land area or trying to find new sources for irrigation water, technologies are emerging to improve the use efficiency of the water from already irrigated lands. For example, from 1984 to 2013, water usage in the 17 westernmost United States remained mostly stable. During those three decades, only half a million acres of land were converted from non-irrigated to irrigated cropland. Despite production and yield increases, water use in those states also remained relatively stable (Hrozencik 2019). Thanks to irrigation and other yield increasing technologies, an estimated 1.69 billion hectares of land have been spared from conversion to agriculture since 1961 (Ritchie and Roser 2019).

Plants simply cannot produce harvestable mass without enough water. Efforts toward making plants more drought tolerant or water efficient (see Chapter 6 in this volume) will go a long way to improving water conservation. Smart digital systems, such as sprinklers that only irrigate when soil is dry, have already improved water use efficiency on many farms worldwide.

### 9.6.3 MONOCULTURE

Another common criticism of the GR system is that when farmers worldwide use the same crop varieties, the global agricultural system becomes more susceptible to catastrophic failure (Anderson 1975). Essentially, while a single field is planted with the same species of crop for multiple seasons, it is easier for diseases to develop and wipe out the crops. One major concern is that the use of similar HYVs across continents could lead to similar collapses on a much larger scale. However, the HYVs produced using the methods pioneered by Norman Borlaug incorporate unprecedented

---

[5] As of the 2014–2018 American Community Survey (ACS) 5-year estimate.

genetic variation. As of 1975, Borlaug's team was performing more than 10,000 wheat crosses every year; every cross involves two wheat plants from separate lines. To evaluate promising varieties, seeds were distributed to more than 100 locations around the world, from deserts and tropical areas to the Arctic Circle, and evaluated for productivity, stress resistance, and pest resistance. Only the variants that performed well in most of those locations were released commercially. Norman Borlaug argued that with all the global genetic variation that went into each variety's development, the resulting plants were robust enough to resist global disease events (Anderson 1975). It is also important to note that new varieties are still developed with similar methods and released regularly, so should a disease begin to take over one variety, a new variety can be distributed and used quickly.

However, disease vulnerability is not the only issue with monoculture. When governments promoted GR techniques in the 20th century, it was often by installing national irrigation systems, subsidizing fertilizers to make them more available to farmers, and generally giving farmers additional incentives to produce HYV staple crops. All these incentives toward the production of staple crops necessarily de-incentivized the cultivation of non-staple crops. With farmers focusing on producing high-value staple crops, cultivation of traditional crops decreased. Traditional crops, including many fruits and vegetables, contain important micronutrients (Pingali 2012); by focusing on HYV staple crops, farmers reduce the availability of micronutrients to consumers, potentially leading to increased malnutrition. Because the GR focused so strongly on staple crops, there are fewer high-yielding varieties for non-staple crops that contribute to a varied, balanced, flavorful, and culturally rich diet. Furthermore, because the 20th century Green Revolution had specific regions of focus (first the Americas, then East Asia), many other regions were neglected. Africa only began receiving attention from international hybrid-developing organizations in the late 1980s (Pingali 2012), putting the region decades behind in the development of locally adapted, high-yielding staple crops.

A Second Green Revolution should build on the success of the 20th century GR by using robust variety development techniques on non-staple, but still vitally important, food crops. This will require conversations between international organizations, local governments, and locals to identify the most culturally and nutritionally important species to focus on in each region. Development of non-staple HYVs may lag behind the release of new staple crops, but non-staples should not be neglected as they were in the 20th century.

## 9.7   ECONOMIC IMPACT

Observing the two Green Revolutions of the 20th century, there is a clear and unfortunate pattern for smallholders. Poor farmers in debt tried to increase their incomes by starting or increasing crop production on marginal land. This often involved removal of lower-demand or non-HYV crops and animals from the farm environment (Pollan 2016). In traditional crop rotation or farm systems, nitrogen-fixing legume crops and cover crops are rotated between seasons of higher-value crops in order to re-stabilize the soil and return nitrogen to it. Varied crops also made it possible for a farm to be self-sustaining, giving the farm family and some of its workers a varied, homegrown diet (Pollan 2016). However, when farmers are incentivized to plant increasing volumes of HYVs and have unrestricted access to synthetic fertilizers, productivity of high-value crops can be maintained without crop rotation or cover crops. When farmers eliminated non-GR plants, reduced biodiversity on the farm made it impossible to feed the farm families and depleted soil fertility, shifting the average farm from self-reliance to reliance on outside food and inputs (Pollan 2016).

As the land became unproductive, farmers either (1) invested in increasingly costly inputs and equipment or (2) defaulted on their debts and sold or abandoned their land. If the costly inputs failed to maintain production, farmers might also default despite their expensive investments. As farmers moved to urban centers in search of employment, they brought their cultural knowledge of how to sustainably farm their native lands with them (Sonnenfeld 1992). With that knowledge lost from the agricultural area, the lands were brought under industrial control and usually converted to HYV monoculture. Increasingly depleted soil increased ecological fragility and, through weather events like dust storms, could even harm the productivity of additional agricultural lands, driving the cycle of farmer poverty and land abandonment.

The pattern of smallholders falling into poverty and abandoning their lands is obviously a strongly negative aspect of the Green Revolution. However, the GR did broadly reduce poverty in the non-farming population. Thanks to the high-yielding varieties produced during the GR, global grain yields tripled in developing countries between 1960 and 2000. The increase in grain supply led to a 35%–65% decrease in food prices and 12%–13% greater food availability in developing countries (Pingali 2012). While these low sale prices for staple grains can incentivize farmers to use marginal lands and unsustainable management practices, on the consumer end, the benefit of having more grain available at lower prices is obvious: more people have access to more food, reducing the instance of starvation. By paying lower prices for staple foods, consumers also have more money available for non-staple foods, like fruits and vegetables that enrich the diet (Pingali 2012).

The agricultural changes at the front end of the 20th-century Green Revolutions also indirectly fought poverty by increasing the value of the products grown on the land and the value of the land itself, which could then be used as collateral against other large loans and investments. During the last century, a 1% increase in value per hectare of agricultural land led to a long-term 1.9% reduction in poverty for the local region (Pingali 2012). This dramatic, lasting reduction in poverty that comes with agricultural improvement is higher than the impacts of growth in other sectors of the economy. For example, in Sub-Saharan Africa, growth in the agricultural sector reduces poverty over four times more than growth in the service sector (Pingali 2012).

For a Second Green Revolution to achieve lasting success over poverty, further increases in the value of agricultural land should come from sustainable techniques. For example, terracing and listing are known methods of soil conservation that also produce a significant increase in the value of the altered land (Hurt 1984). Since irrigation projects have already covered most of the land that can benefit from irrigation, methods like terracing and incorporating local native ecosystems are most likely to give the greatest improvements to land value and poverty rates. It will also be important to ensure that sustainable techniques for increased production are equally available to smallholders and large corporations, and possibly redistribute land to smallholders once again to try and bring land back under the care of experts who can identify and address the region's specific agricultural needs.

## 9.8 GOVERNMENT INTERVENTIONS

Government interventions produced some positive changes for agriculture during the 20th-century Green Revolution. For example, cooperation between governments and private organizations led to the development of high-yielding varieties for staple crops, which can support increased production on land already devoted to agriculture. This can reduce the demand for more marginal lands to be brought under production (Pingali 2012). However, in attempts to support the agricultural sector, many governments implemented policies which incentivized farmers to bring marginal land into production, negating the density benefit of HYVs. When output subsidies encourage high production of commodity crops regardless of market demand, the price of commodities consistently decreases, until the only reliable way for farmers to make a profit is to continue producing those commodities and receiving output subsidies (see Chapter 8 in this volume). Furthermore, input subsidies encourage farmers to use unsustainable techniques like the application of synthetic fertilizers as opposed to sustainable fertility techniques like crop rotation or the use of animal manure (Pollan 2016). As a result, subsidies that are meant to support farmers and maintain food supplies can actually shift the food supply from polyculture to monoculture, increase the supply of staples at the cost of other important foods, and intensify the environmental impacts of the GR technologies discussed above.

Another important aspect to consider is the subsidization of various technologies, inputs, and outputs. Government support via subsidies made new 20th-century technologies like irrigation and

**TABLE 9.1**

**Summarized Impacts of the Green Revolution**

| Subject | Positive Impacts | Negative Impacts |
|---|---|---|
| 9.6 Environmental impacts | Relative to native varieties, HYVs give increased yields in a range of farm biomes | HYV gains are greater for industrially managed lands, limiting their benefits to smallholders |
| | Irrigation and synthetic fertilizers considerably increase yields | Synthetic fertilizer use contributes to air and water pollution; irrigation can reduce soil stability |
| | Green Revolution research created HYVs for important global staple crops | HYVs are most successful when grown in monoculture, reducing the production and availability of important non-staple crops |
| 9.7 Economic impacts | Increased food availability at lower prices reduced malnutrition in the consumer population | Farmers were encouraged to buy expensive inputs, then forced to sell products at a loss, creating a debt cycle |
| | Increases to agricultural land value led to lasting reduction in local poverty | Farmers forced to move to urban centers also remove knowledge of sustainable methods for cultivating their lands |
| 9.8 Government interventions | Government cooperation with private organizations led to creation of HYVs | The effects of input and commodity subsidies reduce the potential environmental benefits of HYVs |
| | Subsidies gave smallholders access to infrastructure improvements and new technologies | Subsidies drive commodity price cycles that bankrupt smallholders |

synthetic fertilizers readily available to smallholders who would be unable to afford such inputs without government assistance. Federal-scale projects like hydroelectric dams and irrigation development benefitted large and small landowners equally, at least initially. Unfortunately, input subsidies tend to disrupt the relationship between producers and the market. Government intervention often includes commodity subsidies for staple crops, creating a vicious cycle by which the sale price of commodities forever decreases (see Chapter 8 in this volume). When smallholders go bankrupt or default on their loans, their land is not taken out of production and returned to a natural, ecologically neutral state. Instead, the land will be purchased by other, more financially successful smallholders or by large farming corporations (Pollan 2016). In the subsidized commodity-driven market, the most financially successful farming operations are those that operate with economically effective but ecologically unstable methods like monoculture and heavy input use. Therefore, commodity subsidies, which are intended to help smallholders, ultimately lead to smallholder bankruptcy and additional environmental issues.

## 9.9 CONCLUSION

The Green Revolutions of the 20th century caused complex social, economic, and environmental shifts. Looking back, we can recognize the harmful patterns of the Green Revolutions and take steps to avoid similar pitfalls in a new, sustainable Second Green Revolution. Table 9.1 summarizes some of the impacts of the previous Green Revolutions.

## REFERENCES

Anderson, Alan Jr. 1975. The green revolution lives. *New York Times*, April 27: 240. Accessed December 5, 2019. https://www.nytimes.com/1975/04/27/archives/the-green-revolution-lives-the-green-revolution.html?login=google&auth=login-google.

Environmental Literacy Council. 2015. *Green Revolution*. Accessed November 5, 2019. https://enviroliteracy.org/food/food-production-supply/green-revolution/.

Folger, Tim. 2014. The next green revolution. *National Geographic*, October. Accessed November 3, 2019. https://www.nationalgeographic.com/foodfeatures/green-revolution/.

Food and Agriculture Organization of the United Nations. 2015. *Urban and Peri-urban Agriculture in Latin America and the Caribbean*. Accessed December 6, 2019. http://www.fao.org/ag/agp/greenercities/en/ggclac/mexico_city.html.

Ganzel, Bill. 2007. *The Mexican Agricultural Program*. Accessed December 5, 2019. https://livinghistoryfarm.org/farminginthe50s/crops_14.html.

Hergert, Gary, Rex Nielsen, and Jim Margheim. 2015. *A Historical Overview of Fertilizer Use*. University of Nebraska-Lincoln Institute of Agriculture and Natural Resources. March. Accessed November 5, 2019.

Hurt, Douglas R. 1981. *The Dust Bowl: An Agricultural and Social History*. Edited by Brian W Blouet and Frederick C Luebke. Nelson-Hall, Chicago, IL, USA, pp. 67–86.

Mid-America Regional Council. 2019. *Know Your Roots*. Accessed January 7, 2020. http://www.cleanwater-kcmetro.org/know-your-roots/.

National Drought Mitigation Center. 2018. *The Dust Bowl*. University of Nebraska. October 3. Accessed November 5, 2019. https://drought.unl.edu/dustbowl/Home.aspx.

Official Data Foundation. 2019. *Inflation Calculator*. Alioth Finance. Accessed December 12, 2019. https://www.officialdata.org/.

Pingali, Prabhu L. 2012. Green revolution: impacts, limits, and the path ahead. *Proceedings of the National Academy of Sciences of the United States of America* 109(31): 12302–12308. www.pnas.org/cgi/doi/10.1073/pnas.0912953109.

Rockefeller Archive Center. n.d. *Mexico*. Rockefeller Foundation. Accessed January 16, 2019. https://rockfound.rockarch.org/mexico.

Russel, D. A., and G. G. Williams. 1977. History of chemical fertilizer development. *Soil Science Society of America Journal* 41(2): 260. doi:10.2136/sssaj1977.03615995004100020020x.

Scherer, Thomas F, and Dean D Steele. 2019. Irrigation *Scheduling by the Checkbook Method. North Dakota State University Extension*. North Dakota State University. February. https://www.ag.ndsu.edu/publications/crops/irrigation-scheduling-by-the-checkbook-method-1/ae792.pdf.

Sonnenfeld, David A. 1992. Mexico's "Green Revolution," 1940–1980: Towards an environmental history. *Environmental History Review (The Forest History Society and American Society for Environmental History)* 16(4): 28–52. Accessed November 22, 2019. https://www.jstor.org/stable/3984948.

United States Geological Survey. n.d. *Land Subsidence*. U.S. Department of the Interior. Accessed January 16, 2020. https://www.usgs.gov/special-topic/water-science-school/science/land-subsidence?qt-science_center_objects=0#qt-science_center_objects.

Zwerdling, Daniel. 2009. *India's Farming 'Revolution' Heading for Collapse*. NPR.org, April 13. Accessed November 5, 2019. https://www.npr.org/templates/story/story.php?storyId=102893816.

# 10 The Second Green Revolution and the Role of Mycoagroecology

*Elizabeth "Izzie" Gall*

## CONTENTS

## 10.1 INTRODUCTION

The Green Revolution (GR) of the 20th century is over, but the human population continues to expand, increasing demand for high volumes and high quality of food. Furthermore, climate change is altering drought patterns and shifting agriculturally favorable zones globally (King et al. 2018), presenting new challenges to agriculture while also demonstrating the need for a holistic, environmentally conscious approach to food production. Most of the growing population projected to exist by 2050 will be living in Sub-Saharan Africa and South and Southeast Asia, some of the areas that will also be the hardest hit by climate change (Folger 2014). There is therefore need for another movement to increase the production and quality of crops worldwide – a Second GR. This time, however, international experts are urging attention to be wider than the crops in the field and profits in the bank (United Nations 2008). To create a successful Second GR, we must build on the successes of the 20th century GR while denouncing, and learning from, its failures. In particular, it will be important to give attention to socioeconomic and global environmental factors in addition to basic increases in crop production.

DOI: 10.1201/9780429320415-10

Because the agricultural industry is focusing more on the products and people left behind by the 20th century GR, this is an excellent time to integrate mycology into agroecology for the betterment of agriculture.

## 10.2 MYCOLOGY: REVEALING THE FUNGAL ROLES OF PARTNERS, PESTS, PEST CONTROL AGENTS, AND PRODUCTS

**Mycology** is the study of fungi, a generally underrepresented kingdom in public discourse. Despite their enormous contributions to nutrient cycling across terrestrial ecosystems, in the author's experience, fungi are often neglected in biology, botany, and ecology courses. Over more than 100 years, mycologists have worked to identify and understand fungi, leading to our current knowledge that they are extremely diverse organisms with equally diverse ecosystem functions.

Fungi are important partners of plants, conferring resistance to environmental stressors and pathogens as well as forming fundamental partnerships that give plants access to important nutrients (see Chapter 11 in this volume). Fungi are also important, abundant members of soil systems, including those in agriculture. Beneficial native fungi have long been excluded from agricultural systems as a side effect of fungal pest control or techniques that disrupt soil structure (see Chapter 12 in this volume). However, soil fungi and plant endophytes have the potential to add tremendous value to agricultural systems. Beneficial fungi should be incorporated and encouraged, not ignored or destroyed. There is considerable value in fungi as agricultural **Partners**.

Mycology can also contribute to the control of harmful fungi. Some fungi are important diseases in agricultural and forestry systems, but understanding these diseases and targeting them specifically will be more beneficial than using blanket strategies (like fungicide application) which also harm beneficial microbes. Furthermore, some beneficial fungi and other members of native ecosystems can be used to biologically control fungal diseases (see Chapter 15 in this volume). Fungi should be understood both as **Pests** and as **Pest Control Agents**.

Finally, many fungi produce edible mushrooms or can be encouraged to create other important agricultural products through fermentation of both solids and fluids (see Chapter 18 in this volume). Farmers who understand the growth requirements of fungi can use them to recycle waste into usable materials and nutritious foods (see Chapters 16 and 17 in this volume). These **Products** can add considerable value to a farm, whether the mushrooms are eaten, sold, or merely used as a waste disposal method.

## 10.3 AGROECOLOGY: INTEGRATING PRODUCTIVITY AND SUSTAINABILITY

**Agroecology** is the study of how human agriculture influences the nutrient and calorie cycles of the natural world. In the past, agroecology was mostly used to make the world more agriculturally productive, with little regard to the environmental impact of that productivity (see Chapter 9 in this volume). In the 21st century, the term is often used to describe farming practices that incorporate ecological understanding and use some techniques meant to help preserve the ecosystem near the farm. In agroecology, there is an emphasis on maintaining a diverse biosphere, using local and natural inputs (as opposed to imported and synthetic inputs), and buffering the farm ecosystem to be resilient to environmental stressors and threats like drought or disease. Practicing agroecology requires a firm understanding of the ecological principles discussed in the introductory section of this text, though the role of microorganisms like fungi is often overlooked.

Integrating ecology into agricultural practices is at once the oldest and newest farming method, as ancient peoples needed to understand the ecology of their fields to get the best harvests but many generational values and techniques were lost or destroyed in the face of artificial fertilizers and other technological advancements of the 20th century GR. Moving back to naturalistic or environmentally friendly agricultural practices after instituting industrial methods may require some upfront costs, but should yield long-lasting benefits for both agricultural yields and the ecosystems around farms.

There are countless agricultural techniques, each of which falls on a spectrum of ecological responsibility. Some techniques focus on the whole farm while others focus on specific parts of the farm, like the areas between fields; some require small changes to established industrial protocols while some require more effort to implement; and some consider the surrounding or global environment while others, like industrial GR agriculture, focus on the productivity of the farm over all other factors.

### 10.3.1 INDUSTRIAL FARMS VALUE PRODUCTION OVER ECOSYSTEM BALANCE

The current industrial model of agriculture, which is used in most large farms, relies on the use of artificial inputs to increase production. This includes the application of fertilizer, pesticide, herbicides against weeds, and water irrigation. In the industrial model, the farm's production is all that matters, and success is measured as the ability of the farm to sell enough products to break even or make a profit.

Over decades of use, this technique has led to numerous environmental crises (as discussed in Chapter 9 of this volume) and economic problems as well. Small farms cannot consistently afford the large amounts of fertilizer and other inputs that are required to make designer hybrid seeds perform at peak efficiency and buying expensive inputs and equipment can put farmers into considerable debt. Modern or 20th-century industrial equipment strips the soil of its natural microbiome and techniques like over-irrigation can drain the soil of nutrients, both of which necessitate the use of additional fertilizer – and if the soil cannot hold the fertilizer, runoff becomes an ecological problem. In addition to expensive installation costs of the sprinklers and pipes themselves, irrigation may require drills and pumps to access the water table, which is an ongoing cost that in many parts of the world is continually increasing as the water level drops. If irrigation proceeds too quickly, the water table can be depleted and underground sources of brackish (salty) water may accidentally be used on fields (Zwerdling 2009). Furthermore, in addition to harming the ecosystems around a farm, application of pesticides and herbicides can also be harmful to human workers (e.g., Bassil et al. 2007; Cohen 2020; Guyten et al. 2015).

The major benefit of high-input GR techniques is in **economies of scale**, where the high upfront cost of expensive inputs is balanced by the sheer volume of product they are used on. Economies of scale are only beneficial to large farms, and industry giants, in particular, may be able to negotiate discounts for buying inputs in bulk volumes that other farms can't match. Similar principles encourage monoculture, where large swaths of land can be planted with a single seed variety and harvested by specialized machinery. Monoculture reduces the biological diversity of a system and makes it more susceptible to diseases and abiotic shocks (see Chapter 6 in this volume); extended monocultures of commodity crops can also lead to market distortions and farmer debt cycles (see Chapter 8 in this volume).

The following sections outline some alternative agricultural philosophies, noting the cost of converting from the high-input food production technique since it was so heavily championed by governments and organizations during the 20th-century GR that it is now the industry standard around the globe.

### 10.3.2 BUFFERING THE ECOSYSTEM FROM THE FARM

Some farms are run conventionally, but buffering zones are added around the farm to try to insulate the environment from the effects of industrial farming. In this philosophy, the first consideration is still the productivity of the farm, but the health of the surrounding ecosystem is also important. For example, a **riparian buffer** is a sequence of trees, shrubs, and other plants installed along the outline of bodies of water on the farm. As farm runoff goes through the roots of the plants, excess nitrogen that could otherwise poison downstream ecosystems is filtered and used. With the buffer in place, farm managers might apply the same amount of nitrogen fertilizer to the field that they used before the buffer was installed, but the downstream impacts will be mitigated. Another example might be putting netting around a field to reduce the likelihood that wildlife will eat crops that have been sprayed with pesticides, reducing the buildup of such toxins in the food chain. The pesticides are still being used, but their effect on the wildlife is reduced.

Changing to this philosophy from the industrial norm requires the installation of buffer zones, fences, and other technology that will reduce the impact of conventional farming on the surrounding ecosystem.

### 10.3.3  CLOSING THE LOOP: THE FARM AS AN ISLAND ECOSYSTEM

With this philosophy, the farm is maintained as separately from the surrounding environment as possible. The goal is to make the farm self-reliant so that minimal inputs are needed, and minimal waste is produced that must be disposed of externally. Crop rotations and co-cropping work this way to some extent, helping the crops use and replenish as much soil nutrition as possible without external inputs. The incorporation of animals often helps achieve this goal, as animal waste can be used in place of synthetic nitrogen fertilizer and the portion of crops that are not eaten by humans can often be nutritious feed for ruminants like cattle and sheep. Use of such techniques is often known as "closing the loop" because it helps the farm be an enclosed system rather than a linear system which must import inputs and export waste (see Figure 16.1).

Turning an industrial farm into a closed loop requires substantial changes in how the farm is run. For a large monocultured field, the first step is diversification of the crops and the addition of crop rotations to help maintain the fertility of the soil in a natural way. Diversification of crops also means diversification of pest and disease control practices, which may represent a large investment of time and effort as well as money. Diversifying a plant-only farm to incorporate livestock may be impractical at large scales, but it may be possible to have livestock and crop farms exchange waste to create a locally closed loop even if each individual farm remains linear.

Many organic farms strive to achieve a closed-loop system. It is also possible to incorporate some artificial inputs as long as they are buffered from the environment, such as by using riparian buffers.

As this technique seems to present a happy medium of high productivity and environmental sustainability, it is probably the most practical technique to use in helping to achieve the Second GR.

### 10.3.4  THE FARM AS A PRODUCTIVE PORTION OF THE ECOSYSTEM

Under this philosophy, the farm blends seamlessly with the ecosystem while still maintaining crop production. Local, native species' **ecosystem services**, like nitrogen deposition into the soil and pest control, are used to reduce the need for industrial or anthropogenic inputs. This can be accomplished by using native plants and animals to the farmer's advantage, such as placing birdhouses in fields so native birds help control bug populations without the need (or as much need) for pesticides or pheromone traps. In sustainably managed forests, leaves and other debris compost in place through the action of native microbes, reducing the need for fertilizers. **Insectary strips** are installations of native plants that encourage local beneficial insect populations, helping reduce insect pests through competition or predation. Instead of fences, trees would be used for field borders to aid the movement of native species like birds and mammals.

Though it is ecologically ideal, this technique is possible only on small scales. It works best for small numbers of coevolved crop plants and trees located near their native regions, such as in subsistence farms. There is currently no known way to support the global human population, nor the global livestock population, with this method unless considerable changes are made to global humans' habits, such as conversion to global vegetarianism or conversion of all livestock to forage-fed (Foley 2014; Muller et al. 2017).

### 10.3.5  INCORPORATING OTHER TECHNIQUES

Ultimately, a combination of techniques will need to be employed to achieve high food production in an ecologically sustainable way. Food production is a vital area that cannot afford to leave any safe technology behind. Therefore, the Second GR must also include the best and most current techniques of producing improved plant varieties, including traditional crossing-based breeding programs as well as the genetic modification of plant lines (see Chapter 6 in this volume). Once a new variety is produced, it can be incorporated into a farm with an ecological philosophy just as easily as it could be used in conventional monoculture.

There are other environmentally friendly techniques that can be incorporated into even conventional farms. For example, large fields can be ravaged by strong winds if there is nothing to interrupt air flow. **Windbreaks** reduce wind damage by breaking up large, flat areas. Windbreaks made of native trees also provide habitat for birds and other wildlife that can reduce pest populations, while the trees' deep root systems can help detoxify runoff before it can escape the field. Riparian buffers require relatively small cropland losses, so these can also be used on large farms to mitigate environmental damage. Using trees or shrubs in place of fences is another technique that large farms can benefit from. Such steps require the upfront cost of tree planting and ongoing maintenance costs and labor to keep the trees healthy, but fences and constructed windbreaks also require upkeep. Ideally, the ecological services of such conversions will outweigh the upfront costs.

## 10.4  MYCOAGROECOLOGY

Mycoagroecology is a proposed practice of recognizing the diverse roles of fungi in agricultural ecosystems and integrating them into agroecological practices. The three major roles of fungi we focus on in this text are as Partners (increasing the production and value of plant crops); as Pests and Pest Control Agents (causing and preventing crop damage); and as Products (adding value to agricultural systems directly).

### 10.4.1  PARTNERS

It is becoming increasingly easy to identify and monitor fungal populations of various kinds, allowing us to better understand their interactions with other organisms and their roles in a given biome. Plant-fungal mutualisms are ancient and widespread (Chapter 11) and can give considerable benefits to farm systems if understood and used appropriately (Chapter 12).

### 10.4.2  PESTS AND PEST CONTROL AGENTS

Some fungi are important diseases in agriculture and forestry (Chapters 14 and 15); understanding their life cycles and ecology will help farmers more effectively control the diseases (Chapter 13). Moreover, certain fungi may be usable as biological control agents against other pests (Chapters 12 and 14) or may even partner with other perceived pests to create value for an agroecosystem (Chapter 15).

### 10.4.3  PRODUCTS

Fungi are key recyclers of organic waste, so they are good candidates for integration into agricultural systems as means of closing farm waste loops (Chapters 16 and 17). They can also be utilized in liquid culture for incorporation into other parts of agricultural production and to create high-value chemical products and flavors (Chapter 18). Furthermore, many fungi can turn waste into nutritious products or be cultivated alongside plant crops to provide high-protein products to smallholders (Chapter 17).

## 10.5   MYCOAGROECOLOGY IS COMPATIBLE WITH THE GOALS OF THE SECOND GREEN REVOLUTION

For the Second GR to succeed, global food productivity and quality will need to continue increasing in an ecologically and socially responsible way. There are several smaller goals that feed into this effort, many of which are meant to patch gaps in the 20th century GR and all of which can be helped along, or directly addressed, by fungi.

### 10.5.1   ADDRESS "ORPHAN" REGIONS AND CROPS

During the intense scientific study and crop development cycles of the 1960s–1980s, the focus was on intensifying crop production on lands already under agricultural management. As a result, areas like Sub-Saharan Africa, where land was relatively abundant, were not given as much attention; the prevailing thought was that to increase production in land-abundant regions, one need only to expand into a new area.[1] Therefore, important staple crops in land-abundant areas were neglected in plant breeding programs of the 20th century GR (Pingali 2012). The first international project focused on improving African crops began in the 1980s and was still mostly focused on generating locally productive varieties of the global staples wheat, maize, and rice. Sorghum and cassava, the latter of which is a staple crop for more than 250 million people, were considered as secondary subjects (Pingali 2012; Folger 2014). Such "orphan" crops and regions, largely ignored by the 20th century GR, deserve to have full focus in the Second GR. Mushrooms are an important foraged part of the diet in many parts of the world, and because the GR focused on plant crops, fungi are also orphan crops that should be given due consideration in the Second GR (e.g., through cultivation on small farms; see Chapter 17 in this volume).

Because the 20th century GR occurred during a time when artificial inputs were becoming abundant and their ecological cost was not yet clear, GR HYVs (high-yield varieties) were developed for use in irrigated, highly fertilized fields (see Chapter 9 in this volume). The Second GR should give equal focus to crops that can be grown in marginal areas, such as in floodplains, drought-stricken areas, and salty soils. Fungi will be very useful in making marginal lands agriculturally productive in an ecologically responsible manner. Rather than using large amounts of fertilizer to try making the land productive, native fungi can naturally mitigate plant stress (Chapter 11 in this volume), bind marginal soils to improve their structure and water capacity, and ameliorate certain problems such as by concentrating salt or heavy metals in soil for removal (see Chapter 17 in this volume). Native arbuscular mycorrhizal and ectomycorrhizal fungi can also be used to improve nutrient uptake in certain plants, especially if those plants are cultivated and developed in native soil without synthetic fertilizers (see Chapter 12 in this volume).

### 10.5.2   RETURN TO POLYCULTURE

Now that the ecological damages of monocultures are clear (see Chapter 9 in this volume), a return to polyculture should be another priority of the Second GR. This will involve the use and breeding of plants that grow well together or in succession, such as rotations of clover and other nitrogen-fixing crops before corn and other crops with high nitrogen requirements. Rotations and co-cropping provide farmers insurance against crop failures, as the failure of one crop can be compensated by the yields of a different crop. These methods also help reduce the instance and severity of diseases in plants by denying disease-causing organisms consistent hosts. Cultivated mycorrhizae can persist between seasons as long as mycorrhizal crops are planted in succession. An established mycorrhizal community can quickly create relationships with newly planted mycorrhizal crops, helping them grow rapidly and produce quality harvests while reducing the instance of microbial diseases (Chapters 11 and 12 in this volume). Fungi that fruit with edible mushrooms can also be incorporated as co-crops in certain fields, providing additional sellable products and further insurance against the failure of another crop.

---

[1]  Indeed, between 2003 and 2019, about 37% of global agricultural land expansion occurred in Africa (Potapov et al. 2022).

### 10.5.3   REDUCE THE NEED FOR SYNTHETIC INPUTS

Herbicides and pesticides, though useful, can harm humans as well as the surrounding ecosystem. Pests and weeds can also develop resistance to commonly used herbicides and pesticides, requiring the development and purchase of new, more aggressive chemicals every few years. A much more environmentally friendly option is to use compatible native species for pest and weed control whenever possible. For example, cassava can be infected with up to two devastating viruses by whiteflies, but whiteflies are also attracted to sunflowers. Instead of spraying insecticides, Tanzanian farmers often plant rows of sunflowers to provide whiteflies with an alternate food source that reduces the instance of viruses in cassava (Folger 2014). Another option is to plant **insectary strips**, rows of plants that naturally harbor insects that prey on herbivorous pest species, or to use trees as fences or windbreaks to encourage native birds to feed on insect pests. Entomopathogenic fungi can also be encouraged to reduce the instance of insect pests (see Chapter 15 in this volume). Meanwhile, a healthy soil microbiome will naturally help exclude disease organisms from the root zone and the presence of beneficial microbes like fungi in the soil can also induce plant resistance to pests in other tissues (see Chapter 11 in this volume).

Synthetic fertilizers are generated from natural gas and release greenhouse gases when applied to fields (see Chapter 9 in this volume). They also commonly leach from fields and cause environmental problems downstream of the farms where they are used. Synthetic fertilizers can be replaced, at least partially, with animal manure and "green manure" (compost and crop residues) for an environmentally responsible method of increasing soil fertility. Compost application restructures the soil, improving water retention. Compost also includes many fungi and other beneficial microbes that can contribute to long-term nutrient cycling in the farm ecosystem. In addition, using compost generated from within the farm can help close the waste loop, making a farm more self-sustaining and ecologically responsible (see Chapter 17 in this volume).

### 10.5.4   MAINTAIN SMALLHOLDER COMPETITIVENESS

The 20th century GR relied on technologies that large-scale farms are best equipped to utilize: specialized machinery, large volumes of expensive inputs, and HYVs that could grow densely in large monocultured fields. However, small farms represent 80% of all farms globally, are culturally important, and are more likely to be managed by families that can use generational knowledge to sustainably manage their lands (Chapter 8). Large farms resulting from consolidation are more likely to be monocultured with commodity crops that are valuable in bulk, while small farms tend to produce higher-value items like vegetables and orchard fruits (Chapter 8). Therefore, it is vital that small farms remain competitive during the Second GR. Rejecting expensive, synthetic inputs in favor of locally produced compost or native species services will certainly help, but newly developed seed varieties will still be costly, especially if they are developed by private companies.

Small farms will need to utilize every possible income stream to reduce the financial impact of implementing environmentally friendly techniques and using the most effective seed stocks. Mushroom cultivation represents a largely untapped, beginner-friendly source of income for smallholders. The setup costs are relatively low, especially for farms that produce considerable lignocellulosic waste or include sustainably managed forests (see Chapter 17 in this volume). Even if the mushrooms are not sold to supplement smallholder income, they can be eaten as a nutritious part of the diet, offsetting costs of food purchases, or at the very least reduce the cost of waste disposal by decreasing the volume of waste that needs to be taken off the farm.

## 10.6   CONCLUSION

Because of population pressure on farm systems globally, fully eliminating the use of artificial fertilizers, pesticides, and other anthropological inputs is not realistic for the Second GR. However, farms that blend modern pest control and irrigation technology with native ecosystem services and techniques that employ ecosystem understanding present an excellent solution to increase food production sustainably. This marriage of practices may also be more affordable and support small, incremental changes, and hence be more attractive to both small-scale farmers and larger industrial farming groups.

A Second GR is needed to increase food production and quality in an ecologically responsible manner. Because of the current shifts in agricultural practices, now is an ideal time to incorporate mycology into agroecology. Fungi serve many ecological roles, but agricultural systems will best benefit from viewing fungi in four main roles: (1) as plant partners able to assist crop plants; (2) as pests dealt with through ecological understanding rather than through blanket application of fungicides; (3) as pest control organisms useful for fending off diseases and other harmful organisms; and (4) as products that can increase farmer nutrition or income while closing a waste loop. The latter half of this textbook investigates these roles in greater detail so that the ecologically minded farm manager can integrate mycology into their practices and help achieve a solid balance between production and ecological responsibility through mycoagroecology.

## REFERENCES

Bassil, K. L., Vakil, C., Sanborn M., Cole, D.C., Kaur, J.S., and K. J. Kerr. 2007. Cancer health effects of pesticides. *Canadian Family Physician* 53:1704–11.
Cohen, P. 2020. Roundup maker to pay $10 billion to SETTLE cancer Suits. *New York Times*. https://www.nytimes.com/2020/06/24/business/roundup-settlement-lawsuits.html (accessed November 3, 2019).
Foley, J. 2014. A five-step plan to feed the world. *National Geographic Magazine*. National Geographic, May 2014. https://www.nationalgeographic.com/foodfeatures/feeding-9-billion (accessed November 4, 2019).
Folger, T. 2014. The Next Green Revolution. *National Geographic*, October 2014. https://www.nationalgeographic.com/foodfeatures/green-revolution/ (accessed November 3, 2019).
Guyton, K. Z., D. Loomis, Y. Grosse, et al. 2015. Carcinogenicity of Tetrachlorvinphos, Parathion, Malathion, Diazinon, and Glyphosate. *Lancet Oncology* 162015: 490–1.
King, M., D. Altdorff, P. Li, et al. 2018. Northward shift of the agricultural climate zone under 21st-century global climate change. *Scientific Reports* 8:7904. https://doi.org/10.1038/s41598-018-26321-8.
Muller, A., C. Schader, N. El-Hage Scialabba, et al. 2017. Strategies for Feeding the World More Sustainably with Organic Agriculture. *Nature Communications* 8:1290. https://doi.org/10.1038/s41467-017-01410-w.
Potapov, P., Turubanova, S., Hansen, M.C., et al. 2022. Global maps of cropland extent and change show accelerated cropland expansion in the twenty-first century. *Nature Food* 3, 19–28. https://doi.org/10.1038/s43016-021-00429-z.
Pingali, P. L. 2012. Green Revolution: Impacts, limits, and the path ahead. *Proceedings of the National Academy of Sciences USA* 109:12302–8.
United Nations. 2008. *At High-Level Segment of Commission on Sustainable Development, Secretary-General Calls for 'Second Green Revolution' to Feed Burgeoning World Population*. May 14, 2008. https://www.un.org/press/en/2008/envdev984.doc.htm (accessed November 4, 2019).
Zwerdling, D. 2009. *Green Revolution Trapping India's Farmers in Debt*. NPR.org. NPR. https://www.npr.org/2009/04/14/102944731/green-revolution-trapping-indias-farmers-in-debt (accessed November 5, 2019).

# Section I

## Partners

# 11 Plant-Fungal Mutualisms

*Jason C. Slot*

## CONTENTS

## 11.1 INTRODUCTION

Mutualistic relationships between plants and fungi have been examined for over a century, ever since the plant pathologist Albert Bernard Frank promoted revolutionary ideas about the function of fungi in plant roots and lichens in 1885 (Trappe 2005). Frank's notions that fungi nourish plant roots with nitrogen and receive carbon from their plant hosts were initially resisted by the scientific community. However, since these initial hypotheses, the essentiality, ubiquity, and diversity of plant-fungal mutualisms have become far clearer. They occur in many species and throughout a plant, performing various roles. Mutualisms are now understood to be central to the complex web of nutrient flows and chemically mediated defenses in ecological niches. Mutualistic interactions between plants and fungi can be **obligate** (necessary) or **facultative** (optional) for either partner. The diversity of mutualistic relationships, the affected plant tissues, and the physical and chemical nature of several interactions are discussed in this chapter. These mutualistic relationships are impacted by agroecosystem management practices. They are also key players in biological invasions.

## 11.2 PLANT-FUNGAL MUTUALISMS VARY ACCORDING TO PLANT "COMPARTMENTS"

Fungi form partnerships throughout the tissues of plants. It is common to distinguish among resident fungi that occur in different compartments of the plant body as different "mycobiomes". Different plant tissues or compartments house distinct groups of microorganisms including fungi with different ecological roles that affect plant fitness. Fungi that are inhabiting, but not causing disease to, the above-ground parts of the plants are typically referred to as **endophytes** or **epiphytes** depending on whether they are within (**endosphere**) or outside (**phyllosphere**) the epidermal boundary. Roots also harbor endophytic fungi in the specialized **endorhiza** compartment, although the endophyte communities above and below ground are usually very different. The complexity of fungal communities increases from the root surface (**rhizoplane**) through the immediately adjacent soil

DOI: 10.1201/9780429320415-12

(**rhizosphere**) to the surrounding bulk soil (Leach et al. 2017). Some endophytic fungi colonize seeds (the **spermosphere**) enabling them to be directly transmitted from parent to offspring plants. Fungal communities of the internal compartments are affected by nutrient availability and internal metabolites, and the plant influences those in the external compartments through chemical exudates (Bell et al. 2019; see Chapter 12 in this volume).

The roots and rhizosphere comprise the most complex mycobiome communities that associate with plants. Highly diverse and dynamic relationships between plants and fungi arise at the interface between the root system, which actively influences microbe composition, and the soil, which is the ultimate repository of microorganisms. Among these diverse interactions, some mutualistic partnerships are understood to be key to plant and fungal fitness, and in some cases act as the foundation for broader functioning of plant-dominated ecosystems. There are multiple types of nutrition-based root-fungus mutualisms, known as **mycorrhizae**, to be considered. In addition to the nutritional exchanges mycorrhizae facilitate, further ways the plant and fungal partners mutually benefit one another and contribute to ecosystem stability are still being investigated.

Mycorrhizae comprise a diverse group of fungi that associate with the roots of plants, but the full scope of fungal associations that warrant this name is not yet settled. In the classification of mycorrhizae, those that form structures within root cells are termed **endomycorrhizae**, and those that form structures outside the root are termed **ectomycorrhizae**, but these terms do not adequately summarize the diversity of mycorrhizal lifestyles. The Arbuscular Mycorrhizal (**AM**) fungi (class Glomeromycetes) are the most pervasive endomycorrhizal plant partners. More than 80% of land plants have AM partners in the soil (Parniske 2008). They are named for the tree shaped structures (**arbuscules**) the majority of these fungi form within the root cortical cells of plants. Ectomycorrhizal (**ECM**) fungi are so named because they form a thick sheath outside the root tips of their host plants. ECM, mostly from the class Agaricomycetes, are most prevalent in temperate forests. ECM fungi grow hyphae between, instead of within, the root cortical cells when exchanging resources. The mushrooms found growing on forest soil are typically the ECM partners of nearby trees. There are other fungi that link plant roots to the soil and appear to provide benefits to plants. Dark septate root endophytes are particularly intriguing root associates that may provide fitness benefits to plants that overlap those of mycorrhizae and endophytic fungi and require more investigation.

## 11.3   ARBUSCULAR MYCORRHIZAL FUNGI COLONIZE DIVERSE PLANT ROOTS WORLDWIDE

AM fungi form intracellular associations with plant roots through both physical and chemical processes. Plants produce abundant root exudates that initiate and shape the association. Root exudates contain a type of plant hormone called **strigolactones** that induce AM fungi to germinate from their resting spores, grow toward the plant roots, and eventually infect root tips by means of a specialized hypha called an **appressorium**. The fungi penetrate the cell walls of root cortical cells, then develop **haustoria** (the arbuscules) that are often highly branched and serve as the interface for mutual exchange between the plant and fungus. The fungi also commonly produce vesicles filled with lipids in other root cortical cells. The mycorrhizal interaction instigates various changes in the development of the plant and the chemistry of its surroundings that result in improved plant fitness. Outside the roots, AM fungi colonize the surrounding soil where they scavenge nutrients and ultimately produce large, characteristic asexual reproductive/resting spores with thousands of nuclei that are resistant to environmental stress.

The relationship between the plant and the AM fungus is considered a nutritional mutualism in which the plant receives phosphorus, nitrogen, and mineral nutrition from the soil-scavenging fungal mycelium, while the fungus benefits from photosynthetically fixed carbon in the form of sugars and lipids transferred from the host (Keymer et al. 2017). The AM hyphae outside the roots are able to more rapidly colonize the soil and can penetrate smaller pores in soil particles than can plant roots (Campos et al. 2018). The fungi also help to make phosphorus more available either by secreting phosphate solubilizing enzymes or interacting with bacteria that do so (Roy-Bolduc and Hijri 2011). The AM fungi are therefore able to supply the growing root with nutrition despite the rapidly diminishing resources in root-adjacent soil. Nutrients are exchanged through the **interfacial apoplast**, the film of extracellular fluid between the plant and fungal membranes that make up the haustorium/arbuscule within the plant cell wall. Transporters in the membranes on the plant and fungus sides of the interfacial apoplast control the release and uptake of sugar and nutrients, driven by a hydrogen ion gradient to which both partners contribute (Moore et al. 2020). The tendency of a plant to form AM mycorrhizae is dependent on the available nutrients in the soil. For example, the high nitrogen and phosphorus found in fertilized soils tends to reduce the number of mycorrhizal associations as plants balance growth and nutrient acquisition, although other plant and fungal genetic factors also affect colonization rates (Venegas et al. 2021).

The establishment and interaction with arbuscular mycorrhiza results in the development of mycorrhizal-induced resistance (**MIR**) against disease and environmental stress. MIR results from a multi-phase progression of signals, secondary chemical changes, and microbiome modification (Cameron et al. 2013). First, plants recognize the presence of AM fungi in root tissues and respond by inducing a localized defense response and sending plant hormone signals (salicylic acid) that bolster system-wide immunity. Fungi then secrete "effectors" that suppress the plant immune response around the fungal hyphae and trigger further defense responses throughout the plant by hormone (jasmonic acid) induced pathways. These events, and the AM fungi themselves, proceed to change the composition of root exudates, causing plants to recruit specific beneficial bacteria to the rhizosphere. The recruited bacteria ultimately trigger additional defenses and conditioning throughout the plant via other hormone (such as abscisic acid and ethylene) pathways. The overall impacts of MIR on plant health and fitness are broad. In addition to reducing disease, colonization by AM fungi can improve tolerance to drought, salt, heavy metals, and temperature stress. AM fungi can also reduce herbivory, in part through inducing signals that recruit predators of herbivorous insects (Meier et al. 2021).

AM associations are widespread, especially throughout tropical forests and global grasslands, and occur with a large portion of terrestrial plant species (Smith and Read 2010). Based on surveys of a small proportion of species, it was inferred that 56% of dicots, primarily in Magnolidae, Rosidae, and the former Asteridae, and 51% of monocots, primarily in Arecidae and Commelinidae (including all grasses) form AM associations exclusively or along with other types of mycorrhizae (Smith and Read 2010). Despite the broad diversity of plants that depend on AM fungi, the fungi themselves are restricted to four orders within the phylum Glomeromycota (Redecker et al. 2013). However, these lineages diversified anciently, and may have been present in the earliest land plants. The fossil record and shared AM-induced genes among diverse land plants both support the ancient foundational nature of this mutualism. Indeed, the partnership between AM fungi and early vascular plants is put forth as an explanation for how multicellular life was able to overcome the challenges of colonizing land (Rimington et al. 2018).

## 11.4 ECTOMYCORRHIZAL FUNGI COLONIZE ROOTS OF WOODY PLANTS IN NORTHERN AND TEMPERATE FORESTS

ECM are the second most widespread mycorrhizal relationship globally, and they contrast with AM in several fundamental ways. Structurally, ECM develop between (rather than inside) root cortical cells, and they form a robust **mantle** of hyphae on the outside of root tips of their hosts. The intercellular hyphae form a network called a **Hartig Net**, which interfaces with the plant apoplast between the root cells for nutrient exchange. These relationships are dominant among forest trees in northern and temperate environments where they associate with 80%–90% of all trees (Brundrett 2004).

ECM fungi are among the most common forest floor mushroom-forming fungi in these locations, including important fungi to human culture and health like *Amanita* species (Fly Agaric and Death Angels) and truffles, boletes, and other prized culinary mushrooms. Unlike AM fungi, which are derived from a single lineage, ECM fungi have arisen independently in multiple lineages from Ascomycota, Basidiomycota, and Mucoromycota, although most ECM fungi are in the class Agaricomycetes (Miyauchi et al. 2020; Tedersoo et al. 2009).

The relationship between ECM fungi and tree roots is thought to mainly entail an exchange of nitrogen for fixed carbon, although additional benefits like nutrient provisioning are being investigated (Becquer et al. 2019). Importantly, ECM fungi are able to access a variety of forms of nitrogen in the soil that are unavailable to the plants themselves. Nitrogen deposition in forests reduces ECM diversity and shifts community composition toward nitrogen tolerant fungal species (Jo et al. 2019; Lilleskov et al. 2019). Some ectomycorrhizal species are adapted to high nitrogen concentrations associated with animal urine and carcasses. For example, the ectomycorrhizal agaric *Hebeloma radicosum* is thought to be part of a three-way symbiosis with underground moles and oak tree roots, cleaning the nitrogen from the moles' latrines and transferring it to the tree roots (Sagara 1995). Temporary high nitrogen deposits like animal carcasses can set the stage for ECM fungal succession as the nitrogen-loving species give way to low-nitrogen species when the deposit is depleted. Individual fungal species may have differently adapted transporters that scavenge very low-concentration nutrients and secrete enzymes and metabolites that make nutrients more available (Lilleskov et al. 2019).

ECM fungal partners are more diverse than their host trees, which often associate with multiple ECM species, presumably under selection from local environmental conditions (Bogar and Peay 2017; Tedersoo et al. 2010). While many ECM fungi are obligate mutualists, some are able to derive nutrition through saprotrophy. Most woody plants are dependent on fungal partners for normal growth and fitness. Establishment and invasion of pine trees in the southern hemisphere requires simultaneous invasion of ECM species to which they are adapted, or gradual co-option of local ECM species (Nuñez et al. 2009). *Amanita phalloides* is an invasive ECM fungus that is expanding its range by adapting to trees in novel environments (Pringle et al. 2009).

Forest soils and ecosystems are largely created and maintained through the networks of ECM fungi that associate with trees. In addition to allowing trees to obtain sufficient nutrition, ECM fungi are thought to reduce soil respiration, resulting in the growth of soil and soil habitats (Stuart and Plett 2020). The mushrooms and truffles of ECM fungi are important food sources for forest mammals and birds (White 2017). ECM fungi can also enhance the drought and salinity tolerance of woody plants.

## 11.5 MYCORRHIZAL RELATIONSHIPS ARE STRUCTURALLY AND ECOLOGICALLY DIVERSE

Many other mycorrhizal types have been described which have overlapping features with AM, ECM, or both. **Ericoid** mycorrhizae form between the roots of Ericaceae plants (like blueberry, cranberry, and rhododendron) and certain fungal species in the Ascomycete order Helotiales. These endomycorrhizae have a coiled, rather than tree-shaped, haustorium. Interestingly, Ericaceae (Ericales) plants form similar mycorrhizae with the basidiomycete family Serendipitaceae (Vohník et al. 2016). These fungi appear to be important for nitrogen provisioning in the acidic and low-nutrient soils to which plants of Ericaceae are adapted (Smith and Read 2010). **Arbutoid** mycorrhizae are "ectendomycorrhizae", having a coiled intracellular haustoria, a Hartig net, and a hyphal mantle around the root. They are associated with species from multiple Ericales families. Similar to ericoid mycorrhizae, arbutoid mycorrhizas can form from either ascomycetes or basidiomycetes (Kühdorf et al. 2015, 2016). Other ectendomycorrhizae form between *Wilcoxina* spp. (Pezizales, Ascomycota) and conifers (Trevor et al. 2001).

### 11.5.1 Plants in a Community Can Benefit from Shared Mycorrhizal Networks

Mycorrhizal networks can associate with multiple plants, creating the opportunity for the fungi to act as conduits of materials and information between plants. Fixed carbon, for instance, may be transferred from taller plants with more access to sunlight to shaded plants and nitrogen from nitrogen fixers to nitrogen receiver plants via fungal hyphae (Simard et al. 1997). Fungi may also relay defense signals from a plant under attack to neighboring plants that would benefit from increasing production of defense compounds (Song et al. 2014). Because diverse mycorrhizal fungi are able to secrete enzymes that liberate nitrogen and phosphorus from organic forms that plants cannot access, they can directly transfer minerals from complex biomass to plants through mycorrhizal networks, accelerating nutrient cycling (Read and Perez-Moreno 2003).

### 11.5.2 Mycoheterotrophs Exploit Plant-Fungal Mutualisms

Some plants have a nutritional strategy called **mycoheterotrophy** wherein the plant derives its carbon nutrition from other plants through a fungal intermediate for at least part of its life cycle. For example, orchids have no energy reserves in their seeds and pair with saprotrophic fungi to provide the resources for germination and early development (Genre et al. 2020). Monotropes (in the Ericaceae) are plants without chlorophyll that form mycorrhizal relationships with ECM fungi to derive their carbon from trees (Trudell et al. 2003). These plants emerge from a dense, convoluted ball of roots coated entirely with the ectomycorrhizae upon which they are fully dependent. Orchids (Orchidaceae) make up one of the two most diverse plant families and almost all are mycoheterotrophs. Many orchids invest in the production of elaborate flowers that attract specific pollinators with which they are co evolved, while investing comparatively little in nutrient scavenging and photosynthesis. Most orchids require a fungal associate for seed germination, and many rely on fungi for most or all of their nutrition throughout their life (Rasmussen and Rasmussen 2009). So-called **orchid mycorrhizae** tend to be saprotrophic or ectomycorrhizal fungi. Rather than forming typical structures like haustoria or Hartig nets with orchids, the fungi appear as densely packed coils called **pelotons** in the root cortical cells that appear to be eventually digested by the orchid root cells. It is not clear how many mycoheterotrophic relationships are mutualisms as opposed to parasitism of the fungi by the plants.

## 11.6 ENDOPHYTIC FUNGI ARE COMMON RESIDENTS OF PLANT TISSUES

The different parts of the plant **endosphere** (interior tissues) host largely distinct communities of fungal endophytes. While the relationships between plants and endophytic fungi are mostly poorly understood, many bear the hallmarks of mutualism. Endophytic fungi may be acquired horizontally (from the environment) or inherited vertically (through reproduction). Both vertically and horizontally transmitted endophytes have members that are deemed to be mutualists. Fungal endophytes may play roles in plant defense against pests and resistance to environmental stressors.

Horizontally transmitted endophytes are extremely diverse, arising from a large number of fungal families, primarily of the filamentous ascomycetes (Arnold 2007). Most of the research on these fungi happens at the community level, either by cultivating and identifying fungi from surface-sterilized plant tissues or by **barcode sequencing**, which involves PCR amplification[1] of taxonomically informative genes shared among fungi. However, these techniques are insufficient for determining whether there are mutual benefits conferred in the relationship. Many of the fungi identified through these methods may be resting saprotrophs, which activate to digest fallen or senesced plant tissues, or similarly latent pathogens, which become aggressive from signals of a

---

[1] A method of investigating gene sequences from very small samples.

weakened or stressed host. Endophyte species often have broad metabolic capabilities enabling them to take multiple nutritional strategies. However, it is rarely clear whether it is the same or a different population of a species that is found in each ecological role.

Individual horizontally transmitted endophytes have been found to increase the diversity of defense compounds in the plant, which is thought to enhance resistance to pathogens and predators (Raguso et al. 2015). Endophytic fungi may also increase chemical diversity by transforming host-produced molecules into new molecules (Ying et al. 2014). Beyond diversifying the molecules in the plant tissue, an interesting phenomenon among these fungi is their production of the same molecules as their host. Important pharmaceuticals like quinine[2] and paclitaxel[3] have been identified in isolated endophytes of *Cinchona* and *Taxus* spp., raising the possibility that some valuable drugs might be produced in culture rather than by harvesting rare plants, thus avoiding ecosystem damage (Kusari et al. 2014; Maehara et al. 2011).

Some endophytic fungi like *Trichoderma* spp. and *Clonostachys* spp. are effective parasites and competitors of plant pathogens, which make them important biological control agents (Pujade-Renaud et al. 2019). These mycoparasitic fungi inhibit fungal pathogens through both antibiotic chemicals and direct attack, coiling around and degrading the plant parasitic fungus (Druzhinina et al. 2011). Notably, these defensive mutualist endophytes are less diverse in crop plantations compared to natural forests, corresponding with increased abundance and diversity of plant pathogenic fungi in plantations (Gazis and Chaverri 2015).

Some horizontally transmitted endophytes are considered "habitat adapted" because they confer resistance to plants growing under specific environmental stresses. For example, salt and heat tolerance are conferred by endophytes isolated from plants in coastal and geothermal soils, respectively (Rodriguez et al. 2008). Habitat adapted endophytes have been used to mitigate salt, heat, and disease stress in agricultural plants. There is probably more complexity to these relationships than we know; for example, an isolate of the fungus *Curvularia protuberata* itself requires a viral symbiont to allow hot springs panic grass (*Dichanthelium lanuginosum*) to grow at 68°C at the surface of the Yellowstone Caldera (Morsy et al. 2010). The virus reprograms the endophyte to express thermotolerance factors that protect the plant.

The interaction between plants and the vertically transmitted **Clavicipitaceous** endophytes has been extensively studied and is one system for which there is good evidence of environmental transformation being driven by plant-fungal mutualism (Saikkonen et al. 2013). Clavicipitaceous endophytes are named for belonging to Clavicipitaceae, which contains the notorious grain pathogen ergot, *Claviceps purpurea*, an important co-traveler of human civilization (Van Dongen and de Groot 1995). Ergot is the source of multiple civilization altering "ergot alkaloids" including the medicines ergometrine[4] and ergotamine[5] as well as the psychoactive drug LSD (lysergic acid diethylamide). Though generally not parasitic, Clavicipitaceous endophytes are similarly endowed with ergot alkaloids, which play an important role in the ecology of the plants they inhabit.

Clavicipitaceous endophytes, e.g. Epichloë and Periglandula, colonize the endospheres of Achnatherum grasses and vines of the morning glory family. The alkaloids these fungi produce are directly responsible for reduction of plant predator and pathogen pressure. For example, insects that eat Epichloë may experience reduced growth and reproduction, so they are deterred from grazing sleepy grass (*Achnatherum robustum*) that harbors Epichloë endophytes (Shymanovich and Faeth 2018). Fescue grasses, which are important for both livestock grazing and green lawns, also harbor Epichloë endophytes that produce toxic alkaloids. Clavicipitaceous endophyte infected grasses are more resistant to disease and predation, which allows them to dominate meadow communities (Saikkonen et al. 2013). However, while endophyte infected grasses improve the health of pastures,

---

[2] An antimalarial drug.
[3] A chemotherapy drug.
[4] A labor-inducing drug.
[5] A migraine drug.

the toxic alkaloids they produce can cause neurological and other health problems for livestock (Hume et al. 2020).

## 11.7 FUNGI CAN PROVIDE BENEFITS TO OTHER PLANT MUTUALISMS

Beyond straightforward bidirectional mutualistic partnerships with plants, some fungi facilitate mutualisms between plants and insects. Yeasts colonize the nectars of flowers and are transmitted to new plant hosts by insect pollinators (Makino 2013). It has been observed that flowers colonized by bacteria instead of these yeasts are less frequently visited by their pollinators (Vannette et al. 2013). Some yeasts may directly enhance the flower-pollinator mutualism by increasing emission of attractive scents (Rering et al. 2018) and warming nectar, thereby facilitating pollination of winter-blooming plants (Herrera and Pozo 2010). The fungus *Trimmatostroma* sp. is part of a three-way association with the plant *Hirtella physophora* and the ant *Allomerus decemarticulatus* (Leroy et al. 2017). At the leaf petioles, the ants build a **domatium** (shelter) from the plant's trichomes (hair-like structures), which become colonized by the fungus. The fungus also grows endophytically and transfers nitrogenous waste from the domatia to the plant tissues.

## 11.8 PLANT-FUNGAL MUTUALISMS REQUIRE AN ADAPTED FUNGAL COMMUNITY

Plants removed from their adapted environment may be negatively impacted by the loss of their fungal associates. This includes crops and ornamentals that have been globally dispersed. Throughout the Green Revolution (see Chapter 9 in this volume), plant genetics and pesticides optimized agricultural productivity in the short term but did not explicitly plan for optimal mutualistic associations with a diverse fungal community in the long term. Defensive endophytes, including those co evolved to best attack their host's enemies, are likely diminished or eliminated by seed treatments and serial monocultures. Understanding and stewarding the fungal associates of plants to help rebuild beneficial microbial communities will be essential for the future sustainability of managed ecosystems (see Chapter 12 in this volume).

## REFERENCES

Arnold, A.E. 2007. Understanding the diversity of foliar endophytic fungi: progress, challenges, and frontiers. *Fungal Biology Reviews* 21:51–66.

Becquer A., C. Guerrero-Galán, J.L. Eibensteiner, et al. 2019. The ectomycorrhizal contribution to tree nutrition. *Advances in Botanical Research* 89:77–126.

Bell, T. H., K.l. Hockett, R. I. Alcalá-Briseño, et al. 2019. Manipulating wild and tamed phytobiomes: challenges and opportunities. *Phytobiomes Journal* 3:3–21.

Bogar, L. M., and K. G. Peay. 2017. Processes maintaining the coexistence of ectomycorrhizal fungi at a fine spatial scale. In *Biogeography of Mycorrhizal Symbiosis*, ed. L. Tedersoo, 79–105. Cham: Springer.

Brundrett, M. 2004. Diversity and classification of mycorrhizal associations. *Biological Reviews* 79:473–95.

Cameron, D. D., A. L., Neal, S. C., van Wees, and J. Ton. 2013. Mycorrhiza-induced resistance: more than the sum of its parts? *Trends in Plant Science* 18:539–45.

Campos, P., F. Borie, P. Cornejo, J. A. López-Ráez, Á. López-García, and A. Seguel. 2018. Phosphorus acquisition efficiency related to root traits: is mycorrhizal symbiosis a key factor to wheat and barley cropping? *Frontiers in Plant Science* 9:752. DOI:10.3389/fpls.2018.00752.

Druzhinina, I. S., V. Seidl-Seiboth, A. Herrera-Estrella, et al. 2011. Trichoderma: The genomics of opportunistic success. *Nature Reviews Microbiology* 10:749–59.

Gazis, R. and P. Chaverri. 2015. Wild trees in the Amazon basin harbor a great diversity of beneficial endosymbiotic fungi: is this evidence of protective mutualism? *Fungal Ecology* 17:18–29.

Genre, A., L. Lanfranco, S. Perotto, and P. Bonfante. 2020. Unique and common traits in mycorrhizal symbioses. *Nature Reviews Microbiology* 18: 649–60.

Herrera, C. M., and M. I. Pozo. 2010. Nectar yeasts warm the flowers of a winter-blooming plant. *Proceedings of the Royal Society B: Biological Sciences* 277:1827–34.

Hume, D. E., A. V. Stewart, W. R. Simpson, and R. D. Johnson. 2020. Epichloë fungal endophytes play a fundamental role in New Zealand grasslands. *Journal of the Royal Society of New Zealand* 50:279–98.

Jo, I., S. Fei, C. M. Oswalt, G. M. Domke, and R. P. Phillips. 2019. Shifts in dominant tree mycorrhizal associations in response to anthropogenic impacts. *Science Advances* 5:eaav6358. DOI:10.1126/sciadv. aav6358.

Keymer, A., P. Pimprikar, V. Wewer, et al. 2017. Lipid transfer from plants to arbuscular mycorrhiza fungi. *elife* 6:e29107. DOI:10.7554/eLife.29107.001.

Kühdorf, K., B. Münzenberger, B. Begerow, J. Gómez-Laurito, and R. F. Hüttl. 2015. *Leotia* cf. *lubrica* forms arbutoid mycorrhiza with *Comarostaphylis arbutoides* (Ericaceae). *Mycorrhiza* 25:109–20.

Kühdorf, K., B. Münzenberger, D., Begerow, J. Gómez-Laurito, and R. F. Hüttl. 2016. Arbutoid mycorrhizas of the genus Cortinarius from Costa Rica. *Mycorrhiza* 26:497–513.

Kusari, S., S. Singh, and C. Jayabaskaran. 2014. Rethinking production of Taxol® (paclitaxel) using endophyte biotechnology. *Trends in Biotechnology* 32:304–11.

Leach, J. E., L. R. Triplett, C. T. Argueso, and P. Trivedi. 2017. Communication in the phytobiome. *Cell* 169587–96.

Leroy, C., A. Jauneau, Y. Martinez, et al. 2017. Exploring fungus–plant N transfer in a tripartite ant–plant–fungus mutualism. *Annals of Botany* 120:417–26.

Lilleskov, E. A., T. W. Kuyper, M. I. Bidartondo, and E.A. Hobbie. 2019. Atmospheric nitrogen deposition impacts on the structure and function of forest mycorrhizal communities: a review. *Environmental Pollution* 246:148–62.

Maehara, S., P. Simanjuntak, C. Kitamura, K. Ohashi, and H. Shibuya. 2011. Cinchona alkaloids are also produced by an endophytic filamentous fungus living in Cinchona plant. *Chemical and Pharmaceutical Bulletin* 59:1073–4.

Meier, A. R. and M.D. Hunter. 2021. Variable effects of mycorrhizal fungi on predator–prey dynamics under field conditions. *Journal of Animal Ecology* 90:1341–52.

Miyauch, I S., E. Kiss, A. Kuo, et al. 2020. Large-scale genome sequencing of mycorrhizal fungi provides insights into the early evolution of symbiotic traits. *Nature Communications* 11: 1–7.

Moore, D., G. D. Robson, and A. P. Trinci. 2020. *21st Century Guidebook to Fungi*. Cambridge: Cambridge University Press.

Morsy, M. R., J. Oswald, J. He, Y. Tang, and M. J. Roossinck. 2010. Teasing apart a three-way symbiosis: transcriptome analyses of *Curvularia protuberata* in response to viral infection and heat stress. *Biochemical and Biophysical Research Communications* 401:225–30.

Nuñez, M. A., T. R. Horton, and D. Simberloff. 2009. Lack of belowground mutualisms hinders Pinaceae invasions. *Ecology* 90:2352–9.

Parniske, M. 2008. Arbuscular mycorrhiza: the mother of plant root endosymbioses. *Nature Reviews Microbiology* 6:763–75.

Pringle, A., R. I. Adams, H. B. Cross, and T. D. Bruns. 2009. The ectomycorrhizal fungus *Amanita phalloides* was introduced and is expanding its range on the west coast of North America. *Molecular Ecology* 18:817–33.

Pujade-Renaud, V., M. Déon, R. Gazis, et al. 2019. Endophytes from wild rubber trees as antagonists of the pathogen *Corynespora cassiicola*. *Phytopathology* 109:1888–99.

Raguso, R. A., A. A. Agrawal, A. E. Douglas, et al. 2015. The raison d'être of chemical ecology. *Ecology* 96:617–30.

Rasmussen, H. N. and F.N. Rasmussen. 2009. Orchid mycorrhiza: implications of a mycophagous life style. *Oikos* 118:334–45.

Read, D. J. and J. Perez-Moreno. 2003. Mycorrhizas and nutrient cycling in ecosystems–a journey towards relevance? *New Phytologist* 157:475–92.

Redecker, D., A. Schüßler, H., Stockinger, S. L. Stürmer, J. B. Morton, and C. Walker. 2013. An evidence-based consensus for the classification of arbuscular mycorrhizal fungi (Glomeromycota). *Mycorrhiza* 23:515–31.

Rering, C. C., J. J. Beck, G. W. Hall, M. M. McCartney, and R. L. Vannette. 2018. Nectar-inhabiting microorganisms influence nectar volatile composition and attractiveness to a generalist pollinator. *New Phytologist* 220:750–9.

Rimington, W. R., S. Pressel, J. G., Duckett, K. J., Field, D. J. Read, and M. I. Bidartondo. 2018. Ancient plants with ancient fungi: liverworts associate with early-diverging arbuscular mycorrhizal fungi. *Proceedings of the Royal Society B.* 285:20181600. DOI:10.1098/rspb.2018.1600.

Rodriguez, R. J., J. Henson, E. Van Volkenburgh, et al. 2008. Stress tolerance in plants via habitat-adapted symbiosis. *The ISME Journal* 2:404–16.

Roy-Bolduc, A., and M. Hijri. 2011. The use of mycorrhizae to enhance phosphorus uptake: a way out the phosphorus crisis. *Journal of Biofertilisers and Biopesticides* 2:1–5.

Sagara, N. 1995. Association of ectomycorrhizal fungi with decomposed animal wastes in forest habitats: a cleaning symbiosis? *Canadian Journal of Botany* 73:1423–33.

Saikkonen, K., K. Ruokolainen, O. Huitu, et al. 2013. Fungal endophytes help prevent weed invasions. *Agriculture, Ecosystems & Environment* 165:1–5.

Shymanovich, T., and S. H. Faeth. 2018. Anti-insect defenses of *Achnatherum robustum* (sleepygrass) provided by two *Epichloë endophyte* species. *Entomologia Experimentalis et Applicata* 166:474–82.

Simard, S. W., D. A. Perry, M. D. Jones, D. D. Myrold, D. M. Durall, and R. Molina. 1997. Net transfer of carbon between ectomycorrhizal tree species in the field. *Nature* 388:579–82.

Smith, S. E., and D. J. Read. 2010. *Mycorrhizal Symbiosis*. London: Academic Press.

Song, Y. Y., M. Ye, C. Li, et al. 2014. Hijacking common mycorrhizal networks for herbivore-induced defence signal transfer between tomato plants. *Scientific Reports* 4:1–8.

Stuart, E. K. and K. L. Plett. 2020. Digging deeper: in search of the mechanisms of carbon and nitrogen exchange in ectomycorrhizal symbioses. *Frontiers in Plant Science* 10:1658. https://doi.org/10.3389/fpls.2019.01658.

Tedersoo, L., T. W. May, and M. E. Smith. 2010. Ectomycorrhizal lifestyle in fungi: global diversity, distribution, and evolution of phylogenetic lineages. *Mycorrhiza* 20:217–63.

Trappe, J. M. 2005 AB Frank and mycorrhizae: the challenge to evolutionary and ecologic theory. *Mycorrhiza* 15:277–81.

Trevor, E. Y., K. N. Egger, and L. R. Peterson. 2001 Ectendomycorrhizal associations–characteristics and functions. *Mycorrhiza* 11:167–77.

Trudell, S. A., P. T. Rygiewicz, and R. L. Edmonds. 2003, Nitrogen and carbon stable isotope abundances support the myco-heterotrophic nature and host-specificity of certain achlorophyllous plants. *New Phytologist* 160:391–401.

Van Dongen, P. W., and A. N. de Groot. 1995. History of ergot alkaloids from ergotism to ergometrine. *European Journal of Obstetrics & Gynecology and Reproductive Biology* 60:109–16.

Venegas, R. A. P., S. J. Lee, M. Thuita, et al. 2021. The phosphate inhibition paradigm: Host and fungal genotypes determine arbuscular mycorrhizal fungal colonization and responsiveness to inoculation in cassava with increasing phosphorus supply. *Frontiers in Plant Science* 12. DOI:10.3389/fpls.2021.693037.

Vohník, M., M. Pánek, J. Fehrer, J., and M. A. Selosse. 2016. Experimental evidence of ericoid mycorrhizal potential within Serendipitaceae (Sebacinales). *Mycorrhiza* 26:831–46.

White, J. F. 2017. Mycophagy and spore dispersal by vertebrates. In *The Fungal Community*, eds. J. Dighton and J. F. White, 401–412. Boca Raton: CRC Press.

Ying, Y. M., W. G. Han, and Z. J. Zhan. 2014. Biotransformation of Huperzine A by a fungal endophyte of *Huperzia serrata* furnished sesquiterpenoid–alkaloid hybrids. *Journal of Natural Products* 77:2054–9.

# 12 Incorporating Microbes into Agricultural Soils

*Elizabeth "Izzie" Gall*

## CONTENTS

## 12.1 INTRODUCTION

Plants can benefit from fungal symbiosis in a variety of ways, including higher yields from the same inputs, disease resistance, and greater defense against abiotic stressors like drought, salinity, and heavy metals. Many fungi also help structure the soil, maintain nutrient cycles, and improve soil retention of important plant nutrients like carbon, nitrogen, and phosphorus (Ellouze et al. 2014). Unfortunately, many common agricultural practices discourage or kill beneficial soil fungi, either by targeting pathogens and incidentally killing beneficial fungi or through soil management practices that disrupt normal microbial community dynamics. Fortunately, there are a few established management methods that actively encourage soil fungi and several techniques for returning healthy microbial communities to agricultural soils. With time, proper management, and microbe cultivation, most soils can regain a healthy microbial ecosystem.

## 12.2 TYPES OF SOIL MICROBES

Most of the microbial activity in the soil occurs within ten centimeters of the surface (Lohman et al. 2010). Only a small number of crop plants do not form mycorrhizal associations. These are chiefly the Brassicaceae, the plant family that includes mustard and canola. For other crop plants, growth

DOI: 10.1201/9780429320415-13

without fungal symbionts should be considered abnormal (Berruti et al. 2016). Where a fungus has colonized a plant, we refer to the plant as "infected" with the fungus. However, infection with beneficial fungi does not lead to disease. Disease-causing fungi are known as **pathogens**.

Mycorrhizal plants recruit partner microbes from the bulk soil into their **rhizospheres** (the area of soil touching the roots and right around the roots) by releasing certain hormones, sugars, and other compounds that appeal to beneficial microbes. Because plants "deposit" these materials into the rhizosphere, the process is known as **rhizodeposition**. Rhizodeposition changes the soil environment around the roots, modifying not just the molecules available to microbes as food sources but the chemical and physical properties of the soil. Some rhizodeposited materials actively encourage the growth and branching of arbuscular mycorrhizal (AM) fungi (Ellouze et al. 2014).

### 12.2.1  AM Fungi

Recall from Chapter 4 that AM fungi form symbioses with most plants, creating arbuscules and sometimes vesicles within the cells of their hosts. Every AM fungus has a unique ability to colonize plant tissues, take nutrients from the soil and transfer them to host plants, bolster plant growth and productivity, and stabilize the soil surrounding the roots (Lohman et al. 2010). AM fungi are obligate symbionts, generally unable to survive without plant hosts. Their hosts are certainly abundant; about 75% of plant families have species that host AM in the wild. Unfortunately, the loss of AM diversity begins with any agricultural changes from the natural state – including any disturbance of soil structure, the removal or addition of certain plant species, tilling, and other methods discussed in Section 12.3. The level of disruption to the AM community correlates with the intensity of management practices, so more heavily managed and disturbed soils have more depleted AM communities (Ellouze et al. 2014).

### 12.2.2  Ectomycorrhizal (ECM) Fungi

Only about 3% of plant species form ECM associations, but their members are common enough that most recognizable mushrooms come from these relationships (see Chapter 4 in this volume). ECM create a two-layered fungal shell, or **mantle**, around growing plant roots, extending hyphae between the cells of the plant's outermost cell layer and sending other hyphae into the soil to scavenge for nutrients and water (see Figure 4.11). Unlike AM fungi and some other microbe groups, ECM fungi are more common deeper in the soil (McGuire et al. 2014).

### 12.2.3  Root Endophytes

Root endophytes are fungi that live almost completely within plant roots. Community ecology and individual function are not known for most endophytic fungi. However, they do seem to be generally beneficial to their plant hosts. Some endophytes are known to fight certain diseases when present in certain plants (Ellouze et al. 2014). In particular, the genus *Trichoderma* contains some potent biocontrol species commonly used against pathogens in such diverse crops as chickpea, onion, and tomato (Ellouze et al. 2014).

### 12.2.4  Other Soil Microbes

Besides the fungi and other organisms that live in the root or the rhizosphere, there are plenty of organisms living in the **bulk soil** (soil not associated with the root of any plant). This is where entomopathogenic and saprotrophic fungi, as well as many bacteria and protists, reside. Bacteria are the most abundant microbes in the rhizosphere, and many genera are known to contain species that have positive effects on crop plants (Rocha et al. 2019). Soil bacteria are also more studied than soil fungi in terms of their presence, community composition, and changes due to shifts in land use.

However, fungi are sensitive to different factors than bacteria, so we cannot predict fungal reactions to agricultural practices based on the patterns seen in bacteria, and vice versa. For example, soil fungi may be more sensitive than soil bacteria to logging activity (McGuire et al. 2014).

## 12.3 INDUSTRIAL AGRICULTURAL PRACTICES DESTROY SOIL MICROBIAL COMMUNITIES

Many conventional agricultural management practices disrupt or destroy natural microbial communities in the soil. Shifting land from its natural state to agricultural use immediately alters the microbial community, either through direct methods like altering the availability of host plants or by changing the abiotic soil factors that the microbes are adapted to. The former change can also lead to the latter, as selecting or introducing certain plants will alter the composition of rhizodeposited materials in the soil. The mix of molecules that a plant releases from its roots depends not only on the plant's species, but its specific genotype, its age, environmental factors such as air temperature and precipitation level, and other factors including the presence of pathogens (Ellouze et al. 2014). Differences in root exudates from crops versus native plants can change the major nutrient sources for microbes and the pH (acidity) of the soil. Rhizosphere fungi are extremely sensitive to changes in pH (Ellouze et al. 2014). Because crop plants are carefully spaced, managed crop systems often create different shade patterns than native ecosystems. The patchiness of sunlight and shade may change the pattern and number of niches available to microbes. Furthermore, continuous monocropping depletes the soil of certain nutrients that might have been abundant when the ecosystem was in its natural state. All these changes can shift the balance of the microbial community and affect the species able to survive in agricultural soils.

Because plants recruit microbes from the bulk soil through rhizodeposition, crop selection may be the single most impactful aspect of agriculture on the soil microbial community (Ellouze et al. 2014). In a study comparing soil fungal communities in forests with communities in formerly forested areas under agricultural use, oil palm plantation soil was almost completely stripped of ECM fungi – even though the forests nearby were dominated by ectomycorrhizal tree species (McGuire et al. 2014). Removing the native trees denied the ECM their hosts, leading to the loss of the ECM species. Likewise, cultivation of crops in the Brassicaceae family can deplete the soil of AM fungi, reducing their ability to colonize and assist other crops rotated in after the Brassicaceae. Recovering the microbes in such an area will prove extremely difficult because, without hosts, AM and ECM fungi will die out, making them unable to establish relationships even if host plants are reintroduced. The missing mycorrhizae can lead to further changes in the soil community since nutrient cycling dynamics in monocropped soil are often completely different from the cycling observed in native ecosystems (McGuire et al. 2014).

Disruptions to soil microbial communities are long-lasting; even after 50 years of recovery, selectively logged lowland tropical forests have different microbial communities than untouched forests. This is partially because an altered plant community can affect soil nutrient cycles and microbiomes, which may in turn affect the types of plants that can re-establish and grow in the disturbed area (McGuire et al. 2014). For example, crops with high nitrogen demands can strip the soil of nitrogen, so that even after a field is reverted to non-agricultural use there is little soil nitrogen available for re-establishing native plants to use. These interactions further complicate land restoration efforts.

### 12.3.1 MONOCULTURE

In addition to selecting for the most virulent pathogenic microbes, monocultures reduce overall soil microbial biomass and diversity (Ellouze et al. 2014). In polyculture or the natural ecosystem state, the rich mix of cultivars and their rhizodepositions creates a range of niches for microbes to occupy, supporting a diverse microbial community. Higher microbial diversity in the soil means that plants have more mycorrhizal partners to select from. It is beneficial for plants to have access to multiple microbes because each microbial variety has the potential to assist the plant during a different type of abiotic or pathogen stress. Therefore, mixing plant cultivars supports a rich microbial community, which in turn stabilizes plant yields in stressful conditions, resulting in higher productivity than monoculture overall (Ellouze et al. 2014).

When a natural ecosystem is shifted to agricultural use or is selectively logged, the removal of certain plants also removes those plants' fallen tissues and root exudates from the ecosystem. That loss reduces the activity and populations of fungi that form specific relationships with those organisms or preferentially digest them. As a result, monocropped fields such as oil palm plantations have significantly lower soil microbial enzyme activity than primary (untouched) forests or transitional ecosystems like selectively logged forests. The reduced enzyme activity considerably alters the nutrient cycles of monocultured plantations for carbon, phosphorus, and nitrogen relative to natural cycles (McGuire et al. 2014).

The ECM population notably shifts when an ecosystem is converted to monoculture, which has important implications for adding ECM products like mushrooms to monocultured fields or orchards. Some families of ECM fungi are more common in old growth forests than in recovering logged forests. For example, the families Amanita, Pezizales, Sordariales, and Lactarius are significantly more abundant in regenerating lowland tropical forests than in primary forest or plantations. Basidiomycetes are known as "late stage" fungi because they grow more slowly than some other fungi and are more sensitive to physical and chemical changes in the soil. Indeed, basidiomycetes are more common in primary forests than transitional forests or monocultures (McGuire et al. 2014). In the comparison study, oil palm plantations had ECM located only in the upper soil, an atypical site for ECM growth; therefore, the ECM in those samples might be spores or dormant structures rather than active ECM fungi (McGuire et al. 2014).

Because oil palm plantations host so few ECM, co-cropping edible fungi with oil palm trees will be nearly impossible. By contrast, cultivating ECM mushrooms would be much more likely in environments that retain healthy native ECM communities (see Chapter 17 in this volume). For example, consider the edible black and white truffles, fungi in the Pezizales family that are highly sought for their use in haute cuisine. Cultivating these fungi would be vastly easier in a regenerating lowland tropical forest which already contains Pezizales species than it would be to cultivate such fungi in the ECM desert of an oil palm plantation. Other edible ECM of interest from lowland tropical forests include the "milk cap" and "candy cap" mushrooms of the genus *Lactarius*, which are more common in recovering logged forests than in primary forests, and chanterelles of the genera *Craterellus* and *Cantharellales*, which are most common in primary forests (McGuire et al. 2014).

### 12.3.2 OTHER MICROBIOME-DAMAGING PRACTICES

Beyond monocropping, many 20th century GR techniques damage the natural microbiome in a converted ecosystem. The effects of some technologies on the microbiome are outlined in this section.

Inorganic (non-biologically based) nitrogen **fertilizers** lower the pH of the soil, affecting the structure and function of the microbial community. A pH that inhibits bacterial function will influence nutrient cycling in the soil, indirectly affecting fungi. A low enough pH (very acidic soil) will also inhibit fungal communities directly. Soil acidification can be countered with lime application, but due to the labor and supply costs, farmers often leave their soil at a low pH (Ellouze et al. 2014). Heavy application of phosphorus fertilizers also reduces the growth and root colonization ability of mycorrhizae (Lohman et al. 2010).

**Fungicides** selectively kill fungi, but do not discriminate between pathogens and beneficial mycorrhizae. Therefore, fungicides applied to reduce disease in a field can also kill off the beneficial soil fungi. Fungicides applied to the upper portions of a plant can also affect soil microbes, because the presence of fungicidal compounds on the leaves influences plant rhizodeposition processes (Ellouze et al. 2014); furthermore, leaves coated with fungicides may fall to the soil, directly affecting the fungi there.

**Pesticides** aimed at controlling insects do not have a direct effect on soil microbes, but by eliminating certain soil insects, they will influence the structure of the soil community and affect some ecological processes. Through these changes, pesticides indirectly alter the microbiome.

Similarly to monoculturing, application of **herbicides** (plant-killing compounds) changes the soil microbiome indirectly by altering the plant tissues and root exudates available to microbes. Over the short term, herbicide application has little effect on the soil fungal community; however, over time, multiple applications lead to reduced diversity in the microbiome. In particular, the herbicide glyphosate is known to negatively alter the composition of the rhizosphere microbial community (Ellouze et al. 2014).

**Removal of leaf litter** and other dropped plant material prevents a layer of organic material from forming in the soil (McGuire et al. 2014), limiting the food source of soil saprobes and altering nutrient cycles, which reduces soil fertility in following seasons (Ellouze et al. 2014).

**Tilling** homogenizes the soil, decreasing the number of distinct niches available for different microbes to occupy. Like removal of leaf litter, tilling also reduces the level of organic material in the soil.

The use of **motorized vehicles** in the field degrades soil fertility by interfering with its natural structure (McGuire et al. 2014). For example, heavy machinery compacts the soil, making it less able to retain resources like water and carbon. Compaction also alters soil pH (McGuire et al. 2014), leading to further disruption of the microbiome.

## 12.4   RESTORING MICROBIAL POPULATIONS IS POSSIBLE AND EFFECTIVE

Fortunately, there are many established methods for **inoculating** (successfully introducing) beneficial microbes into agricultural systems. Beyond re-establishing natural nutrient cycles and plant partnerships, reintroducing soil microbes can be a solid strategy for increasing soil fertility and crop yield without the use of (as many) fertilizers or pesticides (Berruti et al. 2016; Ellouze et al. 2014; Lohman et al. 2010). Therefore, field inoculation methods are sometimes referred to as **biofertilization**. The greatest benefits per effort will probably come from AM fungi and endophytic fungal species (Ellouze et al. 2014), but there is also some promising research into encouraging ECM partners as well.

It is surprisingly easy to reintroduce living fungi to the soil system because soil fungi can regenerate from multiple types of **propagules** (particles that enable the propagation, or spread, of the species). Healthy filamentous fungi can often re-establish colonies from filament fragments or individual hyphal cells, which means that even if we do not know how to encourage sporulation directly, we can encourage certain species to establish in a field simply by spreading fragments of that species throughout the soil. Often, even intensively managed soils have some viable fragments of fungi that can be encouraged with the use of cover crops, diverse crop rotations, and cultivation of mycorrhizal plants (Lohman et al. 2010). Soil inoculation can successfully reintroduce native fungal populations that have been killed off by harmful practices, including the use of fungicides (Lohman et al. 2010).

Each method for reintroducing fungi requires some form of **inoculum**, a mixture that includes viable fungal propagules that are ready to spread and colonize their substrate. Inoculum may include spores, pieces of hyphae, and sometimes plant roots infected with beneficial fungi (Lohman et al. 2010). As long as microbe-friendly management practices are maintained (see Section 12.6), the mycorrhizal community can often thrive and continue spreading to inoculate more soil (Berruti et al. 2016).

While more studies are needed to discern the best possible combinations and methods for soil microbe inoculation, so far studies indicate that the greatest benefits to a field will come from locally native AM fungi (Lohman et al. 2010; Berruti et al. 2016). According to a meta-analysis of more than 160 AM inoculation experiments in more than 40 plant families, inoculating crops with local AM fungi improves fungal colonization of the plant roots, plant root and shoot biomass, crop yield, and plant nutrition. However, successful plant inoculation with foreign (non-native) AM would require knowledge of how the AM will respond to the target environment as well as the target plants, which is not always possible because the factors leading to successful AM-plant symbiosis are still largely unknown (Berruti et al. 2016). In contrast, naturally coevolved relationships allow us to work successfully without complete knowledge of the system. Some studies indicate that native AM fungi are better at promoting plant growth in their native soils than introduced AM because they are adapted to the specific stressors of the environment in which they will be used (Lohman et al. 2010). Using native AM also helps avoid possible problems of introducing invasive species into soil systems, which could cause further disruption to the microbiome instead of repairing it.

While some inoculum formulations are commercially available, and often contain generalist AM that could be expected to colonize a wide variety of plants, the actual utility of such inoculum is limited. One important flaw of commercial inoculum mixtures is that they do not often contain *Gigaspora*, a family which is often found in healthy soils and is known for its significant ability to stabilize soil structure (Lohman et al. 2010). Worse, commercial inoculum does not always contain the species that are listed on the label; sometimes, rather than AM fungi, these mixtures simply contain other additives or fertilizers that encourage plant growth (Berruti et al. 2016). Therefore, the best tactic when trying to encourage microbiome health is for farmers to find cost-effective ways to generate their own inoculum and encourage their native AM communities, rather than to introduce commercial or foreign microbes.

## 12.5  METHODS FOR REINTRODUCING MICROBES TO AGRICULTURAL SOILS

Unfortunately for the creation of inoculum, AM fungi are obligate symbionts. This means they cannot be grown in lab culture in high volumes. Large-scale production of inoculum requires the use of a host plant, either the intended crop or a "trap plant", as discussed below. Furthermore, for microbiome maintenance to be cost-effective, biofertilization methods must be effective at increasing plant yields in difficult conditions but not so expensive that they present a cost burden in years when yields are low (Lohman et al. 2010). Since 90% of global farms are held and run by families, many of them in developing countries and rural areas, it is imperative to find methods of introducing plant-beneficial microbes to the field that are reliable, affordable, and require little specialized technology (Rocha et al. 2019). The amount of inoculum needed depends on the number of plants being cultivated, whether the field or specific plant tissues are inoculated, and when during plant growth inoculation is performed. In general, the creation of inoculum on a farm level using the methods described below will make native AM inoculations viable for smallholders and farmers in developing countries (Berruti et al. 2016) as well as for large, commercial farms.

### 12.5.1  SOIL INOCULATION

The broadest technique for biofertilization is soil inoculation, where inoculum is spread onto field soil or growth medium and allowed to establish. Spreading cultured beneficial soil microbes onto the soil is a "proven" method to increase microbial density and diversity in the soil (Ellouze et al. 2014), though successful establishment of mycorrhizal relationships with plants using this method is not as reliable. The chief benefit of direct soil application is that the introduced microbes will not be deterred by any antimicrobial compounds produced by crop seeds or plant tissues (Rocha et al. 2019).

The simplest method of soil inoculation is to physically transfer some healthy soil with a balanced microbiome into a field with a depleted microbiome. Soils in natural areas which have not been impacted by agricultural practices are the richest in native fungi and provide the best benefit after transplanting (Lohman et al. 2010). If a field contains a known stressor, such as salty soil or low pH, it is best to take soil from similar conditions because the AM fungi in that sample will be adapted to the right type of stress (Berruti et al. 2016). Taking inoculum soil from the rhizosphere of a plant known to associate with AM is another technique that increases the chances of success (Berruti et al. 2016). Because the natural environment has varied microbial niches, AM distribution is naturally patchy, so microbes may more densely inhabit one area of soil than others simply by chance. Therefore, if possible, it is best to pool soil samples together in order to achieve a diverse community in the final inoculum mixture (Lohman et al. 2010).

Mycorrhizal fungi tend to grow within the top 10 centimeters of soil (Lohman et al. 2010), which makes collection and transfer simple even without specialized equipment. It also means that applying AM inoculum to surface soil has a good chance of establishing the right kind of lasting community that farmers need to support their crop plants. However, it also means that transferred soil may contain the wind-dispersed spores of undesired microbes, including pathogens. Therefore, if soil is being moved between agricultural zones, it is important to transfer soil between fields with different species of plants to avoid the spread and establishment of pathogens (Lohman et al. 2010). It is also important to maintain management practices that support the soil microbiome; trying to introduce microbes to a field that lacks their niche will not result in successful inoculation and the AM fungi in the inoculum will not have their intended benefits (Ellouze et al. 2014).

Unfortunately, the expense of fully inoculating an agricultural field with biofertilizer is too great to be cost-effective for large-scale applications (Lohman et al. 2010; Rocha et al. 2019). Devoted equipment for large-scale storage, transport, and application of tonnes of inoculum can also be costly. However, it may be possible to create "fertility islands", where small portions of a field are directly inoculated and the whole field is managed carefully to allow the mycorrhizae to propagate and spread (Berruti et al. 2016). This technique is much less expensive than trying to inoculate entire fields and is also useful for restoring wild spaces like wildlife preserves (Berruti et al. 2016).

Another cost-effective option is to inoculate plants in a greenhouse. This is most practical for plants that are commonly started in the greenhouse before being **outplanted** into outdoor fields (Berruti et al. 2016). Less inoculum is needed for this type of biofertilization since the plants are being inoculated within pots or greenhouse flats and the bulk soil between plants is not being affected. This method has proven effective at not just increasing AM colonization of host plants, but at producing the desired benefits once the hosts have been transferred to the field. For example, in a 2009 study, strawberry plants grown with cultivated AM before outplanting produced 17% more fruit than uninoculated controls (Lohman et al. 2010). Similar yield increases have been observed for a variety of other crops inoculated with AM in the greenhouse, including potatoes, tomatoes, onions, peanuts, watermelons, garlic, and celery (Lohman et al. 2010). For some hosts, the positive effects of AM on shoot biomass, plant nutrition, and yield can be equal whether the plants are inoculated in the greenhouse or via open field biofertilization (Berruti et al. 2016). However, the extent of root infection with beneficial microbes is significantly higher when plants are inoculated in the greenhouse (Berruti et al. 2016). Overall, positive effects of AM inoculation can be obtained with either greenhouse inoculation followed by outplanting or open field inoculation (Berruti et al. 2016).

When inoculating soils in the greenhouse, it is important to use host plants potted in a mixture of compost and nutrient-poor soil. The compost makes nutrients available to the plants and AMF but the nutrient-poor material incentivizes the plants to form partnerships with the available mycorrhizae (Lohman et al. 2010). Generally, phosphorus is more AM-limiting than nutrients like nitrogen or potassium, so a compost rich in phosphorus needs to be further diluted than one enriched in other nutrients. However, the best dilution rate will depend on the target AM community and the type of compost used and must be discovered through testing (Lohman et al. 2010).

## 12.5.2 TRAP PLANTS

Another biofertilization technique is to cultivate AM fungi within host plants and use the propagules within the roots as the inoculum, a technique known as using **trap plants**. Rather than inoculating the intended final host with the AM, a non-crop species known to form mycorrhizal partnerships is grown in a greenhouse and inoculated with AM. AM establish many structures within plant roots, including storage vesicles and the arbuscules that are the sites of nutrient exchange with the plants. The ability of fungi to regenerate from different propagules, like spores, vesicles, arbuscules, or hyphal fragments, differs between fungal families and species. For example, the Glomeraceae AM fungi are known to grow well from mycelial fragments, whereas some other families are known to colonize their substrate better when allowed to germinate from spores (Berruti et al. 2016). Trap plants make it possible to achieve a high concentration of propagules, allowing the storage or movement of large amounts of AM at a time. Trap plant use is the most-studied AM propagation method (Berruti et al. 2016).

Trap plants can be used to house AM safely over winter or between crop rotations. After outplanting, the AM fungi harbored in trap plant roots will be able to quickly establish a mycorrhizal network during the next rotation or growing season, when their intended hosts (crop plants) are present to receive the benefits of AM association (Lohman et al. 2010). If first established in the greenhouse, trap plants can be used as inoculum at any time. The greenhouse method usually involves planting a frost-sensitive trap plant in a bag or pot filled with local compost and soil. The AM fungi in the soil colonize the trap plants, proliferating and increasing the concentration of the propagules in the trap plants' root systems. If trap plants are not outplanted, their roots can be rinsed off, chopped finely, and used as high-concentration inoculum (Berruti et al. 2016) or mixed with soil to increase inoculum volume for larger applications. Outplanting during a cold period will kill the trap plants, at which point the shoots can be discarded or used in compost; meanwhile, the roots will contain AM inoculum ready to spread (Lohman et al. 2010). In field trials, the trap plant method has successfully increased AM populations by 7,000-fold compared with the AM of the initial soil, for all AM fungi tested (Lohman et al. 2010).

Because the trap plants are grown in open soil inside greenhouses, it is possible for contaminant fungi to be present on the surface of the soil or plant tissues. One method for reducing the instance of pathogens is to sterilize the outside of the trap plant roots before they are used as inoculum. Because many AM propagules (e.g., the vesicles and arbuscules) are inside the root, they will be unaffected by surface sterilization (Berruti et al. 2016). However, this method is time and labor intensive. Another method is to ensure that the inoculum source (in this case, the trap plant) does not share any pathogens with the target crop. One example of a trap plant suitable for pairing with vegetable crops is bahiagrass, *Paspalum notatum*, a tropical (frost-sensitive) grass that reliably forms partnerships with most AM fungal species (Lohman et al. 2010).

It is also possible to use target host plants as trap plants under certain conditions. When trees drop seeds and those seeds are permitted to grow, or if the tree can propagate new trunks from the root system, these **self-sown seedlings** are inoculated with all the same mycorrhizae and bacteria as their parent tree (Hall et al. 2003). These seedlings can then be transplanted to new fields, establishing new orchards with partner fungi included. This method, pioneered in the 18th century by Joseph Talon, was the dominant method of establishing new truffle orchards (**truffières**) until the 1970s (Hall et al. 2003). At that time, propagation of truffle hosts in greenhouses became more popular. The greenhouse method of truffle inoculation is similar to the trap plant method; an ECM host, in this case the target host tree, is inoculated with a mixture of propagules. As with AM, ECM inoculum can be made from the finely chopped roots of an infected host. Unlike AM, however, it also is possible to create a pure inoculum of ECM spores by breaking open the fruiting body, in this case the truffle, and using the spores inside to create an inoculum slurry. The greenhouse method is considered more reliable than self-sown seedling inoculation as there is less risk of contamination from insects and pathogens that could reduce the establishment of truffle ECM in new fields (Hall et al. 2003). However, as with AM inoculation, the risk of contamination from the greenhouse itself is still possible. More than half of the world's crop of *Tuber melanosporum* truffles are now sourced from plantation trees that were inoculated in the greenhouse (Hall et al. 2003).

### 12.5.3 SEED COATING

**Seed coating** is a cost-effective method that delivers inoculum directly to the seed-soil interface (Rocha et al. 2019). Seed coating has a low cost relative to soil inoculation and trap plants due to the precise application of small amounts of inoculum to the seed, which promotes the plant-microbial symbiosis from the moment of plant germination. In other inoculation methods, establishment of the symbiosis can be delayed while microbes reach plant roots from the bulk soil or while microbes grow out of trap roots to reach target plants. In seed coating, the relationship is established quickly, increasing the microbial benefits to the plant and ultimately increasing crop yields (Rocha et al. 2019). Seed coating results in the same colonization density of AM fungi in the crop plant for substantially less inoculum input than other methods, as illustrated by a 2019 study which inoculated the mycorrhizal generalist *Rhizophagus irregularis* on maize. In the seed coating treatment, each maize seed received approximately 273 *R. irregularis* propagules; in the soil treatment, soil was inoculated at a concentration of 4,680 AM propagules per plant. Ultimately, the colonization level of the adult plants was equal, despite the much higher inoculation density and cost of the soil treatment (Rocha et al. 2019).

A reliable seed coat includes (1) the inoculum itself, with propagules dispersed in a solid or liquid carrier medium; (2) a binder (adhesive) to hold the inoculum to the seed, and sometimes (3) a filler to bulk up the size, weight, and shape of seeds, making them easier to handle. Common fillers include peat, talc, and lime, which are reliable microbial carriers in addition to having beneficial effects on seed handling. Some binders also act as fillers, including **alginate** (a polysaccharide extracted from brown algae), **biochar** (a carbon-rich substance made by controlled burning of agricultural waste), and **chitosan**[1] (a polysaccharide extracted from the shells of certain shellfish and mollusks). Both the binder and filler can also include micronutrients, compounds that stimulate seed germination, fluorescent dyes to help track seed distribution, and other beneficial additives (Rocha et al. 2019). Some binders can influence the shelf life of the propagules in the inoculum, keeping them viable for years (Rocha et al. 2019).

During seed inoculation, seeds are placed in a mixing drum that rotates and slowly applies the filler (if any), along with any other active ingredients, and finally the binder. When the filler and binder are separate, the binder is added at the end of the mixing process in order to reduce the amount of dust in the final seed mixture (Rocha et al. 2019).

Because they are an intimate part of the mixture, the binder and filler need to be selected almost as carefully as the fungal inoculum. The concentration and makeup of the final coating blend can have substantial effects on seed germination, plant development, and survival of the inoculated fungi. Monetary and ecological cost of the materials must also be considered (Rocha et al. 2019); for example, biochar is produced with controlled burning, which produces much less pollution than uncontrolled burning but still converts almost half of the burned biomass into carbon dioxide, a greenhouse gas of considerable concern. The biochar material could be sourced and processed on the local farm, helping close a disposal loop, or might be sourced from elsewhere, presenting upfront costs at purchase as well as transportation and storage concerns.

Seed coating does of course have limitations. First, the microbes in the inoculum may have a short shelf life, depending on the species' ability to regenerate from different propagules and the makeup of the fillers and binders used. Without testing, it is difficult to predict how long microbes will remain viable, which limits the amount of time that the inoculated seeds can spend in storage and means that seeds might need to be inoculated close to planting, when farmers must also conduct other preparatory tasks. Thus, seed inoculation could present an additional workload during an already busy season. Second, unless small seeds receive enough coating to substantially change their size and weight, they may not receive enough inoculum to make a significant difference in host colonization versus non-inoculated seeds planted in soil with a healthy microbiome (Rocha

---

[1] Pronounced "kite-o-san", from the Greek *chiton* = a type of mollusk.

et al. 2019). Third, seed coating inoculation is incompatible with some other seed treatments, such as the application of fungicides directed at controlling pathogens (Rocha et al. 2019). Finally, seed coat inoculation is not universally successful. Bacteria are the most studied microbes for seed coat application, and in some trials soil inoculation has been more successful at increasing plant colonization and yield than seed coat inoculation (Rocha et al. 2019). However, the seed coat method is promising for many AM fungi. *Trichoderma* has received considerable attention in seed coating trials; when applied to seeds, this fungal genus has led to increased seed germination, plant growth, and pathogen control in both greenhouse and open field experiments (Rocha et al. 2019).

### 12.5.4 MONOSPECIES VS. COMMUNITY INOCULATION

Most experiments of biofertilization study the effects of inoculation with a single microbial species or variety. In particular, the globally distributed generalist AM species *Funneliformis mosseae, Rhizophagus intraradices,* and *Rhizophagus irregularis* are commonly used as inoculants. Unlike many AM, these species have positive effects on a wide variety of host plants, are possible to propagate in the lab, and survive well in long-term storage (Berruti et al. 2016). While the common use of three species is in some ways limiting, different varieties or subspecies of these fungi have different levels of benefits for plant hosts, indicating that there is still considerable genetic variety within these common inoculants and that breeding improvements can be made even within samples of only these three AM species (Berruti et al. 2016).

Monospecies inoculations have been weakly better at increasing shoot biomass than inoculations with multiple AM (a **consortium** of species), though the difference is not statistically significant (Berruti et al. 2016). Many microbial inoculation studies take place in greenhouses, which are more controlled than open fields. Greenhouse studies tend to observe plant-microbe relationships under individual stressors, such as drought stress or salinity stress, but not under combined stresses. It is possible that one species of fungus is sufficient to help the plant through a single stressor, but that such studies do not fully reflect the benefits of a consortium of AM to plants in the field, where multiple stressors may be present at once and a plant has incentive to form partnerships with multiple beneficial fungi (Berruti et al. 2016).

Supposing that multiple species are more beneficial than single species in the field, it can still be difficult to select the AM and other fungi that should be inoculated into field soil. Consortia may help field plants counter multiple simultaneous stressors, but determining the right combination of microbes to create a lasting community in the field can be extremely difficult. For example, if an inoculum mixture contains two microbes that compete for the same niche, competition between them may result in loss of one species and the use of resources that could otherwise be used to benefit the plant. Many fungi and bacteria produce antimicrobial compounds to reduce competition, another factor that must be considered. In some studies, use of a consortium in inoculum can reduce germination of seeds rather than conferring the intended benefits (Rocha et al. 2019).

Fortunately, there are some promising studies of consortium inoculation via seed coating, indicating that while the research to find the right combinations can be time consuming, it is worthwhile. Simultaneous inoculation with *Trichoderma* and *Glomus* fungi can increase the yield and quality of a variety of crop species including winter wheat, and AM fungi can associate with the nitrogen-fixing bacteria in legumes, increasing the yields and protein content of the legumes (Rocha et al. 2019).

## 12.6  MANAGEMENT PRACTICES FOR SOIL MICROBE RETENTION

Once microbe inoculum has been applied to soil, plants, or seeds, it is important to use management practices that help maintain the fledgling microbial community. This increases the chance that soil fertility will be maintained in subsequent seasons, so later crops will have a larger and more diverse soil microbiome to recruit microbial partners from. As crops recruit specific microbes in succession, the community will be able to fill the diverse niches that are required to form a robust system that buffers plants against multiple stressors.

Maintaining a high level of soil organic matter is an important factor in encouraging a dense and diverse soil microbial community (Ellouze et al. 2014). Therefore, it is useful to stop tilling a field entirely (a **no-till** approach) or engage in **conservation tilling**, which leaves at least 30% of the field covered in plant residue. In the absence of tilling that homogenizes the soil, organic matter can build up, increasing the number of niches available for beneficial fungi while helping the soil retain moisture and nutrients. Recovery of once-tilled soil is a long process; it can take decades for microbial communities to recover to the complexity and richness they had before tilling was introduced (Ellouze et al. 2014).

**Crop rotation** fosters a taxonomically diverse microbial community as each plant species in the rotation releases a unique set of compounds through its roots. The breakdown of these unique materials also increases the number of niches available to soil microbes and increases the number of enzymatic functions the community is able to perform (Ellouze et al. 2014). In addition, crop rotation has benefits like reducing the instance of weeds and pathogens, since the major host plant in the field changes regularly. When planning a crop rotation, farmers need to consider water and nutrient levels of the soil, alternate pathogen hosts (such as barberry, which can harbor wheat pathogens), economic factors, and how the microbial community will be affected by the rotation. This additional burden includes transitioning soil pH gently rather than suddenly, growing crops that can handle leaf litter from previous plants, and considering whether certain crops are AM-friendly. For example, canola and mustard are non-mycorrhizal species, but they are highly productive, so farmers like to incorporate them into crop rotations. Because these Brassicaceae are non-mycorrhizal, the population of mycorrhizae in the field soil will decrease during their rotation; the diminished microbial populations might lead to delayed partnerships in the mycorrhizal plants of following rotations (Ellouze et al. 2014). Likewise, the lack of ECM fungi in tree monocultures like oil palm plantations is concerning because the continual culture of non-mycorrhizal trees has depleted the soil of ectomycorrhizae; hence, ectomycorrhizal trees planted in addition to, or after, oil palm trees will have difficulty recruiting beneficial partners from the soil (McGuire et al. 2014). The ecological and economic benefits of various plant species need to be considered to create a holistic agroecological system.

In addition, the best pathogen protection comes from crop rotation systems that include at least three types of crops (including cover crops). Such **long rotations** are also better at encouraging beneficial microbial biomass than short rotations of just two crops (Ellouze et al. 2014). Finding long rotation patterns that are suitably supportive of native AMF and ECM and are economically viable will require additional studies (Ellouze et al. 2014).

It is also important to **select plants that respond well to beneficial soil fungi**. Plants that recruit native AM or ECM and benefit from the interactions will increase nutrient use efficiency in agroecosystems, benefitting most from biofertilization and receiving the greatest benefits while allowing farmers to move away from chemical fertilizer use. It might also be possible to choose crop varieties based on the materials they deposit into the rhizosphere, selectively rearing plants whose rhizodepositions suppress pathogens and improve the density of beneficial microbes (Ellouze et al. 2014).

Finally, it is important to preferentially use **biological amendments** over chemical amendments. In addition to the materials released by their roots, plants produce different organic (biologically made) compounds within their roots, stems, leaves, and flowers. It is possible to use some of these tissues as a biological amendment, known as **green manure**, to alter the compounds and microbial communities in the soil, reducing the density of pathogens and encouraging beneficial microbes. For example, apple trees whose roots are covered with rapeseed meal experience fewer fungal pathogens and nematode infections than trees that do not receive the treatment (Ellouze et al. 2014). Other biological amendments like manure and compost contain a variety of nutrients fungi can readily use and are also excellent at retaining water and binding added nutrients or fertilizers to the soil, further supporting both the plants and fungi in the field (Ellouze et al. 2014). Some biological amendments also trigger fungi to branch, produce more filaments, and secrete sugars that help stabilize the surrounding soil. These changes further increase soil aeration and water retention (Ellouze et al. 2014). However, application should be cautious and based on careful observations; applying too much compost to soil can actually inhibit fungal growth (Ellouze et al. 2014).

## 12.7  CONCLUSION

Despite the damage that conventional agricultural techniques cause to the soil microbiome, reintroducing plant-beneficial fungi to agricultural soils is possible and has definitive benefits for both ecosystem functions and plant productivity. The varied reactions of crop plants to AM, and vice versa, indicate that it is possible to breed and select for AM that are best for crop plants and that crop breeding programs can also select for AM-responsive plants, generating a complete beneficial system that does not require as much fertilizer as modern conventional agriculture.

Biofertilization techniques can also be used to introduce fungi with edible mushrooms into orchards or fields for co-cropping with plants. For example, inoculated saplings have been used to create successful fields of honshimeji mushrooms (*Lyophyllum shimeji*), Shoro mushrooms (*Rhizopogon rubescens*), edible boletes like *Suillus granulatus*, and several truffle species from inoculated saplings (Hall et al. 2003). However, there are still hundreds of edible ECM species that have not been successfully cultivated. Even when an apparently hearty infection is established in a greenhouse sapling, there is no guarantee that the ECM will persist when its host is transplanted into a field (Hall et al. 2003). There are several hypotheses which might explain this difficulty.

First, mycologists may have classified species as mycorrhizal that are actually pathogenic or saprobic. For example, morels (genus *Morchella*) were long considered mycorrhizal fungi, but in fact they are now widely cultivated in China free from plant hosts. Cultivation of morels is possible with bags of nutrient-rich material like wheat, sawdust, or corn, demonstrating that the now-domesticated species *Morchella importuna* is in fact saprobic (Pecchia 2017). Attempting to infect a host plant with a fungus that is not actually mycorrhizal will not result in successful infection! For another example, the "ECM" matsutake mushroom, *Tricholoma matsutake*, initiates a mycorrhizal relationship with its host pine but then pathogenically infects the root's outermost layer, sometimes fully destroying the cortex and living off the dead tissues (Hall et al. 2003). Cultivation of such a mushroom may require the sacrifice of the host tree or reduce its wood production on a normal timeline.

Second, it is possible that we do not understand the ecological conditions that lead to successful establishment of mycorrhizal relationships or to successful fruiting of ECM fungi. Basidiomycetes are known as "late stage" fungi, meaning that they appear most often in well-established forests with high canopy cover and decades-old trees (McGuire et al. 2014). Trying to establish basidiomycetes on saplings may result in poor colonization or only result in mushrooms decades after planting. Some truffles can fruit in a new truffière where there is little to no canopy cover and the day-night cycle brings about severe swings in temperature and moisture levels, while other truffles cannot fruit without 20-year-old hosts and a 75% canopy cover providing consistent temperature and humidity (Hall et al. 2003). Saplings inoculated with species that require considerable canopy cover or other host age-related traits will not be successfully infected or produce mushrooms reliably.

Finally, species interactions may be far more complicated than what we think we observe between just mycorrhizal fungi and host plants. Several ECM are often found associated with other ECM or with saprobic fungi. For example, the porcini mushroom *Boletus edulis* commonly fruits at the same time and in the same locations as *Amanita muscaria,* and in fact the hyphae of these two species have been found colonizing the same host roots. This phenomenon has been noted across Europe, the United States, and New Zealand. If the success of porcini relies in some way on this interaction with *Amanita* species, that could help explain why monospecies inoculations of *Boletus edulis* often fail. Furthermore, several species of bacteria have been isolated from within the hyphae of *Tuber* truffle species. Some *Pseudomonas* bacteria, in particular, release metabolites that influence *Tuber borchii* mycelial growth and may play a role in truffle development (Hall et al. 2003).

If multiple fungal species or bacterial species are required to ensure the success of an inoculation, then it is possible that trap plant techniques and soil transplants are better than sterile, monospecies inoculation for the initial stages of ECM cultivation. As Hall and coauthors have noted, those non-sterile techniques may help farmers spread the correct consortium without necessarily understanding the "intricate biotic interactions" that make mycorrhiza work so well in the wild (Hall et al. 2003).

## REFERENCES

Ellouze, W., A. E. Taheri, L. D. Bainard, et al. 2014. Soil fungal resources in annual cropping systems and their potential for management. *BioMed Research International* 2014:531824. https://doi.org/10.1155/2014/531824.

Hall, I. R., W. Yun, and A. Amicucci. 2003. Cultivation of edible ectomycorrhizal mushrooms. *Trends in Biotechnology* 21:33–8.

Lohman, M., C. Ziegler-Ulsh, and D. Douds. 2010. How to inoculate arbuscular mycorrhizal fungi on the farm Part 1. Rodale Institute. https://rodaleinstitute.org/science/articles/how-to-innoculate-arbuscular-mycorrhizal-fungi-on-the-farm-part-1/.

McGuire, K. L., H. D'Angelo, F. Q. Brearley, et al. 2014. Responses of soil fungi to logging and oil palm agriculture in Southeast Asian tropical forests. *Microbial Ecology* 69:733–47.

Pecchia, J. 2017. *China Trip Unveils Morel Cultivation Mysteries.* Department of Plant Pathology and Environmental Microbiology. Penn State. https://plantpath.psu.edu/news/china-trip-unveils-morel-cultivation-mysteries.

Rocha, I., Y. Ma, P. Souza-Alonso, M. Vosatka, H. Freitas, and R. S. Oliveira. 2019. Seed coating: A tool for delivering beneficial microbes to agricultural crops. *Frontiers in Plant Science* 10:1357. https://doi.org/10.3389/fpls.2019.01357.

# Section II

## Pests and Pest Control Agents

# 13 Fungal Diseases in Agriculture
## *Significance, Management, and Control*

*Noureddine Benkeblia*

## CONTENTS

## 13.1 SIGNIFICANCE OF FUNGAL DISEASES IN AGRICULTURE

The Kingdom Fungi includes millions of species, many with ecological roles as degraders and pathogens. For those purposes they are equipped with complex enzymes such as proteases (which digest proteins), phospholipases (membranes), cellulases (cellulose), hemicellulases (hemicellulose), pectinases (cell walls), ligninases (lignin), and nucleases (DNA and RNA). These enzymes can enhance the virulence of fungal pathogens by hydrolyzing (digesting) the cell wall polymers and enabling pathogen entry (Fisher et al. 2020). Approximately 8,000–10,000 fungal species are known plant pathogens; these are estimated to cause about 100,000 different diseases in plants (Agrios 2005; Hawksworth 1991; Vadlapudi Naidu 2011), or around 80% of the diseases observed in agriculture and horticulture (Agrios 2009; El Hussein et al. 2014). Infection by fungi can cause plant or harvest damage during pre- or postharvest stages, causing heavy losses or even food shortages, disrupting the value chain, and putting economic strain on farmers. Pathogens on commodity crops (see Chapter 8 in this volume) jeopardize the economies of countries who depend upon export revenues to access food grown elsewhere (Fones et al. 2017). Plates 13.1, 13.2, and 13.3 show examples of fungal diseases on plants.

Fungal attacks do not occur randomly but are specific to plant host species or groups (Knogge 1996). In order to access nutrients for their germination, growth, and development, fungi use different strategies to affect their host plants. Initial infection can occur either by forced entry via mechanical and chemical actions (e.g., digesting or breaking the cell wall) or by growing hyphae through host stomata or existing wounds (Horbach et al. 2011; Knogge 1998). Plants have developed different strategies of resistance (structural, innate, or induced) to fungal infection, but abiotic stresses like heat or drought and biotic stresses like insect damage reduce the extent to which a plant can fight off infections, accelerating disease invasion (De Luca 2007).

Plants have pre-existing defense strategies which consist of defense tools endogenously present in the plant before pathogen colonization (Iriti and Faoro 2007; Shittu et al. 2019), for example, by synthesizing peptides such as defensins and oxylipins; by conducting **apoptosis** (programmed cell death) of infected or damaged cells; through superficial structures such as leaf epicuticular layers (e.g., wax); by altering epidermal tissues to be suberized (covered in phenolic or aliphatic

Pods of *Theobroma cacao* in central Costa Rica display symptoms of the fungal disease black pod. This disease causes pods to rot on the cacao tree during the last stages of ripening. (Photo by Christopher J. Saunders. USDA Agricultural Research Service.)

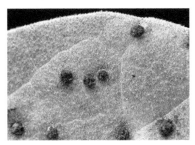

Close-up of pustules (fungal fruiting structures containing thousands of spores) of bean rust fungus on a susceptible bean leaf. (Photo by Peggy Greb. USDA Agricultural Research Service.)

Pecan scab is a fungal disease that can cause blackish lesions on pecan fruit and leaves and can reduce nut quantity and quality. (Photo by Chuck Reilly. USDA Agricultural Research Service.)

**PLATE 13.1**

compounds), cutinized (waterproofed), and/or lignified (strengthened with additional lignins); and by circulating biochemical inhibitors that deter pathogens and pests (Thevissen et al. 2003; Irti et al. 2005; Lam et al. 2001; Dixon et al. 1994; Horsfall and Cowling 1980; Marcell and Beattie 2002; Vance et al. 1980). Induced strategies are mechanisms which become active in response to pathogen attacks. These consist of structure formation (like further suberization) and activation of biochemical reactions by signaling substances like pathogenesis-related proteins and phenolics (Bostock and Stermer 1989; Doughari 2015; Linthorst and Van Loon 1991).

Anthracnose (*Colletotrichum truncatum*)
symptoms. (Photo by Nancy Gregory, University
of Delaware, Bugwood.org.)

Powdery mildew (*Erysiphe flexuosa*) symptoms.
(Photo by Nancy Gregory, University of
Delaware, Bugwood.org.)

Rust (*Uromyces geranii*) sign. (Photo by Cesar
Calderon, Cesar Calderon Pathology Collection,
USDA APHIS PPQ, Bugwood.org.)

**PLATE 13.2**

Extensive literature addressing the significance of fungal diseases in agriculture is readily available, but many aspects of infection and prevention remain unclear; further interdisciplinary research is needed to tackle various issues. First, many of the forces driving the emergence, evolution, and spread of fungi affecting plants are still unknown. Second, it is important to elucidate the mechanisms of how fungi adapt to and interact with their hosts. Third, we need to decipher the mechanisms of evolution and development of fungal resistance to **fungicides** (chemicals that either kill fungi or prevent the growth of fungi and their spores). Fourth, it will be important to implement

Powdery mildew of squash caused by *Podosphaera xanthii.*

Phytophthora crown and root rot (Left) and fruit rot (Right) of squash caused by
*Phytophthora capsici.*

Downy mildew of cucumber caused by *Pseudoperonospora cubensis.*
Left: upper side of leaf; Right: bottom side of leaf.

Late blight of tomato caused by *Phytophthora infestans.*
Left: symptoms on bottom side of the leaf; Right: symptoms on fruit.

**PLATE 13.3**

efficient diagnostic approaches in the identification of fungal diseases in order to thwart fungal threats to agriculture (see Chapter 14 in this volume for examples) (Almeida et al. 2019; Fisher et al. 2012, 2020). Finally, data on how climate change will affect economically important plants are very limited, and the margin for adapting to higher temperatures and changing humidity is reduced by pathogen threat in tropical regions (Paterson et al. 2013).

## 13.2 MANAGEMENT AND CONTROL OF FUNGAL DISEASES IN AGRICULTURE

The management of plant diseases aims to maintain infestation levels below economic thresholds rather than to completely eliminate these diseases (see Chapter 7 in this volume). However, agricultural disease management faces numerous challenges (He et al. 2016; Spadaro and Gullino 2019). Management and mitigation of damage caused by fungal pathogens are most effective when fungi are detected and identified early in the infection process, when distribution, virulence, incidence, and severity might be attenuated, and damages thus reduced (Jain et al. 2019). Although research on fungi has advanced significantly since the 1960s, diagnosing diseases can be difficult because phytopathogenic fungi are classified based on specific traits such as fruiting body and spore structure, while the disease is often classified by plant symptoms (Crous et al. 2015). Nevertheless, progress in biological sciences has made significant contributions to the development of plant pathology. In mycology, studies of plant-microbe interactions, biochemical and physiological studies of fungi and their hosts, and biochemical genetics and DNA-mediated transformation have created a better understanding of fungal biology including their life histories and reproductive cycles (Yarden et al. 2003). In practice, many problems and difficulties are encountered in combating fungal pathogens, and determining their taxonomic groups is of great importance for the identification and hence management of the pathogen (Crous et al. 2015). Advanced technologies, particularly -omics (e.g., genomics, metabolomics, proteomics) and computer models, are increasingly helpful in developing strategic decisions related to agricultural protection. Recently, advances in image processing have resulted in detection, identification, and accurate quantification of the first symptoms of fungal diseases in certain crops from photographs (Arnal Barbedo 2013; Kaur et al. 2019; Pujari et al. 2015). For example, Hyperspectral Imaging (HSI) is used as a small-scale analysis of symptoms caused by different pathogens in sugar beet (Mahlein et al. 2012) and grapes (Knauer et al. 2017), and Deep-Learning-Based Detectors have been used to detect plant diseases and recognize pests in real time in tomato (Fuentes et al. 2022) and maize (DeChant et al. 2017).

At the turn of the 20th century, agriculture was characterized by numerous achievements including good agricultural practices (GAP) aimed at the destruction of potential diseases. However, plant disease control has become heavily dependent on fungicides to control the wide variety of fungal pathogens that are causing heavy losses of agricultural crops worldwide (De Waard et al. 1993; Schwinn 1992). In the field context, plant disease management strategies integrate numerous technologies, but first, it is required to understand the biology of the pathogen and its destructive potential (Fry 1982). It also needed to identify which factors influence the emergence and the increasing incidences of these diseases. Consequently, disease management has become an integral component of crop production and is crucial to avoid crop losses, especially as climatic conditions are being significantly altered by anthropogenic activities.

The tremendous advances in molecular biology in recent years have brought the capacity to compare regions of genomic DNA as well as the ribosomal internal transcribed spacer (ITS) regions of pathogens, which are incredibly useful for taxonomic identification of pathogenic fungi in both the field and the laboratory (see Chapter 14 in this volume) (Martin et al. 2000). Nanotechnology, an innovative and emerging discipline, is also becoming a key approach to target phytopathogens for disease management in agriculture (Khan and Rizvi 2014; Ul Haq et al. 2020). Geographic information systems (GIS), global positioning systems (GPS), and geostatistics techniques are also of burgeoning importance, having been already used in the management of plant diseases like tomato virus, the fungus *Aspergillus flavus*, late blight of tomatoes and potatoes (the oomycete *Phytophthora infestans*), and cotton leaf crumple virus (Nelson et al. 1999). Indeed, studies of the biology and the pathogenicity of plant diseases have revealed, and are still revealing, fundamental principles of cell biology by further investigating virulence of pathogens, susceptibility factors in host plants, and favorable environments for diseases to spread. Such studies will likely reveal additional pathways crucial for the plant immune response, and discoveries on the cell biology of pathogenesis are being translated into the development of efficient methods to control and manage plant diseases (Hajek and Eilenberg 2018; He et al. 2021).

### 13.2.1 CHEMICAL CONTROL

For more than a hundred years, plant protection techniques against fungal diseases have been largely based on fungicide application. Fungicide modes of action vary, although most of them damage fungal cell membranes or interfere with energy production and other metabolic processes within cells of the fungi (Leroux 1996; Yang et al. 2011). The first reported use of fungicides occurred in 1824, when sulfur dust was used against powdery mildew and other pathogens (McCallan 1967; Morton and Staub 2008), though the discovery of Bordeaux mixture (copper sulfate, lime, and water) in 1885 is considered the starting point of widespread chemical disease control (Ragsdale and Sisler 1991, De Waard et al. 1993). In the early 1970s, discovery of **systemic fungicides** (which are absorbed into the plant) revolutionized the chemical control of plant fungal diseases (Cohen 1986; Erwin 1973). Throughout the years, a large number of chemicals have been used to control fungal plant pathogens, and some of these have been advertised as cheaper substitutes for, more effective than, or less hazardous than others on the market. Overall, fungicides are classified into five different groups:

- **Group 1: Fungistats**
  These fungicides temporarily inhibit the growth and development of the fungus. *Example*: Fluconazole®
- **Group 2: Antisporulants**
  These fungicides inhibit spore production without affecting the growth of vegetative hyphae. Because these chemicals do not kill the fungi, they are also called "Fungitoxicants" instead of fungicides. *Example*: Aldehyde traps
- **Group 3: Protectants**
  These fungicides are prophylactic (preventative) in their behavior and are effective only if applied to plants prior to fungal infection. They act by providing a protective barrier that prevents infection. *Example*: Zineb®, Sulfur
- **Group 4: Therapeutants**
  Therapeutants, also called chemotherapeutants, eradicate a fungus after it has caused infection and then cure the plant of infection. These chemicals are systemic in their action. *Example*: Carboxin®, Oxycarboxin®
- **Group 5: Eradicants**
  These fungicides remove pathogenic fungi from an infection and eradicate the dormant or active pathogen from the host by remaining effective on or in the host for some time. *Example*: Organic mercurials, lime sulfur, dodine

Nowadays, more than two hundred fungicides are commercially available from different suppliers and companies (see, for example, University of Tennessee Institute of Agriculture 2016), and their use is subjected to national authorities' approval.

These commercial fungicides are classified by chemical group, by their general or specific mode of action, by their physical properties after being absorbed by plants, or by their persistence on plants and in soils. Some fungicides, such as Captan®, Sulfur, and Mancozeb®, have broad-spectrum activity and are effective against a large variety of fungal pathogens. However, other fungicides have a very narrow spectrum of activity, such as mefenoxam, which is effective only against oomycetes like *Phytophthora* (which are not true fungi).

### 13.2.2 CULTURAL METHODS AND INTEGRATED PEST MANAGEMENT (IPM)

Cultural practices are important for the management of plant pathogens. Cultural pest control includes the techniques outlined in Table 13.1.

In controlled environment crop production (greenhouse production), disease management is more challenging due to the maintained environmental conditions of warmth and humidity, which

**TABLE 13.1**

**Examples of Some Cultural Pest Management Methods**

| Scope of Management | Examples |
|---|---|
| Crop selection | • Crop rotations and timing<br>• Intercropping or separation of crops<br>• Removal and/or chemical control of volunteer plants ("weeds")<br>• Use of pathogen "trap" crops to lure pathogens from vulnerable crops<br>• Use of cover crops<br>• Consideration of plant architecture (three-dimensional shape of the adult crop) when planning |
| Seed selection and planting | • Selection of pathogen-free seeds<br>• Plant spacing<br>• Seeding date and depth<br>• Row direction and spacing |
| Water | • Installing drainage to prevent unwanted flooding<br>• Irrigation methods and timing |
| Soil | • Maintenance of soil fertility, pH, temperature, and soil organic matter<br>• Regular soil testing<br>• Tilling depth and timing<br>• Use of mulch<br>• Removal of debris<br>• Application of fertilizer<br>• Clean fallow practices (periods when fields are empty of crop plants) |
| Crop growth and harvest | • Harvest date<br>• Regular inspections of plant health<br>• Roguing (removing defective or infected plants) |
| Other | • Restricting animal and human movement in fields<br>• Installing barriers<br>• Setting and monitoring storage conditions for seeds and harvest products |

*Source:* Selected from Abawi and Widmer 2000; Awad et al. 1978; Dordas 2008; Hall et al. 1996; Katan 2000; Ogle and Dale 1997; Palti 1981; Sill 1982; Thurston 1992.

are often favorable to disease development (Menzies and Bélanger 1996). Under these conditions, cultural and biological methods are considered better alternatives for disease management than chemical pesticides, and new environmentally friendly technologies to control pathogens have been proposed including soluble silicon, plant extracts, or other biological control agents (see Section 13.2.3 below) (Menzies and Bélanger 1996). For example, silicon application on plants was reported to induce a thicker cellulose membrane, which may help to reduce the severity of disease in plants (Pozza et al. 2015), and foliar application of silicon extended prevention of powdery mildew in grapevine (*Vitis vinifera*), cucumber (*Cucumis sativus*), and melon (*C. melo*) (Menzies et al. 1992; Miyake and Takahashi 1992).

In various crops, foliar sprays of fertilizers have been shown to induce systemic protection against foliar pathogens (Reuveni and Reuveni 1998a) such as powdery mildew (*Leveillula taurica*) (Reuveni and Reuveni 1998a; Reuveni et al. 1998) and *Sphaerotheca fuliginea* (Reuveni et al. 2000). However, heavy agricultural use of fertilizers has been implicated in degradation of the quality of the environment (see Chapter 9 in this volume).

Recent developments in molecular biology regarding small RNAs and the mechanisms of gene silencing provide new opportunities to explore fungal pathogen–host interactions and potential strategies for novel disease control by developing genetically resistant crop species (Nunes and Dean 2012; Punja 2001). Presently, a number of cultivars, varieties, and hybrids resistant to prevailing diseases are commercially available, and many more are being developed by seed companies. Hence, the development and use of genetically resistant plants should be considered one of the most efficient agronomic strategies to control plant diseases caused by fungi and related organisms, like developed hybrid potato cultivars resistant to late blight (*Phytophthora infestans*) (Arora et al. 2014; Halterman et al. 2008; Sun et al. 2016) and soybean cultivars resistant to downy mildew (*Peronospora manshurica*) (Chowdhury et al. 2002; Dong et al. 2018).

Integrated Pest Management (IPM) is another effective and environmentally friendly approach to manage fungal diseases and other pests. Relying on the combination of several common-sense practices, IPM uses comprehensive information on pathogen life cycles and how they interact with the environment, managing fungal diseases using economical means with the least possible hazard to the environment and biodiversity. Rather than a single method, the concept of IPM is a series of management evaluations, decisions, and controls, consisting of (1) setting action thresholds, (2) monitoring and identifying fungal pathogen species, (3) preventing the fungal diseases, and (4) controlling the diseases that do occur (EPA 2021; Kogan 1988). Thus, IPM aims to minimize the impacts of pesticides and maximize the advantages of existing socioeconomic and ecological systems. Karuppuchamy and Venugopal (2016) recommended to carefully consider the following points when implementing IPM:

- Understanding the Agricultural Ecosystem
- Planning of Agricultural Ecosystem
- Conducting a Cost–Benefit Ratio Analysis
- Establishing a Tolerance Level of Pest Damage (see Chapter 7 in this volume)
- Leaving a Pest Residue (contaminating residues in water and soils)
- Timing of Treatments
- Public Understanding and Acceptance of IPM

Initial IPM models focused on ecological aspects of pest management directed at suppression of target pests. However, recent advances in agricultural science have led to the development of an eco-friendlier, systems approach to disease management (Ha 2014). With newly advanced technologies in crop protection using computer modeling and GPS systems, the production of resistant cultivars, use of trap crops, new methods of habitat management, and new insights into consumer preferences for sustainably produced foods, the IPM paradigm needs to be revisited and expanded to include communication and information-based management, business, and sustainability aspects (Dara 2019; Ha 2014). Some alternatives to IPM are outlined in Chapter 7.

### 13.2.3 BIOLOGICAL CONTROL AGENTS (BCAS)

Nowadays, with constraints on plant disease control practices such the use of pesticides, "green" or eco-friendly control methods are capturing the interests of plant pathologists and farmers. During the last few decades, intense research activity on biological control has been carried out in order to adapt the concept of IPM as a sustainable agroecosystem approach to control plant pathogens (Baker and Cook 1974; Carmona et al. 2020; Rashad and Moussa 2020). This biology-based approach is expected to become one of the major components in pest and disease management practices, which will be particularly helpful in the developed countries that are currently major users of chemical pesticides (Baker 1968; Papavizas 1973; Tsao 1977; Wood and Tveit 1955).

Biological control is based on the use of various antagonistic **biological control agents** (BCAs) that suppress diseases by different modes of action (Cook 1993), discussed below. The combination of BCAs with low doses of fungicides is considered a good, integrated approach to control plant diseases, remaining effective at pathogen control while reducing fungicide selection pressure on pathogens and thereby the chances of resistance development (Ons et al. 2020).

All living organisms are subject to predation, parasitism, and/or competition from other organisms (see Chapters 2 and 3 in this volume), and so BCAs can harness these relationships to control fungal diseases. BCA modes of action include mycoparasitism, antibiosis, metabolite (elicitors) production, and competition (Heydari and Pessarakli 2010). Overall, the BCA paradigm consists of investigating relevant modes of action, then introducing the appropriate biological antagonistic organisms to the field (Alabouvette et al. 2006). While the BCA technique is promising, the efficiency of such agents in the field is not always reliable; tests of BCA effectiveness are carried out in pure laboratory culture, which differs considerably from field conditions, so the results of studies do not always translate to the real world (Alabouvette et al. 2006; Wood and Tveit 1955).

Currently, numerous biopesticides such as *Verticillium lecanii* are commercially available for control of fungal diseases (Rodgers 1993; Upadhyay et al. 2014). *V. lecanii* is a particularly interesting biocontrol agent as it can control both several insects and several pathogenic fungi. The action of *V. lecanii* against insects (aphids) involves (1) spore attachment to the host cuticle through a mucilaginous matrix, (2) spore germination and colonization of the cuticle surface, (3) cuticle penetration by germ tubes, (4) active multiplication of blastospores and invasion of host tissues, and (5) release of the fungus from aphid cadavers through the production of conidiospores (St. Leger et al. 1996). It also involves production and diffusion of toxic metabolites possessing insecticidal activity (Claydon and Grove 1982; Gidin et al. 1994). Several studies have also identified other potential control organisms, such as competitive antagonists or hyperparasites of plant diseases (Weltzien 1991). For example, powdery mildews (Erysiphaceae) and rusts (Uredinales) are frequently attacked by *V. lecanii* as well as *Ampelomyces quisqualis*, *Tilletiopsis* spp., and some other fungi (Adams 1990; Hijwegen and Buchenauer 1984).

During the last decades, plant extracts have gained great interest due to their ability to exert biological activity against plant fungal pathogens both *in vitro* (in the lab) and *in vivo* (in real-world conditions), hence demonstrating potential utility as bio-fungicidal products (Fawzi et al. 2009; Romanazzi et al. 2012; Zaker 2016). The use of plant extracts as natural fungicides and pest control agents dates back to ancient Egypt (Loyson 2011; Aboelsoud 2010) but is now becoming much more directed. Plant extracts have been considered as *Generally Regarded As Safe* (GRAS) by the U.S. FDA (Food and Drug Administration), more acceptable to the consumers and farmers, and less hazardous for the environment than fungicides; therefore, they could be an interesting alternative to control plant diseases (Chuang et al. 2007).

More than 50,000 secondary metabolites have been discovered in over 250,000 species in the plant kingdom (Teoh 2015), and more than 50% of these metabolites have been evaluated for their antifungal and anti-plant pathogenic activity. To date, few botanical fungicides have been registered and commercialized, though extensive research is being carried out on the development and isolation of a variety of antifungal plant derivatives (Yoon et al. 2013). Valuable and promising results have been obtained and some commercial formulations have been prepared and marketed (Zaker 2016). As an example, in the United States, there are approximately 245 commercially registered biopesticide-active ingredients; these are used in hundreds of products and 20% of them account for all pesticide-active ingredients registered in the United States. These biopesticides are classified into microbial pesticides, biochemical biopesticides, plant incorporated protectants (-PIPs; e.g., from GM organisms or crops), predators, entomopathogenic nematodes, and parasitoids (Yoon et al. 2013). Among these biopesticides, we can cite Cinnamaldehyde as effective against dry bubble (*Verticillium fungicola*), dollar spot (*Sclerotinia homeocarpa*), and pitch canker disease (*Fusarium moniliforme* var. *subglutinans*) (Copping 2004). Another effective biopesticide is pipernonaline, a piperidine alkaloid derived from long pepper (*Piper longum*), which was reported to be effective against the development of plant diseases caused by barley powdery mildew (*Blumeria graminis*), *Botrytis cinerea*, rice blast (*Magnaporthe oryzae*), powdery mildew (*Phytophtora infestans*), *Puccinia recondite*, and *Rhizoctonia solani* (Copping 2004). Such biopesticides that protect against multiple pathogens are of considerable interest because their development and application is more cost-effective than single-target biopesticides or chemical pesticides.

## REFERENCES

Abawi, G. S., and T. L. Widmer. 2000. Impact of soil health management practices on soilborne pathogens, nematodes and root diseases of vegetable crops. *Applied Soil Ecology* 15:37–47.

Aboelsoud, N. H. 2010. Herbal medicine in ancient Egypt. *Journal of Medicinal Plants Research* 4:82–6.

Adams, P. B. 1990. The potential of mycoparasites for biological control of plant diseases. *Annual Review of Phytopathology* 28:59–72.

Agrios, G. N. 2009. Plant pathogens and disease: general introduction. In *Encyclopedia of Microbiology*, ed. M. Schaechter, 613–646. Amsterdam: Elsevier.

Agrios, G. N. 2005. *Plant Pathology*. 5th ed. Amsterdam: Academic Press.

Alabouvette, C., C. Olivain, and C. Steinberg. 2006. Biological control of plant diseases: The European situation. *European Journal of Plant Pathology* 114:329–41.

Almeida, F., M. L. Rodrigue, and C. Coelho. 2019. The still underestimated problem of fungal diseases worldwide. *Frontiers in Microbiology* 10:214. https://doi.org/10.3389/fmicb.2019.00214.

Arnal Barbedo, J. G. 2013. Digital image processing techniques for detecting, quantifying and classifying plant diseases. *SpringerPlus* 2:660. https://doi.org/10.1186/2193-1801-2-660.

Arora, R. K., S. Sharma, and B. P. Singh. 2014. Light Blight disease of potato and its management. *Potato Journal* 41:16–50.

Awad, M. A., Z. El-Shenawy, A. F. Omran, and M. N. Shatla. 1978. Cultural practices in relation to purple blotch disease of onion. *Scientia Horticulturae* 9:237–43.

Baker, K. F., and R. J. Cook. 1974. *Biological Control of Plant Pathogens*. San Francisco: Freeman.

Baker, R. 1968. Mechanisms of biological control of soil-borne pathogens. *Annual Review of Phytopathology* 6:263–94.

Bostock, R. M., and B. A. Stermer. 1989. Perspectives on wound healing in resistance to pathogens. *Annual Review of Phytopathology* 27:343–71.

Carmona, M., F. Sautua, O. Pérez-Hérnandez, and E. M. Reis. 2020. Role of fungicide applications on the integrated management of wheat stripe rust. *Frontiers in Plant Science* 11:733. https://doi.org/10.3389/fpls.2020.00733.

Chowdhury, A., P. Srinives, P. Saksoong, and P. Tongpamnak. 2002. RAPD markers linked to resistance to downy mildew disease in soybean. *Euphytica* 128:55–60.

Chuang, P. H., C. W. Lee, J. Y. Chou, et al. 2007. Antifungal activity of crude extracts and essential oil of *Moringa oleifera* Lam. *Bioresource Technology* 98:232–6.

Claydon, N., and J. F. Grove. 1982. Insecticidal secondary metabolic products from the entomopathogenous fungus *Verticillium lecanii*. *Journal of Invertebrate Pathology* 40:413–8.

Cook, R. J. 1993. Making greater use of microbial inoculants in agriculture. *Annual Review of Phytopathology* 31:53–80.

Cohen, Y., and M. D. Coffey. 1986. Systemic fungicides and the control of oomycetes. *Annual Review of Phytopathology* 24:311–38.

Copping, L. G. 2004. *The Manual of Biocontrol Agents*. 3rd ed. Alton: BCPC Publications.

Crous, P. W., D. L. Hawksworth, and M. J. Wingfield. 2015. Identifying and naming plant-pathogenic fungi: past, present, and future. *Annual Review of Phytopathology* 53:247–67.

Dara, S. K. 2019. The new integrated pest management paradigm for the modern age. *Journal of Integrated Pest Management* 10:1–9

DeChant, C., T. Wiesner-Hanks, S. Chen, et al. 2017. Automated identification of northern leaf blight-infected maize plants from field imagery using deep learning. *Phytopathology* 107:1426–32.

De Luca A. J. 2007. Harmful fungi in both agriculture and medicine. *Revista Iberoamericana de Micología* 24:3–13.

Dixon, R. A., M. J. Harrison, and C. J. Lamb. 1994. Early events in the activation of plant defence responses. *Annual Review of Phytopathology* 32:479–501.

Dong, H., S. Shi, C. Zhang, et al. 2018. Transcriptomic analysis of genes in soybean in response to *Peronospora manshurica* infection. *BMC Genomics* 19:366. https://doi.org/10.1186/s12864-018-4741-7.

Dordas, C. 2008. Role of nutrients in controlling plant diseases in sustainable agriculture. A review. *Agronomy for Sustainable Development* 28:33–46.

Doughari, J. H. 2015. An overview of plant immunity. *Journal of Plant Pathology and Microbiology* 6:2–11.

El Hussein, A. A., R. E. M. Alhasan, S. A. Abdelwahab, and M. A. El Siddig. 2014. Isolation and identification of *Streptomyces rochei* strain active against phytopathogenic fungi. *British Microbiology Research Journal* 4:1057–68.

EPA. 2021. *Integrated Pest Management (IPM) Principles*. United States Environmental Protection Agency, Washington, DC. https://www.epa.gov/safepestcontrol/integrated-pest-management-ipm-principles [accessed: October 14, 2021].

Erwin, D. C. 1973. Systemic fungicides: Disease control, translocation, and mode of action. *Annual Review of Phytopathology* 11:389–422.

Fawzi, E. M., A. A. Khalil, and A. F. Afifi. 2009. Antifungal effect of some plant extracts on *Alternaria alternata* and *Fusarium oxysporum*. *African Journal of Biotechnology* 8:2590–7.

Fisher, M. C., S. J. Gurr, C. A. Cuomo, et al. 2020. Threats posed by the fungal kingdom to humans, wildlife, and agriculture. *Host-Microbe Biology* 11:e00449–20. https://doi.org/10.1128/mBio.00449-20.

Fisher, M. C., D. A. Henk, C. J. Briggs, et al. 2012. Emerging fungal threats to animal, plant and ecosystem health. *Nature* 484:186–94.

Fones, H. N., D. P. Bebber, T. M. Chaloner, W. T. Kay, G. Steinberg, and S. J. Gurr. 2020. Threats to global food security from emerging fungal and oomycete crop pathogens. *Nature Food* 1:332–42.

Fones, H. N., M. C. Fisher, and S. J Gurr. 2017. Emerging fungal threats to plants and animals challenge agriculture and ecosystem resilience. Eukaryotes: Fungi and parasitology. *Microbiology Spectrum* 5(2). https://doi.org/10.1128/microbiolspec.FUNK-0027-2016.

Fry, W. E. 1982. *Principles of Plant Disease Management*. New York/London: Academic Press.

Fuentes, A., S. Yoon, S. C. Kim, and D. S. Park. 2017. A robust deep-learning-based detector for real-time tomato plant diseases and pests recognition. *Sensors* 17:2–21.

Gindin, G., I. Barash, N. Harari, and B. Raccah. 1994. Effect of endotoxic compounds isolated from *Verticillium lecanii* on the sweet potato whitefly, *Bemisia tabaci*. *Phytoparasitica* 22:189–96.

Ha, T. M. 2014. A review on the development of integrated pest management and its integration in modern agriculture. *Asian Journal of Agriculture and Food Science* 2:336–40.

Hajek, A., and J. Eilenberg. 2018. Biology and ecology of microorganisms for control of plant diseases. In *Natural Enemies: An Introduction to Biological Control*, eds. A. Hajek, and J. Eilenberg, 291–307. Cambridge: Cambridge University Press.

Hall, R., and N. C. B Luiz. 1996. Practice and precept in cultural management of bean diseases. *Canadian Journal of Plant Pathology* 18:176–85.

Halterman, D. A., L. C. Kramer, S. Wielgus, and J. Jiang. 2008. Performance of transgenic potato containing the Late Blight resistance gene plant disease. *Plant Disease* 92:339–43.

Hawksworth, L. 1991. The fungal dimension of biodiversity: magnitude, significance, and conservation. *Mycology Research* 95:641–55.

He, D. C., M. H. He, D. M. Amalin, W. Liu, D. G. Alvindia, and J. Zhan. 2021. Biological control of plant diseases: An evolutionary and eco-economic consideration. *Pathogens* 10:131. https://doi.org/10.3390/pathogens10101311.

He, D. C., J. S. Zhan, and L. H. Xie. 2016. Problems, challenges and future of plant disease management: from an ecological point of view. *Journal of Integrative Agriculture* 15:705–15.

Heydari, A., and M. Pessarakli. 2010. A review on biological control of fungal plant pathogens using microbial antagonists. *Journal of Biological Control* 10:273–90.

Hijwegen, T., and H. Buchenauer. 1984. Isolation and identification of hyperparasitic fungi associated with Erysiphaceae. *Netherlands Journal of Plant Pathology* 90:79–83.

Horsfall, J. G., and E. B. Cowling. 1980. *Plant Disease*. Vol. 5. New York: Academic Press.

Iriti, I., and F. Faoro. 2007. Review of innate and specific immunity in plants and animals. *Mycopathologia* 164:57–64.

Iriti, M., and F. Faoro. 2005. Lipid biosynthesis in Spermathophyta. In *Floriculture, Ornamental and Plant Biotechnology*, Vol. I, ed. A. Teixeira da Silva, 359–372. Global Science Books.

Jain, S. S., Q. Wu, Y. Lu, and J. Shi. 2019. A review on plant leaf fungal diseases and its environment speciation. *Bioengineered* 10:409–24.

Karuppuchamy, P., and S. Venugopal. 2016. Integrated pest management. In *Ecofriendly Pest Management for Food Security*, ed. P. Karuppuchamy, 651–684. Amsterdam: Academic Press.

Katan, J. 2000. Physical and cultural methods for the management of soil-borne pathogens. *Crop Protection* 19:725–31.

Kaur, S., S. Pandey, and S. Goel. 2019. Plants disease identification and classification through leaf images: A survey. *Archives of Computational Methods in Engineering* 26, 507–30.

Khan, M. R., and T. F. Rizvi. 2014. Nanotechnology: Scope and application in plant disease management. *Plant Pathology Journal* 13:214–31.

Knauer, U., A. Matros, T. Petrovic, et al. 2017. Improved classification accuracy of powdery mildew infection levels of wine grapes by spatial-spectral analysis of hyperspectral images. *Plant Methods* 13:47. https://doi.org/10.1186/s13007-017-0198-y.

Knogge, W. 1996. Fungal infection of plants. *Plant Cell* 8:1711–22.

Knogge, W. 1998. Fungal pathogenicity. *Current Opinion in Plant Biology* 1:324–8.

Kogan, M. 1988. Integrated pest management theory and practice. *Entomologia Experimentalis et Applicata* 49:59–70.

Lam, E., N. Kato, and M. Lawton. 2001. Programmed cell death, mitochondria and the plant hypersensitive response. *Nature* 411:848–53.

Leroux, P. 1996. Recent developments in the mode of action of fungicides. *Pesticide Science* 41:191–7.

Linthorst, H. J. M., and L. C. Van Loon. 1991. Pathogenesis-related proteins of plants. *Critical Reviews in Plant Sciences* 10:123–50.

Loyson, P. 2011. Chemistry in the time of the pharaohs. *Journal of Chemical Education* 88:146–50.

Mahlein, A. K., U. Steiner, C. Hillnhütter, et al. 2012. Hyperspectral imaging for small-scale analysis of symptoms caused by different sugar beet diseases. *Plant Methods* 8:3. https://doi.org/10.1186/1746-4811-8-3.

Marcell, L. M., and G. A. Beattie. 2002. Effect of leaf surface waxes on leaf colonization by *Pantoea aglomerans* and *Clavibacter michiganensis*. *Molecular Plant Microbe Interaction* 15:1236–44.

Martin, R. R., D. James, and C. A. Lévesque. 2000. Impacts of molecular diagnostic technologies on plant disease management. *Annual Review of Phytopathology* 38:207–39.

McCallan, S. E. A. 1967. History of fungicides. In *Fungicides, An Advanced Treatise*, Vol. I, eds. V. Morton, and T. Staub, 1–37. New York: Academic Press.

Menzies, J. G., and R. R. Bélanger. 1996. Recent advances in cultural management of diseases of greenhouse crops. *Canadian Journal of Plant Pathology* 18:186–93.

Menzies, J., P. Bowen, D. Ehret, and A. D. M. Glass. 1992. Foliar applications of potassium silicate reduce severity of powdery mildew on cucumber, muskmelon, and zucchini squash. *Journal of the American Society for Horticultural Science* 117:902–5.

Mitchell, J. E. 1973. The mechanisms of biological control of plant diseases. *Soil Biology and Biochemistry* 5:721–8.

Miyake, Y., and E. Takahashi. 1982. Effect of silicon on the growth of solution-cultured cucumber plant comparative studies on silica nutrition in plants. Comparative studies on silica nutrition in plants. *Journal of Soil Science and Plant Nutrition* 53:23–9.

Morton, V., and T. Staub. 2008. *A Short History of Fungicides*. American Phytopathological Society. APSnet Features. https://www.apsnet.org/edcenter/apsnetfeatures/Pages/Fungicides.aspx [accessed: December 13, 2021].

Nelson, M. R., T. V. Orum, R. Jaime-Garcia, and A. Nadeem. 1999. Applications of geographic information systems and geostatistics in plant disease epidemiology and management. *Plant Disease* 83:308–19.

Nunes, C. C. and R. A. Dean. 2012. Host-induced gene silencing: a tool for understanding fungal host interaction and for developing novel disease control strategies. *Molecular Plant Pathology* 13:519–29.

Ogle, H., and M. Dale. 1997. Diseases management: Cultural practices. In *Plant Pathogens and Plant Diseases*, eds. J. F. Brown, and H. J. Ogle, 390–404. Armidale: Rockvale Publication.

Ons, L., D. Bylemans, K. Thevissen, and B. Cammue. 2020. Combining biocontrol agents with chemical fungicides for integrated plant fungal disease control. *Microorganisms* 8:1930. https://doi.org/10.3390/microorganisms8121930.

Palti, J. 1981. *Cultural Practices and Infectious Crop Diseases*. Berlin: Springer-Verlag.

Papavizas, G. C. 1973. Status of applied biological control of soil-borne plant pathogens. *Soil Biology and Biochemistry* 5:709–20.

Paterson, R. R. M., M. Sariah, and N. Lima. 2013. How will climate change affect oil palm fungal diseases? *Crop Protection* 46, 113–20.

Pozza, E. A., A. A. A. Pozza, and D. M. D. S Botelho. 2015. Silicon in plant disease control. *Revista* Ceres 62:323–31.

Punja, Z. K. 2001. Genetic engineering of plants to enhance resistance to fungal pathogens - A review of progress and future prospects. *Canadian Journal of Plant Pathology* 23:216–35.

Pujari, J. D., R. Yakkundimath, and A. S. Byadgi. 2015. Image processing based detection of fungal diseases in plants. *Procedia Computer Science* 46:1802–8.

Ragsdale, N. N., and H. D. Sisler 1991. The nature, modes of action, and toxicity of fungicides. In *Handbook of Pest Management in Agriculture*, ed. D. Pimental, 461–496. Boca Raton: CRC Press.

Rashad, Y. M., and T. A. A. Moussa. 2020. Biocontrol agents for fungal plant diseases management. In *Cottage Industry of Biocontrol Agents and Their Application*, eds. N. El-Wakeil, M. Saleh, and M. Abu-Hashim, 337–363. Cham: Springer.

Reuveni, M., M. Harpaz, and R. Reuveni. 1998. Integrated control of powdery mildew on field-grown mango trees by foliar sprays of mono-potassium phosphate fertilizer, sterol inhibitor fungicides and the strobilurin kresoxym-methyl. *European Journal of Plant Pathology* 104:853–60.

Reuveni, R., G. Dor, M. Raviv, M. Reuveni, and S. Tuzun. 2000. Systemic resistance against *Sphaerotheca fuliginea* in cucumber plants exposed to phosphate in hydroponics system, and its control by foliar spray of mono-potassium phosphate. *Crop Protection* 19:355–61.

Reuveni, R., and M. Reuveni. 1998a. Foliar-fertilizer therapy – a concept in integrated pest management. *Crop Protection* 17:110–8.

Reuveni, R., and M. Reuveni. 1998b. Local and systemic control of powdery mildew (*Leveillula taurica*) on pepper plants by foliar spray of mono-potassium phosphate. *Crop Protection* 17:703–9.

Rodgers, P. B. 1993. Potential of biopesticides in agriculture. *Pesticide Science* 39:117–29.

Romanazzi, G., A. Lichter, F. M. Gabler, and J. L. Smilanick. 2012. Recent advances on the use of natural and safe alternatives to conventional methods to control postharvest gray mold of table grapes. *Postharvest Biology and Technology* 63:141–7.

Schwinn, F. l. 1992. Significance of fungal pathogens in crop production. *Pesticide Outlook* 3:18–25.

Shittu, H. O., E. Aisagbonhi, and O. H. Obiazikwor. 2019. Plants' innate defence mechanisms against phytopathogens. *Journal of Microbiology, Biotechnology and Food Science* 9:314–9.

Sill, W. H. 1982. *Plant Protection: An Integrated Interdisciplinary Approach*. Ames: Iowa State University Press.

Spadaro, D., and M. L. Gullino. 2019. Sustainable management of plant diseases. In *Innovations in Sustainable Agriculture*, eds. M. Farooq, and M. Pisante, 337–359. Cham: Springer.

St. Leger, R. J., L. Joshi, M. J. Bidochka, N. W. Rizzo, and D. W. Roberts. 1996. Characterization and ultrastructural localization of chitinases from *Metarhizium anisopliae*, *M. flavoviride*, and *Beauveria bassiana* during fungal invasion of host (*M. sexta*) cuticle. *Applied Environmental Microbiology* 62:907–12.

Sun, K., A. M. A. Wolters, J. H. Vossen, et al. 2016. Silencing of six susceptibility genes results in potato late blight resistance. *Transgenic Research* 25:731–42.

Teoh, E. S. 2015. Secondary metabolites of plants. *Medicinal Orchids of Asia* 5:59–73.

Thevissen, K, K. K. A. Ferket, I. E. J. A. Francois, and B. P. A. Cammue. 2003. Interaction of antifungal plant defensins with fungal membrane components. *Peptides* 24:1705–12.

Thurston, H. D. 1992. *Sustainable Practices for Plant Disease Management in Traditional Fanning Systems*. Boulder: Westview Press.

Tsao, P. H. 1977. Prospects of biological control of citrus root disease fungi. *Proceedings of the International Society of Citriculture* 3:857–63.

Ul Haq, I., S. Ijaz, and N. A. Khan. 2020. Application of nanotechnology for integrated plant disease management. In *Plant Disease Management Strategies for Sustainable Agriculture through Traditional and Modern Approaches. Sustainability in Plant and Crop Protection*, eds. I. Ul Haq, and S. Ijaz, 173–185. Cham: Springer.

University of Tennessee Institute of Agriculture, 2016. Fungicides by trade name. https://ag.tennessee.edu/cpa/Documents/Pumpkins%20for%20Profit%20Tour/MS.labeled.fung.for.pumpkin.pdf (accessed February 12, 2022).

Upadhyay, U., D. Rai, M. Ran, P. Mehra, and A. K. Pandey. 2014. *Verticillium lecani* (Zimm.): A potential entomopathogenic fungus. *International Journal of Agriculture, Environment & Biotechnology* 7:719–27.

Vadlapudi, V., and K. C. Naidu. 2011. Fungal pathogenicity of plants: molecular approach. *European Journal of Experimental Biology* 1:38–42.

Vance, C. P., T. K. Kirk, and R. T. Sherwood. 1980. Lignification as a mechanism of disease resistance. *Annual Review of Phytopathology* 18:259–88.

Waard, M. A., S. G. Georgopoulos, D. W. Hollomon, et al. 1993. Chemical control of plant diseases: Problems and prospects. *Annual Review of Phytopathology* 31:403–21.

Weltzien, H. C. 1991. Biocontrol of foliar fungal diseases with compost extracts. In *Microbial Ecology of Leaves*, eds. J. H. Andrews, and S. S. Hirano, 430–450. New York: Springer.

Wood, R. K. S., and M. Tveit. 1955. Control of plant diseases by use of antagonistic organisms. *Botanical Review* 21:441–92.

Yang, C., C. Hamel, V. Vujanovic, and Y. Gan. 2011. Fungicide: Modes of action and possible impact on non-target microorganisms. International Scholarly Research Notices, 2011:130289. https://doi.org/10.5402/2011/130289.

Yarden, O., D. J. Ebbole, S. Freeman, R. J. Rodriguez, and M. B. Dickman. 2003. Fungal biology and agriculture: Revisiting the field. *Molecular Plant-Microbe Interactions* 16:859–66.

Yoon, M. Y., B. Cha, and J. C. Kim. 2013. Recent trends in studies on botanical fungicides in agriculture. *Plant Pathology Journal* 29:1–9.

Zaker, M. 2016. Natural plant products as eco-friendly fungicides for plant diseases control – A review. *Agriculturists* 14:134–41.

# 14 Fungal Pathogens in Forested Ecosystems

*Denita Hadziabdic, Aaron Onufrak, and Romina Gazis*

## CONTENTS

## 14.1 INTRODUCTION

As a key component of ecosystem services globally, healthy forests are not only vital for the preservation of biodiversity but an integral part of the socio-economic development of our society. Forest health is inextricably linked to human health and the global economy, contributing an estimated $125–145 trillion per year to the value of global ecosystem services (Costanza et al. 2014; Donovan et al. 2013; Hadziabdic et al. 2021). Healthy forests are not necessarily disease-free but balanced and dynamic systems capable of adapting to natural disturbances (Hadziabdic et al. 2021; Trumbore et al. 2015). Unfortunately, our forests are currently threatened by invasive insect pests and pathogens, largely as a by-product of globalization, climate change, and land-use changes (Eriksson et al. 2019; Hadziabdic et al. 2021). There are several cases in which economically, ecologically, and socially devastating fungal pathogens have caused irreversible changes to natural forest ecosystems. Well-known and extensively studied examples include Chestnut Blight caused by *Cryphonectria parasitica*, Dutch Elm Disease caused by *Ophiostoma ulmi* and *O. novo-ulmi*, and White Pine Blister Rust caused by *Cronartium ribicola* (Anagnostakis 1987; Geils et al. 2010; Gibbs 1978; Kinloch Jr 2003). Here we describe emerging fungal disease complexes that involve diverse plant hosts, multiple vectors, and pathogen(s) with major consequences on our forested ecosystems. We discuss their

DOI: 10.1201/9780429320415-16

basic biology, impact on North American forests, diagnosis, current management measures, and best strategies to mitigate their further spread.

## 14.2 DETECTION TOOLS AND SURVEILLANCE METHODS

Early detection and rapid response are universal recommendations to reduce the establishment and spread of pests and diseases and to prevent outbreaks that cause economic and ecological harm. The timely implementation of control measures depends on the proper identification of the pest(s) and/or the causal agent(s) of the disease. Disease diagnosis often starts with the observation of symptoms in the host as the disease progresses, but accurate identification of the causal agent is crucial as symptoms are frequently nonspecific. Conventional pathogen detection methods are often culture-based and require the **axenic** (pure culture) isolation of the causal agent from symptomatic host tissue. Pathogen isolation can become challenging when dealing with diseases of long-living trees, mainly due to the diversity and abundance of endophytes, secondary pathogens, and/or saprotrophs usually associated with the symptomatic tissue used for pathogen isolation. Unless there is a selective medium in which to solely recover the target pathogen, culture-based detection methods can take up to several days and pathogen isolation might not even be possible. However, selective media are not available for the majority of new and emerging diseases. Besides difficulties in isolation, morphology-based identification in fungal pathogens is usually not recommended due to the lack of diagnostic characteristics and the phenotypic plasticity associated with growth conditions. In the last decade, molecular tools have become widely used in the identification of forest pathogens, speeding the diagnosis process and facilitating the implementation of regulatory policies.

Molecular diagnostic tools use the pathogen's DNA for detection, using general primers that target universal fungal markers such as the Internal Transcribed Spacer (ITS) followed by sequencing of the amplified product (amplicon) or pathogen-specific primers followed by amplicon visualization. Taxon-specific primers need to solely amplify the DNA of the pathogen (be specific) and consistently amplify DNA from strains of the same species from diverse populations (be consistent). When using taxon-specific primers, DNA extraction can be completed from mixed samples (i.e., crude extraction from infected tissues) and amplicon sequencing is not needed. Amplicons can be visualized in an electrophoresis gel or by using fluorescence-based technologies (quantitative PCR (qPCR), as well as loop-mediated isothermal amplification [LAMP]). Target regions can be single copy or multicopy genes, protein-coding or intragenomic regions, nuclear, or mitochondrial, all of which will affect the sensitivity of the assay or the detection threshold.[1]

## 14.3 FUSARIUM DIEBACK – INVASIVE SHOT HOLE BORERS

### 14.3.1 Disease Biology and Impact

Fusarium Dieback – Shot Hole Borer is an invasive pest disease complex affecting numerous woody species across continents. The *Euwallacea fornicatus*, a species complex consisting of multiple cryptic species of ambrosia beetles, is commonly called "invasive shot hole borers (ISHB)". ISHB comprises genetically and ecologically diverse but morphologically identical lineages of emerging pests. There are currently four recognized species in this beetle complex based on molecular evidence (Gomez et al. 2018; O'Donnell et al. 2016; Stouthamer et al. 2017). The natural geographic distribution of the morphological species *E. fornicatus sensu lato* expands from Sri Lanka through South East Asia to Vietnam, China, Taiwan, and the islands of Okinawa, Indonesia, Philippines, and Papua New Guinea (EPPO Global Database 2021). Countries in which this species complex

---

[1] For further details on molecular detection methods for plant pathogenic fungi, we refer readers to Hariharan and Prasannath (2021).

has been introduced are Israel (Mendel et al. 2012), the United States (Eskalen et al. 2012), South Africa (Paap et al. 2018), Mexico (García-Avila et al. 2016), Costa Rica, and Panama (Atkinson 2015; Kirkendall and Ødegaard 2007). In their native region in southern Asia, these beetles mostly colonize dead or declining species (Hulcr et al. 2007), but they are also known as destructive pests of several economically important woody plants, including tea (*Camellia sinensis*), avocado (*Persea americana*), citrus (*Citrus* spp.), rubber tree (*Hevea brasiliensis*), and cacao (*Theobroma cacao*), in which the pest-disease complex can cause extensive dieback and even death (Danthanarayana 1968; Walgama and Pallemulla 2005).

In managed settings (i.e., forestry plantings, orchards) and urban landscapes, these beetle species attack live exotic tree species, and in invaded areas they also infest naturally occurring hosts. For instance, the ISHB-fungus complex has had a detrimental impact on fragile riparian ecosystems in Southern California, USA (Lynch 2019). Unlike most bark and ambrosia beetles, species in the ISHB group infest healthy, well-watered trees (Boland and Woodward 2019; Swain et al. 2017). Therefore, species in this group represent a significant threat to naïve forests as well as urban landscapes and to wood, fruit, and nut production worldwide (Coleman et al. 2019; O'Donnell et al. 2016; Paap et al. 2020). Species in this complex are nearly morphologically identical but have distinct preferred host ranges, levels of pathogenicity, and geographic distributions. Several studies have been conducted to delineate the species within this complex, as some lineages are known to be primary borers attracted to living, apparently healthy trees, whereas others colonize dead hosts (Gomez et al. 2018). Therefore, knowing which lineage within the complex is present in an area has important ecological, economical, and biosecurity implications.

Adult female beetles bore into trees and build galleries in the sapwood where they inoculate a suite of symbiotic fungi, including *Fusarium euwallaceae*, *Graphium euwallaceae*, and *Paracremonium pembeum* (Lynch et al. 2014). *Fusarium euwallaceae* (causal agent of Fusarium Dieback, FD) infection causes disruption of water and nutrient movement in the host phloem and xylem and, coupled with extensive beetle boring, leads to tree injury and mortality (Coleman et al. 2019). High levels of beetle infestation and fungal infection result in high levels of tree mortality (Umeda 2017). Symptoms of beetle attack and FD vary by host species. With a wide reproductive host range that includes over 65 tree species, from which 20 are California, U.S. natives, this beetle complex has caused significant damage to natural and urban forests as well as fruit orchards in California (Umeda et al. 2016). Additionally, at least 260 tree species in 64 families can be attacked but do not support beetle reproduction (Lynch 2019). It is important to note that not all susceptible species are threatened with rapid tree mortality; only species in which the beetle can reproduce succumb to the disease. The non-reproductive tree species do not sustain the growth of the symbiont fungus (Mendel et al. 2021). Nevertheless, some non-reproductive species might become reproductive hosts under certain environmental conditions that favor disease development (Paap et al. 2020).

## 14.3.2 Detection and Identification of the Pathogen and Vector

External signs associated with members of the ISHB complex include entrance holes (attacks) on main stems and larger limbs to small diameter (<2.5cm) branches and exposed root flares (Eskalen et al. 2013; Mendel et al. 2012). Adult entrance holes are round and about 0.85mm in diameter. Strands of **frass** (insect feces mixed with wood particles) or boring dust, resembling tubes or packed cylindrical columns, can protrude from the holes; frass can also be found at the base of infested trees. Dieback symptoms include wilting branches, discolored leaves, and breaking of heavy branches (Mendel et al. 2012). High levels of beetle attacks on a tree can cause severe branch and crown dieback; **epicormic** growth (from previously dormant buds) along the stem and at the base of the tree; stem or branch failure; and eventual tree death (Coleman et al. 2013). Depending on the tree species attacked, disease symptoms may also include sap oozing (gumming) from the holes and wet staining and discoloration on the bark (Eskalen et al. 2013). Internal symptoms (those observed underneath the bark) include brown to black staining of the vascular tissue surrounding the entrance holes and beetle galleries; depending on host susceptibility, vascular staining can extend beyond the beetle gallery (Eskalen et al. 2013).

Members of the ISHB can be distinguished from other borers in the beetle complex present in the United States by molecular detection, using conventional PCR and Sanger sequencing (Gomez et al. 2018) or rapid high-resolution melt curve analysis with real-time PCR. Both methods target the cytochrome oxidase I locus in the insect's DNA (Rugman-Jones and Stouthamer 2017).

Most ambrosia fusaria share a unique spore morphology, **clavate** (club-shaped) macroconidia, which is different from the classic banana-shaped conidia present in other species of *Fusarium* (Kasson et al. 2013; Short et al. 2017). Nevertheless, species within the ambrosia fusaria group can't be identified based solely on morphological characteristics. Although selective media for *F. euwallaceae* is currently not available, the pathogen can be easily isolated using non-selective media, such as potato dextrose agar (PDA) or malt extract agar (MEA) amended with antibiotics (i.e., tetracycline), since it is the most abundant symbiont in the beetle's mycangia and galleries (Eskalen et al. 2013; Na et al. 2018; Paap et al. 2020; see Section 15.1.1.3). The internal transcribed spacer (ITS) region of ribosomal DNA (rDNA) is the most used region for identifying fungal pathogens because it amplifies across fungal taxa, it is present in multiple copies per genome which increases detection sensitivity, and there is a robust publicly available database to use for comparisons (Schoch et al. 2012). However, the ITS region is not informative for species-dense *Fusarium* lineages, especially in certain clades or species complexes such as those associated with *Euwallacea* spp. Other loci like translation elongation factor 1-α (TEF1-α) and the largest (RPB1) and second-largest (RPB2) subunits of DNA-directed RNA polymerase II are more useful in distinguishing species within this species complex (Freeman et al. 2013). More recent detection methods include a **multiplex** conventional PCR, which targets multiple genomic regions at once (Short et al. 2017), and multiplex real-time quantitative PCR assays using hydrolysis probes targeting the beta-tubulin gene (Carrillo et al. 2020).

## 14.4  LAUREL WILT

### 14.4.1  Disease Biology and Impact

Laurel Wilt (LW), caused by the fungus *Raffaelea lauricola* (RL), is a deadly vascular disease affecting numerous hosts in the Lauraceae family, including many important North American native forest species and avocado trees (Harrington et al. 2008). The Ophiostomatalean fungus is a nutritional symbiont of the ambrosia beetle *Xyleborus glabratus* (Curculionidae: Scotylinae), native to Southeast Asia (Rabaglia et al. 2006). Both the fungus and its vector were introduced to the United States in 2002 through Port Wentworth, Georgia, USA, but the disease was not reported until 2004 (Fraedrich et al. 2008). Even though *X. glabratus* is the most efficient vector of the pathogen, nine other native and naturalized beetles are known to carry the pathogen (externally or within the mycangia) and at least two species have been confirmed to efficiently transmit the disease into healthy avocado trees (Carrillo et al. 2014; Cruz et al. 2021; Ploetz et al. 2017). The extent to which other ambrosia species act as vectors in natural areas remains unknown. Besides beetle transmission, the pathogen can move to adjacent trees through root grafts (Ploetz et al. 2017). Currently, the disease has been detected in 12 U.S. states, affecting multiple forest species native to North America (SREF Forest Health Program 2021). LW has been found as far west as eastern Texas, USA and continues to move through natural forests across the Northeast and the Midwest. This disease represents not only a threat to natural ecosystems but also to the major avocado-producing areas in California, USA and Mexico. The broad geographic range of the multiple susceptible hosts, in addition to the abundance and generalist habit of the beetle vectors, has promoted the unprecedented rapid expansion of LW across ecosystems.

LW is a lethal disease that compromises the functionality of the host's vascular system. When the pathogen establishes and multiplies in the host's tissues, the tree responds by producing gums and **tyloses** (bulbous cellular outgrowths), which ultimately obstruct the sap flow (Inch et al. 2012). Therefore, external symptoms associated with the disease resemble those caused by drought stress. Symptoms first appear where the beetle attack has occurred. When pathogen transmission is root-to-root, symptoms tend to appear on the side of the new host that faces the infected tree. In avocado, symptoms start with **turgor** (water pressure) loss and consequent foliar wilting ("green wilting"), which turns into brown-wilting within two to three weeks (Crane et al. 2020). Following this initial discoloration, brown desiccated leaves can stay attached to the branches for months (Kendra et al. 2013). In deciduous trees such as sassafras (*Sassafras albidum*), LW symptomatology varies. For instance, instead of "green wilting", the foliage of the infected sassafras turns reddish to purplish brown (similar to when the winter season is approaching) and defoliates rapidly as trees wilt and die (Smith et al. 2009). Internal symptoms include a blackish-blue-stained streaking of the sapwood which can be also observed surrounding the entry hole leading to the beetle's galleries (Kendra et al. 2013). Depending on the inoculum concentration and tree diameter, infected trees can succumb to the disease within two to eight weeks (Cameron et al. 2015; Ploetz et al. 2016).

The potential economic impact of LW on the avocado industry could be devastating, as the disease has led to the destruction of $42 million worth of fruit-bearing trees and an annual production loss of $4 million in Florida, USA alone. To date, this disease has caused the loss of more than 140,000 trees, threatening not only the livelihood of farmers but also the state's economy, as the avocado industry has an overall economic impact of about $100 million per year in Florida, USA (Evans and Nalampang 2010). In forested ecosystems, LW has decimated entire populations of dominant and ecologically important tree species, likely changing ecosystem functioning and compromising important reservoirs of biodiversity.

### 14.4.2 Detection and Identification of the Pathogen and Vector

Differences in symptoms and disease progress among orchards with different cultivation practices (distance among trees, cultivars, initial health status, etc.) are the biggest challenges for avocado farmers. These irregularities also cause confusion for determining which, when, and how many trees should be removed to contain the spread of the disease within the orchard. In natural forests, where susceptible hosts are not dominant but of scattered distribution (i.e., sassafras) and the canopy is composed of multiple tree species, identifying infected trees becomes even more challenging. Since the early diagnosis of LW is critical in facilitating the timely implementation of proper management practices and therefore containing the spread of the disease within orchards and urban and natural forests, having a rapid and accurate diagnostic method is a priority. The conventional detection method can take up to ten days and requires the isolation of the pathogen RL from symptomatic (stained) sapwood chips or slivers (for detailed sampling guidelines see Crane et al. 2020), using a semi-selective medium (CSMA: 1% malt extract and 1% agar, supplemented with 200 ppm cycloheximide and 100 ppm of streptomycin sulfate), followed by DNA extraction and the amplification of two RL-specific microsatellite markers, IFW and CHK (Dreaden et al. 2014).

A PCR-based method to detect RL directly from symptomatic sapwood was developed by Parra et al. (2020), reducing the diagnosis time to 24 hours. In this direct test, DNA is isolated from sawdust collected from drilled symptomatic sapwood or sapwood slivers and used as a template to amplify the two RL-specific microsatellite primers, IFW and CHK (Dreaden et al. 2014), through conventional PCR and electrophoresis visualization. Parra et al. (2020) assessed the specificity of both primers by testing cross-amplification against a diverse set of fungal strains, representing close relatives of the pathogen and fungi that are often found colonizing avocado sapwood and phloem in commercial orchards. The amplification of the IFW region was found to be more consistent and reliable, with a detection rate close to 90%, when symptomatic tissue was tested. Although not taxon-specific, the primers LWD3 and LWD4, based on the 18s SSU rDNA of RL and closely related species, have been used in quantitative PCR (qPCR) assays for detection of RL (Dreaden et al. 2014). Species-specific primers based on the 28S rRNA have also been designed for use in TaqMan qPCR detection assays

from host tissue (Jeyaprakash et al. 2014). Another approach based on a LAMP assay, using primers targeting the beta-tubulin gene of RL, was developed by Hamilton et al. (2020). This assay
detected the pathogen in artificially inoculated redbay trees as early as ten days after inoculation
when high-quality DNA was tested. The authors were also able to detect the pathogen 12 days after
inoculation by using high-quality or crude extract DNA, as well as utilizing beetle crude extracts
(Hamilton et al. 2020). The same assay has been reported to detect RL in symptomatic samples
from naturally infected redbay and sassafras trees using benchtop equipment in the lab, and using a
portable device in the field (Hamilton et al. 2021). So far, there is not a method sensitive enough to
detect the pathogen at the early stages of infection, when the **titer** (concentration in solution) of the
pathogen is low and the tissue remains asymptomatic (no vascular damage). Therefore, there is still a
need to develop technologies able to detect molecules secreted by the pathogen or by the host in the
early stages of pathogen colonization that can be used under field conditions.

## 14.5 THOUSAND CANKERS DISEASE

### 14.5.1 DISEASE BIOLOGY AND IMPACT

Thousand Cankers Disease (TCD) outbreaks are the product of complex interactions among (1) the
fungal pathogen *Geosmithia morbida* M. Kolarik, E. Freeland, C. Utley, & N. Tisserat (Hypocreales:
Bionectriaceae), (2) the primary insect vector, *Pityophthorus juglandis* Blackman (Coleoptera:
Curculionidae), also known as the walnut twig beetle (WTB) (Kolařík et al. 2011; Rugman-Jones
et al. 2015; Seybold et al. 2019; Tisserat et al. 2009), and (3) the tree hosts, *Juglans* (walnuts) and
*Pterocarya* (wingnuts) spp. (Fagales: Juglandaceae) (Hishinuma et al. 2016; Serdani et al. 2013;
Tisserat et al. 2009). Disease signs and symptoms include wilting and **chlorosis** (loss of chlorophyll)
of the upper canopy, branch dieback, appearance of epicormic shoots, numerous entrance and exit
holes, gallery formation by WTB, and ultimate tree mortality within three to four years after initial
symptoms are observed (Daniels et al. 2016; Kolařík et al. 2011; Tisserat et al. 2009). Initial walnut
decline was observed in the mid-1990s in the Western United States. (Willamette Valley, Oregon),
and records associated with WTB-induced mortality of Eastern black walnut were first reported in
New Mexico, USA in 2001 (Daniels et al. 2016; Tisserat et al. 2009). Seven years later, TCD was
reported in Colorado, USA (Tisserat et al. 2009; Tisserat et al. 2011), and by 2010, it was confirmed
in the native range of Eastern black walnut in Tennessee, USA (Grant et al. 2011).

It was not until 2011 that the fungal symbiont, *G. morbida*, was identified as the causal agent of
TCD (Kolařík et al. 2011). Since its initial discovery, TCD has spread across 15 different states in
the United States (Daniels et al. 2016; Hadziabdic et al. 2014; Juzwik et al. 2016; Rugman-Jones
et al. 2015; Tisserat et al. 2009; Zerillo et al. 2014) and reached Italy in 2013 (Montecchio and
Faccoli 2014; Montecchio et al. 2014, 2016; Moricca et al. 2019). The common name of the disease,
"thousand cankers", comes from the fact that the pathogen leads to the formation of numerous dark
brown to black cankers underneath the bark that often coalesce and **girdle** twigs and branches.
Fungal colonization eventually destroys the vascular tissues of the branches and main stem, resulting in nutrient depletion. For the past two decades, this devastating disease has spread on multiple
occasions and from different sources across the United States and Italy, which include the native
ranges of Eastern black (*J. nigra*) and English (*J. regia*) walnuts, respectively (Hadziabdic et al.
2014; Montecchio and Faccoli 2014; Montecchio et al. 2014; Zerillo et al. 2014).

*Geosmithia morbida* is the first species in the *Geosmithia* genus to be considered pathogenic
(Huang et al. 2019; Kolařík et al. 2011). Most *Geosmithia* spp. are bark beetle symbionts and
are not considered important forest pests. *Geosmithia morbida* fungal colonies can be grown on
non-selective media (e.g., PDA or MEA). Sporulating isolates produce cylindrical to ellipsoid conidia
that are cream to tan in color and can form chains. Recent work has indicated that *G. morbida* is
more likely native to the United States as the pathogen populations are complex, present high genetic
diversity, and potentially have coevolved with native walnut species (Daniels et al. 2016; Sitz et al.

2021; Zerillo et al. 2014). Although no evidence of sexual reproduction exists, Zerillo et al. (2014) provided strong support suggesting that *G. morbida* has evolved in close association with WTB and its spread has been widely documented to coincide with the range and spread of WTB. The high genetic diversity found in both the pathogen and the vector of TCD further confirmed the original hypothesis of multiple introductions from multiple sources as a result of anthropogenic movement (Hadziabdic et al. 2014; Kolařík et al. 2011; Rugman-Jones et al. 2015; Zerillo et al. 2014).

The WTB is a small (~2 mm), phloem-feeding insect, yellowish to reddish-brown, that is native to the Southwestern United States and Northern Mexico (Blackman 1928; Bright 1981; Kolařík et al. 2011; LaBonte and Rabaglia 2012; Seybold et al. 2013; Wood and Bright 1992). This vector of TCD has three generations per year in Tennessee, Eastern United States, whereas two overlapping generations were identified in Colorado, Western United States WTB overwinters as both larvae and adults within the walnut galleries. In the late spring, adults either resume their reproductive activity by flying to new branches, mating, and developing new tunnels for egg galleries or remain in the existing, infested branches to expand overwintering tunnels (Nix 2013). Similarly to its fungal companions, WTB populations have high levels of genetic diversity, presence of **genetic structure** (distribution of genetic variation within and among the populations), and evidence of two genetic lineages in the United States, suggesting the presence of two morphologically indistinguishable beetle species (Oren 2016; Rugman-Jones et al. 2015; Seybold et al. 2019).

TCD poses a major threat to urban, natural, and forested landscapes and their ecological sustainability, with damage estimates over $1.2 billion for wood product businesses and up to $3.4 billion for community trees by 2030 in the United States (Feeley 2010; Randolph et al. 2013; Treiman and Tuttle 2009; USDA-FS-PPQ 2021). Furthermore, the incidence and severity of TCD are higher in the host's introduced (Western United States) range of disease distribution (Griffin 2015). Geographical differences in TCD severity may be partially driven by differences in host and genotype susceptibility (Sitz et al. 2017, 2021) and the current level of genetic diversity of the host, the pathogen (Hadziabdic et al. 2014; Sitz et al. 2021; Zerillo et al. 2014), and the vector(s) (Oren 2016; Rugman-Jones et al. 2015). Differences in disease severity may also be partially driven by geographical variation in the walnut **phytobiome** (plant host, environment, and their associated microbial communities) as a result of various levels of disease pressures (Gazis et al. 2018; Onufrak et al. 2020). Walnut soil and **caulosphere** (branch and stem tissues) microbial communities differ between regions where the host is native vs. introduced, harboring different mutualistic and pathogenic microorganisms (Onufrak et al. 2020). TCD-affected trees from introduced regions had an abundance of *Ophiostoma* spp., cosmopolitan fungi associated with both conifers and hardwood trees, whereas TCD-compromised trees in the native region were dominated by *Trichoderma* spp. This included several *Trichoderma* spp. that are antagonistic to *G. morbida* (Gazis et al. 2018) and are currently being explored as potential biological control agents (see Section 14.6 below*)*. These findings suggest that the absence of mutualists in the walnut phytobiome in its introduced range and/or the presence of microbial antagonists of *G. morbida* in the native range may affect disease severity.

## 14.5.2   DETECTION AND IDENTIFICATION OF THE PATHOGEN AND VECTOR

Early detection of TCD using symptomology is not reliable due to the cryptic nature of TCD cankers that can only be revealed after bark removal. In addition, TCD external symptoms such as wilting and yellowing resemble those caused by drought-induced stressors (Randolph et al. 2013). WTB can be identified using a morphological key and a descriptive guide (Bright 1981; Seybold et al. 2013), as well as species-specific microsatellite DNA markers (Hadziabdic et al. 2015; Oren et al. 2018; Stackhouse et al. 2021). The pathogen is slow-growing under laboratory conditions and is quickly outcompeted by faster-growing fungi, making traditional identification very difficult. There is no selective medium for *G. morbida* and morphological characters are not diagnostic, making traditional confirmation challenging, time-consuming, and often unreliable (Oren et al. 2018). Challenges in pathogen identification create obstacles for timely and efficient regulatory responses. To mitigate this issue, a number of research groups have developed rapid, accurate, and cost-effective molecular detection tools for TCD diagnosis (Lamarche et al. 2015; Oren et al. 2018; Rizzo et al. 2020, 2021; Stackhouse et al. 2021).

Moore et al. (2019) developed primers for *G. morbida* detection based on the beta-tubulin region with the visualization of results in conventional electrophoresis gel. However, the group reported only 86% accuracy and the need to use Sanger sequencing in some cases. Using species-specific microsatellite markers for both the pathogen and the vector of TCD, Oren et al. (2018) designed a molecular detection protocol to identify the presence of TCD complex members directly from infected host tissues. This protocol reduced the time for disease diagnosis from five weeks to eight to ten hours by eliminating the lengthy and laborious isolation and culturing steps. This protocol has a high degree of sensitivity and specificity (Oren et al. 2018). However, the protocol requires specialized equipment and highly trained personnel. An improved, simplified, and cost-effective protocol by Stackhouse et al. (2021) utilizes fluorescent molecular probes (TaqMan technology) designed to elicit a specific light spectrum, simplifying the visualization process by using an inexpensive flashlight. Recently, Rizzo et al. (2021) developed a diagnostic protocol based on LAMP using the 28S ribosomal RNA gene (Rizzo et al. 2021). Two additional detection approaches based on the gold standard in molecular diagnoses, qPCR assays, have been developed for TCD confirmation, based on the beta-tubulin genes for the identification of the pathogen and COI genes for the vector (Lamarche et al. 2015; Rizzo et al. 2020). The protocols utilize environmental samples from both symptomatic and asymptomatic host phloem tissue, insect frass, insect tissues, and pathogen mycelia (Lamarche et al. 2015; Rizzo et al. 2020). To further reduce the resource cost, the protocol can be applied to both insect and fungal samples in a single reaction (multiplex processing).

## 14.6 MANAGEMENT OF FUNGAL DISEASES IN FORESTED AREAS

Although it is widely recognized that prevention and early detection are key strategies to prevent outbreaks of forest pests and diseases, they are often not applied and alternative approaches are needed. The management of forest fungal diseases requires a multi-pronged approach that uses combinations of cultural practices, chemical fungicides, and biological control organisms. These science-based practices are often distributed by universities and extension agents to increase community awareness of forest diseases and can be used by land managers and growers to limit the ecological and economic impacts associated with forest fungal diseases. While not always entirely effective at preventing diseases from occurring, these management strategies can significantly reduce disease incidence and severity. Below we describe cultural practices, chemical management, and biological control strategies employed to manage some of the most devastating and threatening forest fungal diseases including Chestnut Blight (CB), Dutch Elm Disease (DED), Butternut Canker Disease (BCD), Oak Wilt (OW), Laurel Wilt, and Thousand Cankers Disease.

### 14.6.1 CULTURAL PRACTICES

Identifying cultural practices that can be used by land managers and growers to reduce the susceptibility of their managed tree stands is often one of the first steps to limit the economic and ecological impact of diseases. These cultural practices can include quarantine and sanitation, resource management such as water and nutrient availability, and plant breeding. Below we will review these cultural practices in more detail and provide examples of how they are employed to manage the impacts of forest diseases.

#### 14.6.1.1 Quarantine and Sanitation

In an effort to slow the anthropogenic mediated spread of emerging forest diseases, federal and state agencies have historically enacted quarantine procedures that restrict the movement of host plant materials across state and county lines. These quarantine procedures aim to limit disease spread into regions where the disease has yet to be detected and to reduce disease-associated damages. For example, to limit the interstate spread of TCD in *Juglans* spp., the state of Minnesota, USA which was TCD free as of 2021, restricted the import of live trees and plant materials from TCD positive U.S. states such as Arizona, California, and Colorado (Minnesota Department of Agriculture 2015). Furthermore, Tennessee, U.S., which has only detected TCD in the southeastern portion of the state as of 2021, has enacted county-level quarantines, in addition to state-level quarantines, to limit the spread of TCD within the state (Tennessee Department of Agriculture 2014). As of 2021, a total of ten counties were under quarantine in Tennessee, with restrictions placed on the transport of live trees and harvested wood products such as logs, untreated lumber, firewood, and hardwood mulch between TCD positive counties and TCD negative counties (Tennessee Department of Agriculture 2014). Similarly, to control the spread of Rapid 'Ōhi'a Death (ROD) within the state of Hawaii, U.S., the Hawaii Department of Agriculture restricted the movement of plant materials and soils between islands where ROD has been detected and ROD-free islands (State of Hawaii Department of Agriculture 2016).

While quarantine procedures limit the spread of diseases into previously disease-free regions, the use of sanitation practices by federal and state land managers and private growers can limit disease spread within individual tree stands where disease has been detected. Sanitation recommendations are often disease-specific and take routes of pathogen transmission into consideration to limit vector-mediated transmission and tree-to-tree spread. For instance, recommended sanitation practices for trees infected with LW include the removal of the entire tree, including the tree's root system, and chipping the uprooted tree (Crane et al. 2020). Uprooting the tree prevents below-ground pathogen transmission via root grafting and chipping the tree interferes with the lifecycle of ambrosia beetle vectors, limiting vector-mediated spread (Crane et al. 2020). Similar sanitation recommendations have been used for other insect- and root graft-transmitted diseases, such as OW (Koch et al. 2010). For both OW and LW, it is recommended that both symptomatic and asymptomatic trees within a given infection center be removed to limit vector-mediated spread and that root graft disruption be carried out using vibratory plows, backhoes, and trench inserts to prevent belowground transmission (Koch et al. 2010). It should be noted that the removal of trees can be economically costly depending on the size and location of the tree. For instance, the removal of a single TCD-affected walnut tree in Boulder, Colorado, U.S. ranged from $150 to $1,200, illustrating the importance of quarantine and early disease detection for limiting disease spread and reducing economic costs associated with tree removal and disease containment (Seybold et al. 2019).

Following the removal of trees, plant material must be further sanitized to prevent disease spread. For instance, following the removal of trees infected with ISHB-FD, it is recommended that trees be chipped and the chips be solarized, composted, or burned to kill beetle vectors (Chen et al. 2020; Jones and Paine 2015; Lynch 2019). Timber that will be used for lumber and other wood-based products can be treated with chemicals or heat to kill fungal pathogens and their associated vectors (Mayfield III et al. 2014; Schmidt et al. 1997; Seabright et al. 2019). These treatments allow for the continued export of harvested wood products while reducing the risk of disease spread into previously uninfected regions. For example, to kill the fungal pathogen of TCD, *G. morbida*, and its insect vector, *P. juglandis*, logs of Eastern black walnut (*J. nigra*) can be fumigated with methyl bromide or steam heated until sapwood temperatures reach 56°C for 30–40 minutes (Juzwik et al. 2021; Seabright et al. 2019). Similar chemical and heat treatment protocols have been developed for other fungal forest diseases such as OW and ISHB-FD (Juzwik et al. 2019; Lynch 2019; Schmidt et al. 1997).

### 14.6.1.2 Resource Management

The effects of forest fungal pathogens tend to be more pronounced in trees experiencing abiotic stressors such as drought (Griffin 2015; Oliva et al. 2014). Abiotic stressors can hinder host tree defenses, increasing their susceptibility to fungal pathogens (Anderegg et al. 2015; Oliva et al. 2014). Additionally, stressed trees produce higher concentrations of volatiles such as ethanol, which can increase host attractiveness to insect vectors and encourage the growth of the fungal mutualists of insect vectors (Cavaletto et al. 2021; Kelsey et al. 2014; Lehenberger et al. 2021). Regulating resource availability is key in disease prevention. Resource management recommendations for diseases such as BCD, DED, TCD, and LW include supplemental irrigation and fertilization practices to prevent drought stress or nutrient limitations, respectively (Brazee 2020; Crane et al. 2020; Leisso and Hudelson 2008; Lynch 2019; Teviotdale 2017). In addition to resources such as water and nutrients, the management of light availability is also recommended due to its relationship with the flight activity of insect vectors (Crane et al. 2020). For instance, to reduce flight activity of ambrosia beetle vectors of LW, it is recommended to regularly prune branches in commercial avocado orchards to increase light penetration throughout the canopy (Crane et al. 2020).

### 14.6.1.3  Plant Breeding

In addition to resource management and sanitation, land managers and growers are encouraged to use disease-resistant or tolerant cultivars produced by tree breeding programs. The primary goal of many forest tree breeding programs is to produce cultivars with tolerance or resistance to fungal pathogens for the restoration of natural and urban forests (Brennan et al. 2020; Clark et al. 2014; Hughes et al. 2015; Knight et al. 2017). These breeding programs address diseases of coniferous (King et al. 2010) and hardwood tree species (Brennan et al. 2020; Clark et al. 2014; Hughes et al. 2015; Knight et al. 2017). Breeding programs propagate genetically resistant individuals within a species (Sniezko et al. 2014; Townsend and Douglass, 2001; Townsend et al. 2005) or hybridize a susceptible species with a resistant **congeneric** (closely related) species (Brennan et al. 2020; Clark et al. 2019). Plant breeding programs for emerging diseases such as TCD (Sitz et al. 2021) and LW (Hughes and Smith 2014) are currently being developed through the identification, propagation, and evaluation of disease resistant individuals.

LW tolerance and/or resistance has not been formally observed in commercial cultivars; however, a few "escape trees" in commercial orchards that exhibit tolerance to natural infections have been documented (Navia-Urrutia and Gazis 2021). These surviving escape trees were confirmed to be infected with the pathogen but only developed mild LW symptoms, and were able to recover and remained productive. Fruit yield and quality has not been assessed in these surviving trees, and more research is needed to determine if the infection remains latent and can induce symptoms again. Based on field observations and small-scale greenhouse experiments it is hypothesized that a genetic component could be involved, such as the ability of the plant to quickly compartmentalize the pathogen (Navia-Urrutia and Gazis 2021). Previous studies indicated that LW progresses more slowly in avocado cultivars of Mexican genetic background compared with West Indian derived cultivars, due to the smaller xylem vessel diameter in the Mexican varieties (Beier et al. 2020; Castillo-Argaez et al. 2021). In forest species such as redbay and sassafras, tolerance to LW has been observed (Cameron et al. 2015; Hughes et al. 2015) but further research is necessary to better understand the mechanism behind these observations. In greenhouse experiments, tolerance has only been observed in certain avocado cultivars of small diameter size (potted trees) which can produce new sapwood even when infected with the pathogen. This "juvenile tolerance" has been observed in LW-infected avocado stumps which re-sprout. These sprouts remain externally asymptomatic for months or sometimes years, but ultimately succumb to the disease (Crane et al. 2015; Navia-Urrutia and Gazis 2021). Lauraceae native to Asia may have genetically based tolerance to RL (Shih et al. 2018), and such hosts could be used to understand host tolerance mechanisms, identify genes for resistance, or breed LW-resistant hybrids. Although some research efforts have been made to preserve breeding material for locally adapted trees at both botanical gardens and public lands, it can take years to identify tolerant or resistant genotypes that could be used for future restoration efforts of native species in the United States. (Smith et al. 2020).

Some of the most promising forest tree breeding programs have developed cultivars of American elm (*Ulmus americana*) (Griffin et al. 2017), American butternut (*Juglans cinerea*) (Brennan et al. 2020), and American chestnut (*Castanea dentata*) (Clark et al. 2019) with tolerance to DED,

BCD, and CB, respectively. While not entirely resistant to the disease [i.e., symptoms still develop (Roy and Kirchner 2000)], these disease-tolerant trees develop less severe symptoms compared to susceptible trees. Below we review in more detail these three breeding programs, each with a slightly different approach to developing disease tolerant trees.

### 14.6.1.4 DED-Resistant American Elm

Breeding for the DED-resistant American elm has largely been dependent upon the clonal propagation of disease-tolerant individuals found in natural settings (Griffin et al. 2017; Townsend et al. 2005; Townsend and Douglass 2001). One of the most successful American elm cultivars in the ten-year National Elm Trial was the commercially available "New Harmony" cultivar, which is a hybrid of the Siberian elm (*U. pumila*) and the Japanese elm (*U. japonica*) (Smalley and Guries 1993; Griffin et al. 2017). More recent breeding efforts for the American elm have crossed cultivars such as "New Harmony" with mature American elms from the midwestern and northeastern United States. that have escaped DED to develop locally adapted cultivars (Pinchot et al. 2017; Slavicek and Knight 2012). This hybridization approach has also been employed in the butternut tree breeding program, which aims to develop BCD-resistant trees.

### 14.6.1.5 BCD-Resistant Butternut

Current breeding programs for BCD-resistant butternut trees use a hybridization approach, crossing the American butternut with the Japanese walnut (*J. ailantifolia*) (Boraks and Broders 2014; Brennan et al. 2020). Hybridization was adopted because butternut grafted clones from disease-free individuals were found to be susceptible to BCD (Ostry and Moore 2008), but naturally occurring, infected butternut hybrids (*J. cinerea×J. ailantifolia*) developed fewer cankers than naturally infected pure butternuts (Boraks and Broders 2014). Brennan et al. (2020) conducted a field trial evaluating disease severity in artificially inoculated and naturally infected butternut and butternut hybrid trees. In this study, artificially induced cankers were smaller and naturally induced cankers were less frequent in hybrid butternut compared to pure butternut trees, illustrating the potential of the butternut hybrids for use in restoration programs (Brennan et al. 2020). However, it should be noted that while butternut hybrids display tolerance to BCD, they may not be able to fill the ecological niche of pure American butternut because these two lineages differ in drought, flooding, and temperature tolerances (Brennan et al. 2021; Crystal and Jacobs 2014). To circumvent this challenge, the American chestnut breeding program has incorporated a series of backcrosses after hybridization (mating hybrids with *J. cinerea*) to capture the phenotype of the original species (Hebard 2012).

### 14.6.1.6 CB-Resistant American Chestnut

The approach employed by the American chestnut breeding program involves first hybridizing American chestnut with the Chinese chestnut (*C. mollisma*) and then performing a series of backcrosses with American chestnut (Hebard 2012). This allows for the incorporation of disease resistance genes from the Chinese chestnut while preserving the desirable growth habit and phenology of the American chestnut (Hebard 2012). In artificially inoculated field trials, these hybrid backcrosses develop significantly smaller cankers than the pure American chestnut (Steiner et al. 2017). However, forest trials assessing resistance to natural CB development indicated that tolerance to disease was site-dependent (Clark et al. 2019). For instance, backcrossed hybrid and Chinese chestnut trees from one of the three study sites displayed similar levels of CB susceptibility as American chestnut trees (Clark et al. 2019). This highlights the importance of evaluating new genotypes across a variety of environments in breeding programs.

### 14.6.2 CHEMICAL MANAGEMENT

The use of chemical fungicides can be preventative or therapeutic. **Preventative** fungicide applications are used to protect trees from infection or to prevent disease development should an infection take place (Haugen and Stennes 1999; Koch et al. 2010). **Therapeutic** fungicide applications are used on symptomatic or infected trees to limit disease severity or prevent mortality (Haugen and Stennes 1999; Koch et al. 2010). Fungicide treatments are applied in a variety of ways including, but not limited to, foliar sprays (Hagan and Arkidge 2013), basal bark sprays (Ploetz et al. 2011), and soil drenches (Arjona-López et al. 2020; Ploetz et al. 2011; Thomidis and Exadaktylou 2012). Furthermore, fungicides can be directly injected into the host vasculature as microinjections or macroinfusions. **Microinjections** are injections of small volumes of undiluted fungicide at multiple locations in the trunk or branches (Haugen and Stennes, 1999; Ploetz et al. 2017). In a **macroinfusion**, dilute concentrations of fungicide are injected into the tree in large volumes, often at the root flare (Ploetz et al. 2017). Depending on the formulation and the nature of the active ingredient, macroinfusions can distribute the fungicide uniformly throughout the host tree, ensuring that the entire active vascular system is protected from the pathogen (Ploetz et al. 2011, 2017). The method of fungicide application will be dependent upon the disease of interest, the host, the mobility of the fungicide, and the costs associated with the treatment.

One fungicide commonly used to manage forest fungal diseases is the triazole fungicide, **propiconazole**. Propiconazole is a xylem-mobile sterol biosynthesis inhibitor with **fungistatic** (prevents fungal growth) properties that hinders fungal cell wall formation (Amiri and Schnabel 2012; Beckerman 2018; Grosman et al. 2019; Haugen and Stennes 1999; Koch et al. 2010; Mayfield III et al. 2008). It is largely translocated throughout the plant upwards (**acropetal** movement) from the site of injection, but some downwards (**basipetal**) movement has been documented (Amiri and Schnabel 2012; Blaedow et al. 2010). As many xylem-mobile chemical fungicides are only capable of upward movement throughout the plant, the multidirectional movement of propiconazole makes it particularly useful for treating root rot diseases, such as Armillaria root rot in peach (Amiri and Schnabel 2012; Amiri et al. 2008), or root transmitted diseases, such as OW in Live, White, and Red oaks (Appel and Kurdyla 1992; Blaedow et al. 2010; Eggers et al. 2005). The compound is also used to manage DED, though basipetal movement in DED hosts has yet to be documented (Haugen and Stennes 1999; Mayfield et al. 2008; Ploetz et al. 2011).

Perhaps the newest chemical management strategy for fungal forest diseases is the use of propiconazole for the management of LW. Both basal bark sprays and soil drenches of propiconazole reduced the severity of LW in pot-grown trees (Ploetz et al. 2011). However, the use of propiconazole to treat larger trees in orchard settings is still being researched. Field studies using mature trees have found that the xylem concentrations of propiconazole in basal bark applications were insignificant, indicating that basal bark spray application is likely ineffective in orchard settings (Ploetz et al. 2017). Other potential methods for propiconazole application to manage LW include macroinfusions and microinjections (Mayfield et al. 2008; Ploetz et al. 2011). Macroinfusions, while shown to reduce LW symptom development in mature redbay trees (Mayfield et al. 2008), are considered to be too costly for property owners and land managers to be a viable disease management strategy (Ploetz et al. 2011). Some evidence exists that microinjections are capable of delaying LW symptom onset, but additional research is needed to determine the efficacy of microinjection treatments for long-term management of LW (Crane et al. 2020; Ploetz et al. 2017).

### 14.6.3 Biological Control

Biological control (**biocontrol**) organisms are organisms such as fungi, bacteria, and viruses that can limit pathogen damage (Dawe and Nuss 2001; Dumas 1992; Gazis et al. 2018; Mousseaux et al. 1998). Some biocontrol organisms can directly act to antagonize fungal pathogens through mycoparasitism and the production of antifungal metabolites (antibiosis) (Gazis et al. 2018; Mesanza et al. 2016; Mohan et al. 2015). Additionally, biocontrol organisms can inhibit pathogen activity indirectly through stimulation of plant defense cascades or by competing with the pathogen for resources and physical space (Amira et al. 2017; Chen et al. 2019; Marx and Davey 1969). In the case of insect vectored forest fungal diseases, biocontrol organisms with entomopathogenic activity can be used to control insect vectors and limit pathogen transmission (Carrillo et al. 2015, 2017; Mayfield et al. 2019; Zhou et al. 2018). Organisms used for biocontrol can come from commercially available formulations or newly identified strains, such as endophytes isolated from susceptible but surviving host trees (Castrillo et al. 2017; Gazis et al. 2018). One system in which biocontrol organisms are being actively pursued as a management tool is TCD which we will explore in more detail below.

In an effort to explain the differential TCD severity between the Eastern black walnut's native and introduced ranges, endophytic fungi were isolated from *Juglans* spp. in the Eastern United States (native range) and the Western United States (introduced range) (Gazis et al. 2018). From the native range, several *Trichoderma* spp. were recovered that demonstrated antagonistic activity toward the fungal pathogen, *G. morbida*, in culture, suggesting their potential use as biocontrol agents for TCD (Gazis et al. 2018). Despite their antagonistic activity *in vitro*, additional *in planta* studies are needed to determine the best treatment application method and the ability of these isolates to control *G. morbida* in host trees and across different environments. Other potential biocontrol methods for the control of TCD include entomopathogenic fungi to limit vector-mediated spread. For the control of *P. juglandis*, commercially available strains of *Beauveria bassiana* GHA and *Metarhizium ansiopilae* F52 were evaluated *in vitro* and in large walnut bolts (branch sections) (Castrillo et al. 2017; Mayfield et al. 2019). Both entomopathogens caused 100% mortality in *P. juglandis* when the adult beetles were dipped directly into suspensions of entomopathogen conidia (Castrillo et al. 2017). It should be noted that a prior study found that *B. bassiana* application did not reduce *P. juglandis* emergence from bolts but did increase *B. bassiana* infection rate in emerged adults, potentially reducing future population sizes (Castrillo et al. 2017). These results highlight the potential utility of entomopathogens in limiting the vector-mediated spread of fungal pathogens; however, there are still obstacles that must be overcome before these treatments can be utilized by growers and land managers, including the optimization of treatment application methods. The use of entomopathogenic fungi has also been explored for the control of vectors of RL to manage the spread of LW (Carrillo et al. 2015; Zhou et al. 2018).

## REFERENCES

Amira, M. B., Lopez, D., Mohamed, A. T., et al. 2017. Beneficial effect of *Trichoderma harzianum* strain Ths97 in biocontrolling *Fusarium solani* causal agent of root rot disease in olive trees. *Biological Control*, 110, 70–78.

Amiri, A., Bussey, K. E., Riley, M. B., and G. Schnabel. 2008. Propiconazole inhibits *Armillaria tabescens* in vitro and translocates into peach roots following trunk infusion. *Plant Disease*, 92(9), 1293–1298.

Amiri, A., and G. Schnabel. 2012. Persistence of propiconazole in peach roots and efficacy of trunk infusions for Armillaria root rot control. *International Journal of Fruit Science*, 12(4), 437–449.

Anagnostakis, S. L. 1987. Chestnut blight: The classical problem of an introduced pathogen. *Mycologia*, 79(1), 23–37.

Anderegg, W. R., Hicke, J. A., Fisher, R. A., et al. 2015. Tree mortality from drought, insects, and their interactions in a changing climate. *New Phytologist*, 208(3), 674–683. https://doi.org/10.1111/nph.13477.

Appel, D., and T. Kurdyla. 1992. Intravascular injection with propiconazole in live oak for oak wilt control. *Plant Disease*, 76(11), 1120–1124.

Arjona-López, J., Capote, N., Melero-Vara, J. M., and C. López-Herrera. 2020. Control of avocado white root rot by chemical treatments with fluazinam in avocado orchards. *Crop Protection*, 131, 105100. https://doi.org/10.1016/j.cropro.2020.105100.

Atkinson, R. 2015. *Euwallacea fornicatus* (Eichhoff 1868). *Bark and Ambrosia Beetles of North and Central America*. Online via: http://www.barkbeetles.info/regional_chklist_target_species.php.

Beckerman, J. 2018. Fungicide mobility for nursery, greenhouse, and landscape professionals. *Disease Management Strategies for Horticultural Crops*, BP-70-W.

Beier, G. L., Lund, C. D., Held, B. W., Ploetz, R. C., Konkol, J., and R. Blanchette, 2020. Variation in xylem characteristics of botanical races of *Persea americana* and their potential influence on susceptibility to the pathogen *Raffaelea lauricola*. *Tropical Plant Pathology*, 46(2), 232–239. https://doi.org/10.1007/s40858-020-00397-y.

Blackman, M. 1928. *The Genus Pityophthorus eichhoff in North America: A Revisional Study of the Pityophthori, with Descriptions of Two New Genera and Seventy-One New Species.* New York State College of Forestry at Syracuse University.

Blaedow, R. A., Juzwik, J., and B. Barber. 2010. Propiconazole distribution and effects on *Ceratocystis fagacearum* survival in roots of treated red oaks. *Phytopathology,* 100(10), 979–985.

Boland, J. M., and D.L. Woodward. 2019. Impacts of the invasive shot hole borer (*Euwallacea kuroshio*) are linked to sewage pollution in southern California: The Enriched Tree Hypothesis. *PeerJ,* 7, e6812. https://doi.org/10.7717/peerj.6812.

Boraks, A., and K. Broders. 2014. Butternut (*Juglans cinerea*) health, hybridization, and recruitment in the northeastern United States. *Canadian Journal of Forest Research,* 44(10), 1244–1252.

Brazee, N. 2020. *Dutch Elm Disease.* UMass Amherst Center for Agriculture, Food, and the Environment. https://ag.umass.edu/landscape/fact-sheets/dutch-elm-disease.

Brennan, A. N., McKenna, J. R., Hoban, S. M., and D.F. Jacobs. 2020. Hybrid breeding for restoration of threatened forest trees: Evidence for incorporating disease tolerance in *Juglans cinerea. Frontiers in Plant Science,* 11, 1511. https://doi.org/10.3389/fpls.2020.580693.

Brennan, A. N., Uscola, M., Joly, R. J., and D.F. Jacobs. 2021. Cold and heat tolerances of hybrids for restoration of the endangered *Juglans cinerea* L. *Annals of Forest Science,* 78(2), 1–11.

Bright, D. E. 1981. Taxonomic monograph of the genus *Pityophthorus* Eichhoff in North and Central America (Coleoptera: Scolytidae). *Memoirs of the Entomological Society of Canada,* 113(S118), 1–378.

Cameron, R. S., Hanula, J., Fraedrich, S., and C. Bates. 2015. Progression and impact of laurel wilt disease within redbay and sassafras populations in southeast Georgia. *Southeastern Naturalist,* 14(4), 650–674.

Carrillo, D., Duncan, R., Ploetz, J., Campbell, A., Ploetz, R., and J. Peña. 2014. Lateral transfer of a phytopathogenic symbiont among native and exotic ambrosia beetles. *Plant Pathology,* 63(1), 54–62.

Carrillo, D., Dunlap, C., Avery, P., et al. 2015. Entomopathogenic fungi as biological control agents for the vector of the laurel wilt disease, the redbay ambrosia beetle, *Xyleborus glabratus* (Coleoptera: Curculionidae). *Biological Control,* 81, 44–50.

Carrillo, J. D., Mayorquin, J. S., Stajich, J. E., and A. Eskalen. 2020. Probe-based multiplex real-time PCR as a diagnostic tool to distinguish distinct fungal symbionts associated with *Euwallacea kuroshio and Euwallacea whitfordiodendrus* in California. *Plant Disease,* 104(1), 227–238.

Castillo-Argaez, R., Vazquez, A., Konkol, J. L., et al. 2021. Sap flow, xylem anatomy and photosynthetic variables of three Persea species in response to laurel wilt. *Tree Physiology,* 41(6), 1004–1018.

Castrillo, L. A., Mayfield III, A. E., Griggs, M. H., et al. 2017. Mortality and reduced brood production in walnut twig beetles, *Pityophthorus juglandis* (Coleoptera: Curculionidae), following exposure to commercial strains of entomopathogenic fungi *Beauveria bassiana* and *Metarhizium brunneum. Biological Control,* 114, 79–86.

Cavaletto, G., Faccoli, M., Ranger, C. M., and D. Rassati. 2021. Ambrosia beetle response to ethanol concentration and host tree species. *Journal of Applied Entomology,* 145(8): 800–809.

Chen, L., Bóka, B., Kedves, O., et al. 2019. Towards the biological control of devastating forest pathogens from the genus *Armillaria. Forests,* 10(11), 1013. https://doi.org/10.3390/f10111013.

Chen, Y., Coleman, T. W., Poloni, A. L., Nelson, L., and S.J. Seybold. 2020. Reproduction and control of the invasive polyphagous shot hole borer, *Euwallacea nr. fornicatus* (Coleoptera: Curculionidae: Scolytinae), in three species of hardwoods: Effective sanitation through felling and chipping. *Environmental Entomology,* 49(5), 1155–1163.

Clark, S. L., Schlarbaum, S. E., Pinchot, C. C., et al. 2014. Reintroduction of American chestnut in the national forest system. *Journal of Forestry,* 112(5), 502–512.

Clark, S. L., Schlarbaum, S. E., Saxton, A. M., and R. Baird. 2019. Eight-year blight (*Cryphonectria parasitica*) resistance of backcross-generation American chestnuts (*Castanea dentata*) planted in the southeastern United States. *Forest Ecology and Management,* 433, 153–161.

Coleman, T. W., Eskalen, A., and R. Stouthamer. 2013. New pest complex in California: The polyphagous shot hole borer, *Euwallacea* sp., and *Fusarium dieback, Fusarium euwallaceae. USDA Forest Service Pest Alert, R5-PR-032,* 5. https://www.fs.usda.gov/Internet/FSE_DOCUMENTS/stelprdb5441465.pdf.

Coleman, T. W., Poloni, A. L., Chen, Y., et al. 2019. Hardwood injury and mortality associated with two shot hole borers, *Euwallacea* spp., in the invaded region of southern California, USA, and the native region of Southeast Asia. *Annals of Forest Science,* 76(3), 1–18.

Costanza, R., De Groot, R., Sutton, P., et al. 2014. Changes in the global value of ecosystem services. *Global Environmental Change,* 26, 152–158.

Crane, J., Carrillo, D., Ploetz, R., Evans, E., Palmateer, A., and D. Pybas. 2015. Current status and control recommendations for laurel wilt and the ambrosia beetle vectors in commercial avocado (*Persea americana* L.) orchards in south Florida. *ACTAS Proceedings of the VIII Congreso Munidal de la Palta*, 240–244.

Crane, J. H., Carrillo, D., Evans, E. A., et al. 2020. Recommendations for control and mitigation of laurel wilt and ambrosia beetle vectors in commercial avocado groves in Florida: HS1360. *EDIS*, 2020(2), 1–8.

Crane, J. H., Gazis, R., Wasielewski, J., et al. 2020. Sampling guidelines and recommendations for submitting samples for diagnosing Laurel Wilt in avocado trees (*Persea americana* L.): HS1394. *EDIS*, 2020(6), 1–3.

Cruz, L. F., Menocal, O., Kendra, P. E., and D. Carrillo. 2021. Phoretic and internal transport of *Raffaelea lauricola* by different species of ambrosia beetle associated with avocado trees. *Symbiosis*, 84(2), 151–161.

Crystal, P. A., and D.F. Jacobs. 2014. Drought and flood stress tolerance of butternut (*Juglans cinerea*) and naturally occurring hybrids: Implications for restoration. *Canadian Journal of Forest Research*, 44(10), 1206–1216.

Daniels, D., Nix, K., Wadl, P., et al. 2016. Thousand cankers disease complex: A forest health issue that threatens *Juglans* species across the US. *Forests*, 7(11), 260.

Danthanarayana, W. 1968. The distribution and host range of the shot-hole borer (*Xyleborus fornicatus* Eichh.). *Tea Quarterly*, 39(3), 61–69.

Dawe, A. L., and D.L. Nuss. 2001. Hypoviruses and chestnut blight: Exploiting viruses to understand and modulate fungal pathogenesis. *Annual Review of Genetics*, 35(1), 1–29.

Donovan, G. H., Butry, D. T., Michael, Y. L., et al. 2013. The relationship between trees and human health: Evidence from the spread of the emerald ash borer. *American Journal of Preventive Medicine*, 44(2), 139–145.

Dreaden, T. J., Davis, J. M., Harmon, C. L., et al. 2014. Development of multilocus PCR assays for *Raffaelea lauricola*, causal agent of laurel wilt disease. *Plant Disease*, 98(3), 379–383.

Dumas, M. 1992. Inhibition of *Armillaria* by bacteria isolated from soils of the boreal mixedwood forest of Ontario. *European Journal of Forest Pathology*, 22(1), 11–18.

Eggers, J., Juzwik, J., Bernick, S., and L. Mordaunt. 2005. Evaluation of propiconazole operational treatments of oaks for oak wilt control. *Res. Note NC-390*. St. Paul, MN: US Department of Agriculture, Forest Service, North Central Research Station. 6. p. 390.

Eriksson, L., Boberg, J., Cech, T. L., et al. 2019. Invasive forest pathogens in Europe: Cross-country variation in public awareness but consistency in policy acceptability. *Ambio*, 48(1), 1–12.

Eskalen, A., Gonzalez, A., Wang, D., Twizeyimana, M., Mayorquin, J., and S. Lynch. 2012. First report of a *Fusarium* sp. and its vector tea shot hole borer (*Euwallacea fornicatus*) causing Fusarium dieback on avocado in California. *Plant Disease*, 96(7), 1070.

Eskalen, A., Stouthamer, R., Lynch, S. C., et al. 2013. Host range of Fusarium dieback and its ambrosia beetle (Coleoptera: Scolytinae) vector in southern California. *Plant Disease*, 97(7), 938–951.

Evans, E. A., and S. Nalampang. 2010. Sample avocado production costs and profitability analysis for Florida: FE837. *EDIS*, 2010(4), 1–6.

Feeley, T. 2010. *Emerging Threats to Iowa's Forests, Communities, Wood Industry, & Economy Thousand Cankers Disease*. Iowa Department of Natural Resources Forestry Bureau, p. 6.

Fraedrich, S., Harrington, T., Rabaglia, R., et al. 2008. A fungal symbiont of the redbay ambrosia beetle causes a lethal wilt in redbay and other Lauraceae in the southeastern United States. *Plant Disease*, 92(2), 215–224.

Freeman, S., Sharon, M., Maymon, M., et al. 2013. *Fusarium euwallaceae* sp. nov.—a symbiotic fungus of *Euwallacea* sp., an invasive ambrosia beetle in Israel and California. *Mycologia*, 105(6), 1595–1606.

García-Avila, C. D. J., Trujillo-Arriaga, F. J., López-Buenfil, J. A., et al. 2016. First report of *Euwallacea nr. fornicatus* (Coleoptera: Curculionidae) in Mexico. *Florida Entomologist*, 99(3), 555–556.

Gazis, R., Poplawski, L., Klingeman, W., et al. 2018. Mycobiota associated with insect galleries in walnut with thousand cankers disease reveals a potential natural enemy against *Geosmithia morbida*. *Fungal Biology*, 122(4), 241–253. https://doi.org/https://doi.org/10.1016/j.funbio.2018.01.005.

Geils, B. W., Hummer, K. E., and R.S. Hunt. 2010. White pines, Ribes, and blister rust: A review and synthesis. *Forest Pathology*, 40(3–4), 147–185.

Gibbs, J. N. 1978. Intercontinental epidemiology of Dutch elm disease. *Annual Review of Phytopathology*, 16(1), 287–307.

Gomez, D. F., Skelton, J., Steininger, M. S., et al. 2018. Species delineation within the *Euwallacea fornicatus* (Coleoptera: Curculionidae) complex revealed by morphometric and phylogenetic analyses. *Insect Systematics and Diversity*, 2(6), 2. https://doi.org/10.1093/isd/ixy018.

Grant, J. F., Windham, M. T., Haun, W. G., Wiggins, G. J., and P.L. Lambdin. 2011. Initial assessment of thousand cankers disease on black walnut, *Juglans nigra*, in eastern Tennessee. *Forests*, 2(3), 741–748.

Griffin, G. 2015. Status of thousand cankers disease on eastern black walnut in the eastern United States at two locations over 3 years. *Forest Pathology,* 45(3), 203–214.

Griffin, J. J., Jacobi, W. R., McPherson, E. G., et al. 2017. Ten-year performance of the United States national elm trial. *Arboriculture & Urban Forestry,* 43(3), 108–121.

Grosman, D. M., Eskalen, A., and C. Brownie. 2019. Evaluation of emamectin benzoate and propiconazole for management of a new invasive shot hole borer (*Euwallacea* nr. *fornicatus,* Coleoptera: Curculionidae) and symbiotic fungi in California sycamores. *Journal of Economic Entomology,* 112(3), 1267–1273.

Hadziabdic, D., Bonello, P., Hamelin, R., et al. 2021. The future of forest pathology in North America. *Frontiers in Forests and Global Change,* 4(737445). https://doi.org/10.3389/ffgc.2021.737445.

Hadziabdic, D., Vito, L. M., Windham, M. T., Pscheidt, J. W., Trigiano, R. N., and M. Kolarik. 2014. Genetic differentiation and spatial structure of *Geosmithia morbida,* the causal agent of thousand cankers disease in black walnut (*Juglans nigra*). *Current Genetics,* 60(2), 75–87. https://doi.org/10.1007/s00294-013-0414-x.

Hadziabdic, D., Wadl, P. A., Staton, M. E., et al. 2015. Development of microsatellite loci in *Pityophthorus juglandis,* a vector of thousand cankers disease in *Juglans* spp. *Conservation Genetics Resources,* 7(2), 431–433.

Hadziabdic, D., Windham, M., Baird, R., et al. 2014. First report of *Geosmithia morbida* in North Carolina: The pathogen involved in thousand cankers disease of black walnut. *Plant Disease,* 98(7), 992–992.

Hagan, A., and J. Arkidge. 2013. Instrata fungicide evaluated for control of Cercospora Leaf Spot on crapemyrtle. *Journal of Environmental Horticulture,* 31(1), 21–26.

Hamilton, J., Fraedrich, S., Nairn, C., Mayfield, A., and C. Villari. 2021. A field-portable diagnostic approach confirms Laurel Wilt Disease diagnosis in minutes instead of days. *Arboriculture & Urban Forestry,* 47(3), 98–109.

Hamilton, J. L., Workman, J. N., Nairn, C. J., Fraedrich, S. W., and C. Villari. 2020. Rapid detection of *Raffaelea lauricola* directly from host plant and beetle vector tissues using loop-mediated isothermal amplification. *Plant Disease,* 104(12), 3151–3158.

Hariharan, G. and K. Prasannath. 2021. Recent advances in molecular diagnostics of fungal plant pathogens: A mini review. *Frontiers in Cellular and Infection. Microbiology,* 10, 600234. https://doi.org/10.3389/fcimb.2020.600234.

Harrington, T., Fraedrich, S., and D. Aghayeva. 2008. *Raffaelea lauricola,* a new ambrosia beetle symbiont and pathogen on the Lauracea. *Mycotaxon,* 104, 399–404.

Haugen, L., and M. Stennes. 1999. Fungicide injection to control Dutch elm disease: Understanding the options. *Plant Diagnosticians Quarterly,* 20(2), 29–38.

Hebard, F. 2012. The American chestnut foundation breeding program. In *Proceedings of the Fourth International Workshop on the Genetics of Host-Parasite Interactions in Forestry: Disease and Insect Resistance in Forest Trees,* Tech. Coords: R.A. Sniezko; A.D. Yanchuk; J.T. Kliejunas; K.M. Palmieri; J.M. Alexander; S.J. Frankel. Gen. Tech. Rep. PSW-GTR-240. Albany, CA: Pacific Southwest Research Station, Forest Service, US Department of Agriculture. pp. 221–234.

Hishinuma, S. M., Dallara, P. L., Yaghmour, M. A., et al. 2016. Wingnut (Juglandaceae) as a new generic host for *Pityophthorus juglandis* (Coleoptera: Curculionidae) and the thousand cankers disease pathogen, *Geosmithia morbida* (Ascomycota: Hypocreales). *Canadian Entomologist,* 148(1), 83–91. https://doi.org/10.4039/tce.2015.37.

Huang, Y.-T., Skelton, J., Johnson, A. J., Kolařík, M., and J. Hulcr. 2019. *Geosmithia* species in southeastern USA and their affinity to beetle vectors and tree hosts. *Fungal Ecology,* 39, 168–183.

Hughes, M. A., Smith, J., Ploetz, R., et al. 2015. Recovery plan for laurel wilt on redbay and other forest species caused by *Raffaelea lauricola* and disseminated by *Xyleborus glabratus. Plant Health Progress,* 16(4), 173–210.

Hughes, M. A., and J.A. Smith. 2014. Vegetative propagation of putatively laurel wilt-resistant redbay (*Persea borbonia*). *Native Plants Journal,* 15(1), 42–50.

Hulcr, J., Mogia, M., Isua, B., and V. Novotny. 2007. Host specificity of ambrosia and bark beetles (Col., Curculionidae: Scolytinae and Platypodinae) in a New Guinea rainforest. *Ecological Entomology,* 32(-6), 762–772.

Inch, S., Ploetz, R., Held, B., and R. Blanchette. 2012. Histological and anatomical responses in avocado, *Persea americana,* induced by the vascular wilt pathogen, *Raffaelea lauricola. Botany,* 90(7), 627–635.

Jeyaprakash, A., Davison, D., and T. Schubert. 2014. Molecular detection of the laurel wilt fungus, *Raffaelea lauricola. Plant Disease,* 98(4), 559–564.

Jones, M. E., and T.D. Paine. 2015. Effect of chipping and solarization on emergence and boring activity of a recently introduced ambrosia beetle (*Euwallacea* sp., Coleoptera: Curculionidae: Scolytinae) in Southern California. *Journal of Economic Entomology,* 108(4), 1852–1859.

Juzwik, J., McDermott-Kubeczko, M., Stewart, T., and M. Ginzel. 2016. First report of *Geosmithia morbida* on ambrosia beetles emerged from thousand cankers-diseased *Juglans nigra* in Ohio. *Plant Disease,* 100(6), 1238. https://doi.org/10.1094/PDIS-10-15-1155-PDN.

Juzwik, J., Yang, A., Chen, Z., White, M. S., Shugrue, S., and R. Mack. 2019. Vacuum steam treatment eradicates viable *Bretziella fagacearum* from logs cut from wilted *Quercus rubra*. *Plant Disease,* 103(2), 276–283.

Juzwik, J., Yang, A., Heller, S., et al. 2021. Vacuum steam treatment effectiveness for eradication of the Thousand cankers disease vector and pathogen in logs from diseased walnut trees. *Journal of Economic Entomology*, 114(1), 100–111.

Kasson, M. T., O'Donnell, K., Rooney, A. P., et al. 2013. An inordinate fondness for *Fusarium*: Phylogenetic diversity of fusaria cultivated by ambrosia beetles in the genus *Euwallacea* on avocado and other plant hosts. *Fungal Genetics and Biology,* 56, 147–157.

Kelsey, R. G., Gallego, D., Sánchez-García, F., and J. Pajares. 2014. Ethanol accumulation during severe drought may signal tree vulnerability to detection and attack by bark beetles. *Canadian Journal of Forest Research,* 44(6), 554–561.

Kendra, P. E., Montgomery, W. S., Niogret, J., and N.D. Epsky. 2013. An uncertain future for American Lauraceae: A lethal threat from redbay ambrosia beetle and laurel wilt disease (a review). *American Journal of Plant Sciences,* 4(3), 727–738.

King, J., David, A., Noshad, D., and J. Smith. 2010. A review of genetic approaches to the management of blister rust in white pines. *Forest Pathology,* 40(3–4), 292–313.

Kinloch Jr, B. B. 2003. White pine blister rust in North America: Past and prognosis. *Phytopathology,* 93(8), 1044–1047.

Kirkendall, L. R., and F. Ødegaard. 2007. Ongoing invasions of old-growth tropical forests: Establishment of three incestuous beetle species in southern Central America (Curculionidae: Scolytinae). *Zootaxa,* 1588(1), 53–62.

Knight, K. S., Haugen, L. M., Pinchot, C. C., Schaberg, P. G., and J.M. Slavicek. 2017. American elm (*Ulmus americana*) in restoration plantings: a review. In *Proceedings of the American Elm Restoration Workshop 2016,* eds. C.C. Pinchot; K.S. Knight, L.M. Haugen; C.E. Flower; J.M. Slavicek; 2016 October 25–27; Lewis Center, OH. Gen. Tech. Rep. NRS-P-174. Newtown Square, PA: US Department of Agriculture, Forest Service, Northern Research Station: 133–140.

Koch, K. A., Quiram, G. L., and R.C. Venette. 2010. A review of oak wilt management: A summary of treatment options and their efficacy. *Urban Forestry & Urban Greening,* 9(1), 1–8.

Kolařík, M., Freeland, E., Utley, C., and N. Tisserat. 2011. *Geosmithia morbida* sp. nov., a new phytopathogenic species living in symbiosis with the walnut twig beetle (*Pityophthorus juglandis*) on Juglans in USA. *Mycologia,* 103(2), 325–332. https://doi.org/10.3852/10-124.

LaBonte, J. R., and R. Rabaglia. 2012. *A Screening Aid for the Identification of the Walnut Twig Beetle, Pityophthorus juglandis Blackman.* Available online: http://www.nyis.info/wp-content/uploads/files/labonte%20and%20Rabaglia%202010%20WTB%20Key.pdf.

Lamarche, J., Potvin, A., Pelletier, G., et al. 2015. Molecular detection of 10 of the most unwanted alien forest pathogens in Canada using real-time PCR. *PLoS One,* 10(8), e0134265. https://doi.org/10.1371/journal.pone.0134265.

Lehenberger, M., Benkert, M., and P.H. Biedermann. 2021. Ethanol-enriched substrate facilitates ambrosia beetle fungi, but inhibits their pathogens and fungal symbionts of bark beetles. *Frontiers in Microbiology,* 11, 3487. https://doi.org/10.3389/fmicb.2020.590111.

Leisso, R., and B. Hudelson. 2008. Butternut canker. *University of Wisconsin Garden Facts, XHT1142.*

Lynch, S. 2019. A statewide strategic initiative to control Fusarium dieback – invasive shot hole borers in California. *Invasive Species Council of California.* http://www.iscc.ca.gov/docs/ISHB_Final_Report0909019.pdf.

Lynch, S., Wang, D., Mayorquin, J., Rugman-Jones, P., Stouthamer, R., and A. Eskalen. 2014. First report of *Geosmithia pallida* causing foamy bark canker, a new disease on coast live oak (*Quercus agrifolia*), in association with *Pseudopityophthorus pubipennis* in California. *Plant Disease,* 98(9), 1276–1276.

Marx, D. H., and C.B. Davey. 1969. Influence of ectotrophic mycorrhizal fungi on the resistance of pine roots to pathogenic infections. III. Resistance of aseptically formed mycorrhizae to infection by *Phytophthora cinnamomi*. *Phytopathology,* 59(4), 411–417.

Mayfield, A. E., Juzwik, J., Scholer, J., Vandenberg, J. D., and A. Taylor. 2019. Effect of bark application with *Beauveria bassiana* and permethrin insecticide on the walnut twig beetle (Coleoptera: Curculionidae) in black walnut bolts. *Journal of Economic Entomology,* 112(5), 2493–2496.

Mayfield III, A. E., Fraedrich, S., Taylor, A., Merten, P., and S. Myers. 2014. Efficacy of heat treatment for the thousand cankers disease vector and pathogen in small black walnut logs. *Journal of Economic Entomology,* 107(1), 174–184.

Mayfield III, A. E., Barnard, E. L., Smith, J. A., Bernick, S. C., Eickwort, J. M., and T. J. Dreaden. 2008. Effect of propiconazole on laurel wilt disease development in redbay trees and on the pathogen in vitro. *Arboriculture & Urban Forestry*, 34, 317–324.

Mendel, Z., Lynch, S. C., Eskalen, A., Protasov, A., Maymon, M., and S. Freeman. 2021. What determines host range and reproductive performance of an invasive ambrosia beetle *Euwallacea fornicatus*; lessons from Israel and California. *Frontiers in Forests and Global Change*, 4, 29. https://doi.org/10.3389/ffgc.2021.654702.

Mendel, Z., Protasov, A., Sharon, M., et al. 2012. An Asian ambrosia beetle *Euwallacea fornicatus* and its novel symbiotic fungus Fusarium sp. pose a serious threat to the Israeli avocado industry. *Phytoparasitica*, 40(3), 235–238.

Mesanza, N., Iturritxa, E., and C.L. Patten. 2016. Native rhizobacteria as biocontrol agents of *Heterobasidion annosum* ss and *Armillaria mellea* infection of *Pinus radiata*. *Biological Control*, 101, 8–16.

Minnesota Department of Agriculture. 2015. *Minnesota Department of Agriculture State Exterior Quarantine Thousand Cankers Disease of Walnut (Juglans SP) (Version 1)*. https://agriculture.mo.gov/plants/pests/MinnesotaTCDQuarantine.pdf.

Mohan, V., Nivea, R., and S. Menon. 2015. Evaluation of ectomycorrhizal fungi as potential bio-control agents against selected plant pathogenic fungi. *JAIR*, 3(9), 408–412.

Montecchio, L., and M. Faccoli. 2014. First record of thousand cankers disease *Geosmithia morbida* and walnut twig beetle *Pityophthorus juglandis* on *Juglans nigra* in Europe. *Plant Disease*, 98(5), 696–696.

Montecchio, L., Fanchin, G., Simonato, M., and M. Faccoli. 2014. First record of thousand cankers disease fungal pathogen *Geosmithia morbida* and walnut twig beetle *Pityophthorus juglandis* on *Juglans regia* in Europe. *Plant Disease*, 98(10), 1445–1445.

Montecchio, L., Vettorazzo, M., and M. Faccoli. 2016. Thousand cankers disease in Europe: An overview. *EPPO Bulletin*, 46(2), 335–340.

Moricca, S., Bracalini, M., Benigno, A., Ginetti, B., Pelleri, F., and T. Panzavolta. 2019. Thousand cankers disease caused by *Geosmithia morbida* and its insect vector *Pityophthorus juglandis* first reported on *Juglans nigra* in Tuscany, Central Italy. *Plant Disease*, 103(2), 369. https://doi.org/https://doi.org/10.1094/PDIS-07-18-1256-PDN.

Mousseaux, M. R., Dumroese, R. K., James, R. L., Wenny, D. L., and G.R. Knudsen. 1998. Efficacy of *Trichoderma harzianum* as a biological control of *Fusarium oxysporum* in container-grown Douglas-fir seedlings. *New Forests*, 15(1), 11–21.

Na, F., Carrillo, J. D., Mayorquin, J. S., et al. 2018. Two novel fungal symbionts *Fusarium kuroshium* sp. nov. and *Graphium kuroshium* sp. nov. of Kuroshio shot hole borer (*Euwallacea* sp. nr. *fornicatus*) cause Fusarium dieback on woody host species in California. *Plant Disease*, 102(6), 1154–1164.

Navia-Urrutia, M., and R. Gazis. 2021. Líneas de investigación enfocadas al desarrollo de estrategias para el manejo integral de la marchitez del laurel en aguacate. Barrientos-Priego, A. F. (Ed.), *Memorias del VI Congreso Latinoamericano del Aguacate*, p. 9.

Nix, K. A. 2013. *The Life History and Control of Pityophthorus juglandis Blackman on Juglans nigra L. in Eastern Tennessee*. Master's Thesis, University of Tennessee. Available Online: https://trace.tennessee.edu/utk_gradthes/1656/.

O'Donnell, K., Libeskind-Hadas, R., Hulcr, J., *et al.* 2016. Invasive Asian *Fusarium–Euwallacea* ambrosia beetle mutualists pose a serious threat to forests, urban landscapes and the avocado industry. *Phytoparasitica*, 44(4), 435–442.

Oliva, J., Stenlid, J., and J. Martínez-Vilalta. 2014. The effect of fungal pathogens on the water and carbon economy of trees: Implications for drought-induced mortality. *New Phytologist*, 203(4), 1028–1035. https://doi.org/https://doi.org/10.1111/nph.12857.

Onufrak, A. J., Williams, G. M., Klingeman III, W. E., et al. 2020. Regional differences in the structure of *Juglans nigra* phytobiome reflect geographical differences in thousand cankers disease severity. *Phytobiomes Journal*, 4(4), 388–404. https://doi.org/10.1094/PBIOMES-05-20-0044-R.

Oren, E. 2016. *Rapid Molecular Detection and Population Genetics of Pityophthorus juglandis, a Vector of Thousand Cankers Disease in Juglans spp*. Master's Thesis, University of Tennessee. Available Online: https://trace.tennessee.edu/utk_gradthes/4270/.

Oren, E., Klingeman, W., Gazis, R., et al. 2018. A novel molecular toolkit for rapid detection of the pathogen and primary vector of thousand cankers disease. *PLoS One*, 13(1), e0185087. https://doi.org/10.1371/journal.pone.0185087.

Ostry, M., and M. Moore. 2008. Response of butternut selections to inoculation with *Sirococcus clavigignenti-juglandacearum*. *Plant Disease*, 92(9), 1336–1338.

Paap, T., De Beer, Z. W., Migliorini, D., Nel, W. J., and M.J. Wingfield. 2018. The polyphagous shot hole borer (PSHB) and its fungal symbiont *Fusarium euwallaceae*: A new invasion in South Africa. *Australasian Plant Pathology,* 47(2), 231–237.

Paap, T., Wingfield, M. J., Wilhelm de Beer, Z., and F. Roets. 2020. Lessons from a major pest invasion: The polyphagous shot hole borer in South Africa. *South African Journal of Science,* 116(11–12), 1–4.

Parra, P. P., Dantes, W., Sandford, A., et al. 2020. Rapid detection of the laurel wilt pathogen in sapwood of Lauraceae hosts. *Plant Health Progress,* 21(4), 356–364.

Pinchot, C., Flower, C., Knight, K., et al. 2017. Development of new Dutch elm disease-tolerant selections for restoration of the American elm in urban and forested landscapes. In *Gene Conservation of Tree Species—Banking on the Future. Proceedings of a Workshop*, Tech Coords. R.A. Sniezko; G. Man; V. Hipkins; K. Woeste; D. Gwaze; J.T. Kliejunas; B. A. McTeague. 2017. Gen. Tech. Rep. PNW-GTR-963. Portland, OR: US Department of Agriculture, Forest Service, Pacific Northwest Research Station, pp. 53–63.

Ploetz, R., Hughes, M., Kendra, P., et al. 2016. Recovery plan for laurel wilt of avocado, caused by *Raffaelea lauricola*. *Plant Health Progress,* 18(2), 51–77.

Ploetz, R. C., Konkol, J., Pérez-Martínez, J., and R. Fernandez. 2017. Management of laurel wilt of avocado, caused by *Raffaelea lauricola*. *European Journal of Plant Pathology,* 149(1), 133–143.

Ploetz, R. C., Konkol, J. L., Narvaez, T., et al. 2017. Presence and prevalence of *Raffaelea lauricola*, cause of laurel wilt, in different species of ambrosia beetle in Florida, USA. *Journal of Economic Entomology,* 110(2), 347–354.

Ploetz, R. C., Pérez-Martínez, J. M., Evans, E. A., and S.A. Inch. 2011. Toward fungicidal management of laurel wilt of avocado. *Plant Disease,* 95(8), 977–982.

Rabaglia, R. J., Dole, S. A., and A.I. Cognato. 2006. Review of American Xyleborina (Coleoptera: Curculionidae: Scolytinae) occurring north of Mexico, with an illustrated key. *Annals of the Entomological Society of America,* 99(6), 1034–1056.

Randolph, K. C., Rose, A. K., Oswalt, C. M., and M.J. Brown. 2013. Status of black walnut (*Juglans nigra* L.) in the eastern United States in light of the discovery of thousand cankers disease. *Castanea,* 78(1), 2–14.

Rizzo, D., Da Lio, D., Bartolini, L., et al. 2020. A duplex real-time PCR with probe for simultaneous detection of *Geosmithia morbida* and its vector *Pityophthorus juglandis*. *PLoS One,* 15(10), e0241109. https://doi.org/10.1371/journal.pone.0241109.

Rizzo, D., Moricca, S., Bracalini, M., et al. 2021. Rapid Detection of *Pityophthorus juglandis* (Blackman) (Coleoptera, Curculionidae) with the Loop-Mediated Isothermal Amplification (LAMP) Method. *Plants,* 10(6), 1048. https://doi.org/10.3390/plants10061048.

Roy, B., and J. Kirchner. 2000. Evolutionary dynamics of pathogen resistance and tolerance. *Evolution,* 54(1), 51–63.

Rugman-Jones, P. F., Seybold, S. J., Graves, A. D., and R..Stouthamer. 2015. Phylogeography of the walnut twig beetle, *Pityophthorus juglandis*, the vector of thousand cankers disease in North American walnut trees. *PLoS One,* 10(2), e0118264. https://doi.org/10.1371/journal.pone.0118264

Rugman-Jones, P. F., and R. Stouthamer. 2017. High-resolution melt analysis without DNA extraction affords rapid genotype resolution and species identification. *Molecular Ecology Resources,* 17(4), 598–607.

Schmidt, E., Juzwik, J., and B. Schneider. 1997. Sulfuryl fluoride fumigation of red oak logs eradicates the oak wilt fungus. *Holz als Roh-und Werkstoff,* 55(5), 315–318.

Schoch, C. L., Seifert, K. A., Huhndorf, S., et al. 2012. Nuclear ribosomal internal transcribed spacer (ITS) region as a universal DNA barcode marker for Fungi. *Proceedings of the National Academy of Sciences,* 109(16), 6241–6246.

Seabright, K. W., Myers, S. W., Fraedrich, S. W., Mayfield, A. E., Warden, M. L., and A. Taylor. 2019. Methyl bromide fumigation to eliminate thousand cankers disease causal agents from black walnut. *Forest Science,* 65(4), 452–459.

Serdani, M., Vlach, J. J., Wallis, K. L., et al. 2013. First report of *Geosmithia morbida* and *Pityophthorus juglandis* causing thousand cankers disease in butternut. *Plant Health Progress,* 14(1), 38.

Seybold, S. J., Dallara, P. L., Hishinuma, S. M., and M.L. Flint. 2013. Detecting and identifying the walnut twig beetle: monitoring guidelines for the invasive vector of thousand cankers disease of walnut. *UC IPM Program*, University of California Agriculture and Natural Resources. 13 p. www.ipm.ucdavis.edu/thousandcankers.

Seybold, S. J., Haugen, L., O'Brien, J., and A.D. Graves. 2013. Thousand cankers disease. *USDA Forest Service, Northeastern Area State and Private Forestry Pest Alert, NA-PR-02-10.* Available Online: http://na.fs.fed.us/pubs/palerts/cankers_disease/thousand_cankers_disease_screen_res.pdf.

Seybold, S. J., Klingeman, W. E., III, Hishinuma, S. M., Coleman, T. W., and A.D. Graves. 2019. Status and impact of walnut twig beetle in urban forest, orchard, and native forest ecosystems. *Journal of Forestry,* 117(2), 152–163. https://doi.org/10.1093/jofore/fvy081

Shih, H., Wuest, C. E., Fraedrich, S. W., Harrington, T. C., and C. Chen. 2018. Assessing the susceptibility of Asian species of Lauraceae to the laurel wilt pathogen, *Raffaelea lauricola. Taiwan Lin Ye Ke Xue,* 33, 173–184.

Short, D. P., O'Donnell, K., Stajich, J. E., et al. 2017. PCR multiplexes discriminate *Fusarium* symbionts of invasive *Euwallacea ambrosia* beetles that inflict damage on numerous tree species throughout the United States. *Plant Disease,* 101(1), 233–240.

Sitz, R. A., Luna, E. K., Caballero, J. I., Tisserat, N. A., Cranshaw, W. S., and J.E. Stewart. 2017. Virulence of genetically distinct *Geosmithia morbida* isolates to black walnut and their response to coinoculation with *Fusarium solani. Plant Disease,* 101(1), 116–120. https://doi.org/https://doi.org/10.1094/PDIS-04-16-0535-RE.

Sitz, R. A., Luna, E. K., Ibarra Caballero, J., et al. 2021. Eastern black walnut (*Juglans nigra* L.) originating from native range varies in their response to inoculation with *Geosmithia morbida. Frontiers in Forests and Global Change,* 4, 12. https://doi.org/10.3389/ffgc.2021.627911.

Slavicek, J. M., and K.S. Knight. 2012. Generation of American elm trees with tolerance to Dutch elm disease through controlled crosses and selection. In *Proceedings of the Fourth International Workshop on the Genetics of Host-Parasite Interactions in Forestry: Disease and Insect Resistance in Forest Trees,* Tech coords. R.A. Sniezko; A.D. Yanchuk; J.T. Kliejunas; K.M. Palmieri; J.M. Alexander; S.J. Frankel. Gen. Tech. Rep. PSW-GTR-240. Albany, CA: Pacific Southwest Research Station, Forest Service, US Department of Agriculture. pp. 342–346.

Smalley, E., and R. Guries. 1993. Breeding elms for resistance to Dutch elm disease. *Annual Review of Phytopathology,* 31(1), 325–354.

Smith, J., Dreaden, T., Mayfield III, A., Boone, A., Fraedrich, S., and C. Bates. 2009. First report of laurel wilt disease caused by *Raffaelea lauricola* on sassafras in Florida and South Carolina. *Plant Disease,* 93(10), 1079. https://doi.org/10.1094/PDIS-93-10-1079B.

Smith, K., Zhang, J., Hughes, M., et al. 2020. Restoration after the laurel wilt disease epidemic. In *General Technical Reports SRS-252.* Asheville, NC: US Department of Agriculture Forest Service. Southern Research Station, pp. 169–169.

Sniezko, R. A., Smith, J., Liu, J.-J., and R.C. Hamelin. 2014. Genetic resistance to fusiform rust in southern pines and white pine blister rust in white pines—a contrasting tale of two rust pathosystems—current status and future prospects. *Forests,* 5(9), 2050–2083.

Southern Regional Extension Forestry (SREF) – Forest Health Program, 2021. http://southernforesthealth. net/diseases/laurel-wilt/distribution-map.

Stackhouse, T., Boggess, S. L., Hadziabdic, D., Trigiano, R. N., Ginzel, M. D., and W.E. Klingeman. 2021. Conventional gel electrophoresis and TaqMan probes enable rapid confirmation of thousand cankers disease from diagnostic samples. *Plant Disease,* 105(10), 3171–3180.

State of Hawai'i Department of Agriculture. 2016. *Amendments to Chapter 4–72 Hawaii Administrative Rules.* Hawaii Department of Agriculture. https://hdoa.hawaii.gov/wp-content/uploads/2012/12/PI-ROD-admin-rules.pdf.

Steiner, K. C., Westbrook, J. W., Hebard, F. V., Georgi, L. L., Powell, W. A., and S.F. Fitzsimmons. 2017. Rescue of American chestnut with extra specific genes following its destruction by a naturalized pathogen. *New Forests,* 48(2), 317–336.

Stouthamer, R., Rugman-Jones, P., Thu, P. Q., *et al.* 2017. Tracing the origin of a cryptic invader: Phylogeography of the *Euwallacea fornicatus* (Coleoptera: Curculionidae: Scolytinae) species complex. *Agricultural and Forest Entomology,* 19(4), 366–375.

Swain, S., Eskalen, A., Lynch, S., and S. Latham. 2017. Scolytid beetles and associated fungal symbionts threaten California hardwoods. *Western Arborist,* 43, 54–60.

Tennessee Department of Agriculture. 2014. *Thousand Cankers Disease Regulations in Plain Language.* Tennessee Department of Agriculture. https://www.tn.gov/agriculture/businesses/plants/plant-pests--diseases-and-quarantines/ag-businesses-tcd.html.

Teviotdale, B. L. 2017. *UC Pest Management Guidelines: Walnut.* UC ANR Publication 3471.

Thomidis, T., and E. Exadaktylou. 2012. Effectiveness of cyproconazole to control Armillaria root rot of apple, walnut and kiwifruit. *Crop Protection,* 36, 49–51.

Tisserat, N., Cranshaw, W., Leatherman, D., Utley, C., and K. Alexander. 2009. Black walnut mortality in Colorado caused by the walnut twig beetle and thousand cankers disease. *Plant Health Progress,* 10(1), 10. https://doi.org/10.1094/PHP-2009-0811-01-RS.

Tisserat, N., Cranshaw, W., Putnam, M. L., et al. 2011. Thousand cankers disease is widespread in black walnut in the western United States. *Plant Health Progress,* 12(1), 35. https://doi.org/10.1094/PHP-2011-0630-01-BR.

Townsend, A., Bentz, S., and L. Douglass. 2005. Evaluation of 19 American elm clones for tolerance to Dutch elm disease. *Journal of Environmental Horticulture,* 23(1), 21–24.

Townsend, A., and L. Douglass. 2001. Variation among American elm clones in long-term dieback, growth, and survival following *Ophiostoma* inoculation. *Journal of Environmental Horticulture,* 19(2), 100–103.

Treiman, T., and J. Tuttle. 2009. Thousand cankers disease of black walnut: How much will it hurt Missouri's pocketbook? *Notes for Forest Managers,* 16.

Trumbore, S., Brando, P., and H. Hartmann. 2015. Forest health and global change. *Science,* 349(6250), 814–818.

Umeda, C., Eskalen, A., and T.D. Paine. 2016. Polyphagous shot hole borer and Fusarium dieback in California. In *Insects and Diseases of Mediterranean Forest Systems,* eds. T. Pain, and F. lieutier, pp. 757–767. Springer.

Umeda, C. Y. 2017. *Environmental Effects on Polyphagous Shot Hole Borer.* Doctoral Dissertation. University of California Riverside. Available Online: https://escholarship.org/uc/item/243646pn.

USDA-FS-PPQ. 2021. *Thousand Cankers Disease Survey Guidelines for 2021.* United States Department of Agriculture, Forest Service (FS) and Plant Protection and Quarantine (PPQ).

Walgama, R., and R. Pallemulla. 2005. The distribution of shot-hole borer, *Xyleborus fornicatus* Eichh (Coleoptera: Scolytidae), across tea-growing areas in Sri Lanka. A reassessment. *Sri Lanka Journal of Tea Science,* 70(2), 105–120.

Wood, S. L., and D.E. Bright. 1992. A catalog of Scolytidae and Platypodidae (Coleoptera), Part 2: Taxonomic Index. Volume B. *Great Basin Naturalist Memoirs,* 13, 835–1557. https://www.biodiversitylibrary.org/part/144043.

Zerillo, M. M., Ibarra Caballero, J., Woeste, K., et al. 2014. Population structure of *Geosmithia morbida,* the causal agent of thousand cankers disease of walnut trees in the United States. *PLoS One,* 9(11), e112847. https://doi.org/10.1371/journal.pone.0112847.

Zhou, Y., Avery, P. B., Carrillo, D., et al. 2018. Identification of the Achilles heels of the laurel wilt pathogen and its beetle vector. *Applied Microbiology and Biotechnology,* 102(13), 5673–5684.

# 15 Fungal Interactions with Other Pests

*Guillermo E. Valero David and Jason C. Slot*

## CONTENTS

## 15.1 INSECTS

Insects comprise the most diverse class of animals worldwide, with over 900,000 species and representatives on every continent, including Antarctica. There are therefore tremendous opportunities for fungi and insects to interact and to exploit one another for nutrition. Fungi and insects can prey on and parasitize each other, and they can also cooperate, directly or indirectly, to utilize plant biomass. These interactions have a variety of demonstrable impacts on global agroecosystems, no doubt leaving many impacts to be discovered.

### 15.1.1 FUNGI ARE AN IMPORTANT FOOD SOURCE FOR INSECTS AND THEIR RELATIVES

Fungi make up a major portion of surface and below-ground biomass, presenting diverse nutritional targets for insect foragers. Many important **fungivorous** insects are generalist fungal grazers. For instance, the fungus gnats (Diptera: Sciaroidea) feed broadly on both mushroom-producing genera such as *Armillaria* and *Hypholoma* and on wood-decaying mold genera like *Botrytis*. Alternatively, some fungivores like the Malaysian ant, *Euprenolepis procera*, are highly specialized, feeding primarily on oyster mushrooms (*Pleurotus* spp.). Some insects may consume fungi without seeking them out; for example, tussock moth caterpillars (*Lymantria dispar*) incidentally consume the spores of the rust fungus *Melampsora larici-populina* while grazing on the aspen leaves that host the rust. Extreme specialization allows some insect species to enjoy a largely competitor-free food source. For example, some Drosophilid species have evolved resistance to *Amanita* mushroom toxins, which are deadly to most other organisms (Bunyard 2018). These few, brief examples of insect grazing on both microscopic and macroscopic fungi outline the ubiquity and diversity of these interactions.

DOI: 10.1201/9780429320415-17

Fungi are certainly not passive victims of grazing. As with many evolutionary relationships, fungal responses to grazers range from aggressive defenses to mutualisms. Fungal defenses can be direct, such as when insect grazing induces fungi to produce toxic chemicals that protect against further predation (Caballero Ortiz et al. 2013; Döll et al. 2013). Insect grazing can also cause changes in fungal lifecycles to mitigate the costs of predation. For example, when grazed by Collembola, which are hexapods closely related to insects (see Figure 15.1a), the mold *Aspergillus nidulans* undergoes early sexual reproduction to produce thick-walled spores that are more resistant to predation than vegetative mycelia and asexual conidia (Döll et al. 2013). By contrast, some fungi have evolved to take advantage of grazing, even attracting organisms that might not otherwise ingest fungi. The Stinkhorn mushrooms (*Phallus* sp.) release volatile chemicals that mimic the smell of a decaying corpse to attract insects that normally feed on flesh. Diverse fly species then consume the smelly, nutritious slime coating the stinkhorn, as well as the spores embedded inside. By taking advantage of insect grazers, stinkhorn spores may travel further than they would by wind alone.

Some insects even cultivate fungi as a food source. Multiple species of fungivorous insects "farm" plant-decaying fungi in highly coordinated mutualisms. Fungal agriculture is observed in the Attine ants (Mueller et al. 2001), Macrotermitini termites (Aanen et al. 2002), and seven different groups of "ambrosia" beetles (Farrell et al. 2001). In each of these mutualistic relationships, the insects sequester a specific fungus from the environment, prepare a substrate of fresh or decaying plant matter, and provide an environment with the suitable temperature and humidity for fungal growth. In exchange the fungus provides the insects with a reliable, nutritious food source.

### 15.1.1.1 Attine Ants

In the Western hemisphere, the most prominent fungus farmers are the **attine ants**, which are common in the wet forests of South America. Attine ants prepare one or more fungal "garden" chambers inside their nests, which can reach over 100 square feet and contain millions of worker ants. Before the ants "plant" their fungal crops, they fill the garden chambers with plant substrates (De Fine Licht and Boomsma 2010). Attine lineages are defined by the type of plant material they use to cultivate fungi. The "lower attines" use wood fragments, plant debris, seeds, and flower parts in their gardens, while the "higher attines" or "leafcutter ants" use freshly cut leaves and flowers. When a daughter queen ant sets off to start a new nest, she brings a bit of the cultivated fungus in her mouth. This transmission from parent to offspring has caused the domesticated fungi to have evolved separately from their wild relatives and rely on attine ants for growth and reproduction. Different species of attine ants may cultivate coral fungi (Pterulaceae), yeasts, or mushroom-forming fungi in the Lepiotaceae.

As with many human crops, the ants' fungal crop is affected by pests and pathogens such as the mycoparasitic fungi *Escovopsis* and *Trichoderma*, fungal "weeds" including *Xylaria* spp., and mites. Attine ants are meticulous in the care of their fungal crops and actively remove diseased tissue. The ants even use fungicides to keep their crops healthy. Actinobacteria that grow in specialized pits and cavities on the ant exoskeleton produce antimicrobial agents that stifle the fungal pathogen *Escovopsis* (Cafaro et al. 2011; Poulsen et al. 2009).

The immense population size of leaf cutting ant nests can impart tremendous herbivory pressure on the surrounding agroecosystem. Some wet forest agroecosystems, like coffee and cacao plantations, are severely impacted by leaf cutting ants that forage leaves to prepare their gardens. Lewis (1975) reported that the leaf cutting ants *Acromyrmex octospinosus* can defoliate and destroy 6%–17% of a cacao plant during the first year after planting and can collect over 3000 flowers per hectare of cacao each day. Varón et al. (2007) also observed that a single large *Atta cephalotes* nest killed 20 coffee bushes due to herbivory and damage to roots adjacent to the nest. Together these factors indicate major economic and yield losses to different crops impacted by leaf cutting ants, although these losses are yet to be formally quantified. At the same time, attine ants can provide some benefits to the agricultural systems they inhabit. The massive underground ant nests can favorably change soil structure and nutrient availability. Organic matter that is incorporated into

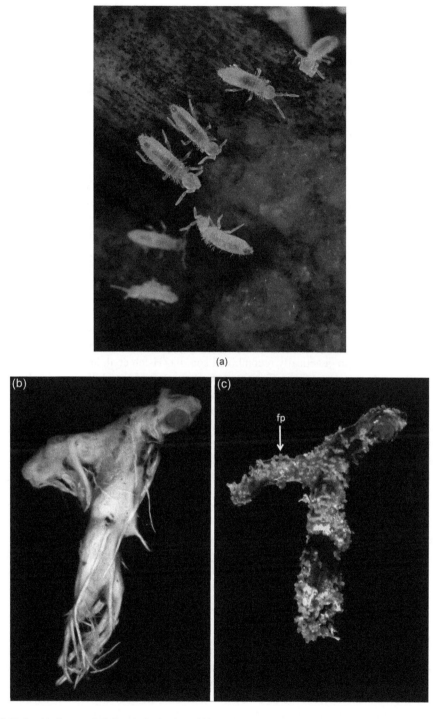

**FIGURE 15.1**    (a) Group of Collembola (springtails) on the organic component of the soil. (b and c) Example of the effects of Collembola (*Folsomia candida*) grazing: (b) Fresh specimen of ECM fungus *Piloderma bicolor* on a root. (c) The same specimen after two weeks of grazing by *F. candida*. The fungal mantle around the root and scavenging mycelia have been consumed, replaced by many Collembola fecal pellets ("fp"). [Image (a) reproduced with permission from Shutterstock: Holger Kirk. Images (b, c) reproduced with permission of Springer Nature: LeFait, A., Gailey, J., and G. Kernaghan. 2019. Fungal species selection during ectomycorrhizal grazing by Collembola. *Symbiosis*. 78:87–95, doi:10.1007/s13199-018-00596-x.]

the soil during the building and disposal of ant gardens can potentially improve the nutrition of nearby plants. In some attine ant nests, symbiotic bacteria fix atmospheric nitrogen, serving as an important source of soil nitrogen in these agroecosystems (Pinto-Tomas et al. 2009) and potentially reducing the amount of fertilizer inputs needed for the crops. Furthermore, coffee monocultures are more heavily foraged by *A. cephalotes* ants than coffee plantations shaded by a complex canopy (Varón et al. 2007), suggesting that diversifying agroecosystems may also help mitigate the economic impacts of attine ant activity.

### 15.1.1.2 Macrotermitini Termites

**Macrotermitini termites** are the chief fungus-farming insects of the Eastern hemisphere, occurring in the tropics of Africa and southern Asia. These termites prepare gardens out of partially dead and decomposed plant material, as well as termite fecal pellets. The termites cultivate a mushroom-forming fungus from the genus *Termitomyces* in the prepared substrate. Within the garden chambers, the fungal crop is arranged in "combs" that produce nodules of conidia that the termites consume. Like the gardens of attine ants, the combs are cleaned and tended. Depending on the termite species, the fungus in a new colony may be grown from newly foraged spores or from spores brought by termites from a previous colony (Korb and Aanen 2003). Some Macrotermitini fungal cultivars produce large mushrooms that generate abundant spores in synchrony with the establishment of new daughter colonies. In other termite species, conidia from active combs are carried in the guts of termites when establishing new colonies (Johnson et al. 1981). Unlike most mushroom-forming fungi, which undergo sexual recombination in the mushrooms themselves, one species of *Termitomyces* has been observed to undergo a type of sexual recombination in the mycelia which then gave rise to genetically recombined conidia (Hsieh et al. 2017). It is interesting that although both are domesticated fungi, Termitomyces has retained the ability to sexually reproduce while no sexual state has been observed for the attine ants' cultivars.

Termite-fungal mutualism results in a highly successful collective with the potential to be beneficial to surrounding agroecosystems. Some of these termite nests could have a significant impact on the local environment through soil restructuring and nutrient cycling. However, the termites can quickly become harmful when their populations are unchecked. Nurseries and plantations of eucalyptus, coconuts, palms, mangos, and other tropical crops are sometimes damaged by heavy termite foraging, with serious economic impacts to local communities (Rouland-Lefèvre 2010). Control of the termites can create additional revenue for farmers through the sale of Termitomyces mushrooms at market, which is an important source of income in some developing countries. *Termitomyces titanicus* is the largest known edible mushroom in the world, with caps growing up to a meter in diameter. However, overharvesting due to market demand may put the sensitive Microtermitini-fungal mutualisms at risk of local extinction (Koné et al. 2013), eliminating the benefits of the termite nests and of general biodiversity, including other local species that rely on the termites as a food source.

### 15.1.1.3 Ambrosia Beetles

While the attine ants and Macrotermitini termites bring plant matter to the nests where fungi can be cultivated, "**ambrosia**" **beetles** bring their farmed fungi to a stationary plant substrate, which becomes the beetle's nest. In this case, a female ambrosia beetle carries the spores of "ambrosia fungi" in specialized pockets on her body called **mycangia**. "Ambrosia" is an allusion to the food of the gods in Greek mythology, said to be brought by doves to the heavenly Mount Olympus. Rather than flying ambrosia fungi to a mountaintop in the clouds, the beetle drops her spores or mycelia as she excavates a complex network of tunnels called a **gallery** in the bark and phloem of a tree. Ambrosia beetles of different species may create galleries in weakened or freshly dead tissue, dead wood, or living trees. As the female digs, the fungus colonizes the wood of her gallery, transforming undigestible, toxic, insect-repelling wood into nutritious mycelium. She lays her eggs in the colonized gallery so that when the larvae hatch, they can feed on specialized spores or modified hyphae from the fungal crop.

The ambrosia crop is generally not a pure fungal culture; instead, multiple fungi are transported in the mycangia and grow in the galleries (Kostovcik et al. 2015). Mites, nematodes, bacteria, and other fungi can also be present. Some species of ambrosia beetles have social structures that communally maintain the fungal gardens (Biedermann and Taborsky 2011). Ambrosia symbioses are diverse and globally distributed, and different beetle species may cultivate *Raffaelea* (order Ophiostomatales), *Ambrosiella* (order Microascales), *Fusarium* (order Hypocreales), or other fungi (Kasson et al. 2013; Mayers et al. 2015). While many ambrosia beetles cause little damage or prefer dead or dying trees as hosts, others are devastating tree pathogens (Figure 15.2; see Chapter 14 in this volume).

## 15.1.2 Insects Are Important Vectors of Plant Pathogenic Fungi

The Asian ambrosia beetle (*Xyleborus glabratus*) cultivates the fungal crop *Raffaelea lauricola*. Together this pair causes laurel wilt disease, which has caused significant losses in both cultivated and wild areas of Florida – killing millions of native trees in the Everglades (Snyder 2014) and over 120,000 commercial avocado trees. Along with the ecological impact of the Everglade losses, the economic cost to the avocado industry has been estimated at $100 million. Another Asian ambrosia

**FIGURE 15.2**   Different stages (eggs, larva, adult) of the Asian ambrosia beetle (*Euwallacea fornicatus*) in galleries made on an avocado tree. [Images reproduced with permission from Shutterstock: Protasov AN.]

beetle, *Euwallacea* sp., and its fungal partner *Fusarium euwallaceae* form an invasive species mutualism that is endangering the avocado industry from California to Israel (Freeman et al. 2013). Other bark-boring beetles cause major tree mortality as well. Urban and wild forests in Europe and the United States have been massively transformed by the deaths of millions of native elms caused by Dutch elm disease and by the Mountain Pine Beetle, which partners with a blue-stain fungus that protects the beetle from tree defenses while together they destroy the tree's vascular system, starving it of its own sugars.

Perhaps more often, insects passively transmit fungal diseases by feeding or walking on infected plant material and moving to an uninfected plant. Sometimes fungi directly exploit their unwitting accomplices. Bees carry and propagate spores when they visit infected flowers, spreading such diseases as *Ustilaginomycetes* smuts and *Microbotryomycetes* anther smuts (Altizer et al. 1998; Shykoff and Bucheli 1995). In a sinister twist, some fungal pathogens induce the growth of false flowers on plant tissues that attract insects with colors, scents, or nectars (Slot and Kasson 2021).

### 15.1.3 Fungi Are Important Pathogens of Insects

We have described many examples of insects consuming fungi, but there are also many fungi that prey on insects. Fungi that derive their nutrition from insect hosts are known as **entomopathogens**. Of approximately 100,000 described species of fungi, around 1,000 species in over 100 genera are pathogens of insects (Araújo and Hughes 2016; Vega et al. 2012). Insect parasitism occurs in five of the eight fungal phyla: Microsporidia, Chytridiomycota, Ascomycota, Basidiomycota, and Zygomycota (Vega et al. 2012; Araújo and Hughes 2016). Entomopathogenic fungi are most diverse in tropical forests but are also common and widespread in temperate regions (Vega et al. 2012).

Infection of insects is typically a multistage process. The entomopathogenic life cycle begins when a spore adheres to the exoskeleton of an insect. After germination, hyphae penetrate weak or damaged points in the exoskeleton and enter the insect's body, a rich source of nitrogen and carbon nutrition. Infection spreads to new tissues via the **haemolymph** (insect "blood"), often in a yeast-like form called hyphal bodies or blastospores. Eventually, there is a switch to filamentous growth, which completes the colonization of the insect and emerges through the exoskeleton to reproduce. The fruiting bodies erupting dramatically from insect exoskeletons can be eerily beautiful (Figure 15.3).

Some entomopathogenic fungi manipulate insect behavior to optimize spore dispersal. *Ophiocordyceps unilateralis* broadly colonizes the bodies of ants, including muscle fibers, but interestingly leaves the host brain intact (Fredericksen et al. 2017). Infected ants exhibit convulsions, walk in zigzags, wander around, and neglect their daily work routines. In time, the fungus causes the ant to climb to an elevated plant tissue and attach itself to a leaf or twig by biting prior to death (Araújo and Hughes 2016, de Bekker et al. 2014, Loreto et al. 2017). de Bekker et al. (2014) found evidence that neurologically active chemicals are involved in the manipulation of the host insect, and others have suggested that toxins may affect the insect's odor response, foraging, and other behaviors (Will et al. 2020). These behavioral changes are known as "**summit disease**" and are exhibited in insects infected by multiple lineages of entomopathogenic fungi, including the Entomophthoralean fly pathogens. In some Entomophthoralean pathogens, insects are induced to actively transmit infectious spores to new hosts (Lovett et al. 2020). When a cicada is infected by a species of *Massospora*, the still-living insect's abdomen is replaced by a plug of spores (see Figure 15.4). Even though the sexual organs are replaced, the cicada can still act sexually. Boyce et al. (2019) found that the fungus produces either psilocybin or an amphetamine-like stimulant, depending on the fungal species, and suggested these chemicals contribute to active disease transmission through the insect host's hyperactive, hypersexual behavior.

Social insects like ants and termites can sense the presence of pathogenic fungi and actively remove them in a "social immune response". The social immune system is described in Cremer et al. (2007) as "the cooperation of social group members resulting in avoidance, control or elimination

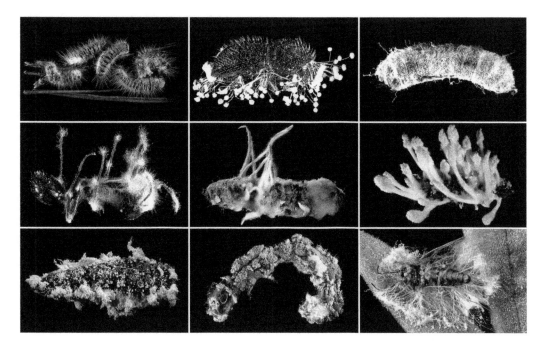

**FIGURE 15.3** Insects infected with diverse entomopathogenic fungi, showing different spore bearing arrangement and colors. [Images reproduced with permission from Shutterstock: Protasov AN.]

of parasitic infection". This behavioral response consists of mutual grooming, removal of spores and of diseased young, and in some instances the application of protective chemicals. Mutual grooming behavior is triggered in ants and termites exposed to *Metarhizium anisopliae* fungal spores, resulting in resistance to the pathogen (Hughes et al. 2002; Yanagawa and Shimizu 2007). In addition to grooming the brood and fellow workers, the ant *Lasius neglectus* secretes a poison composed of formic acid and other chemicals that reduces the viability of *Metarhizium* spores (Tragust et al. 2013) (Figure 15.4).

### 15.1.3.1 Entomopathogenic Fungi in Soils

In agricultural ecosystems, geographical location, soil type farming type, and other abiotic and biotic conditions affect the occurrence and distribution of entomopathogenic fungi (Goble et al. 2009; Klingen et al. 2002; Kumar Sain et al. 2019; Quesada-Moraga et al. 2007). While soils can harbor a great diversity of entomopathogenic fungi, *Beauveria bassiana* and *Metarhizium anisopliae* are the most commonly detected species across agricultural and non-agricultural ecosystems globally (Quesada-Moraga et al. 2007; Sanchez-Peña et al. 2011; Tkaczuk et al. 2014; Wakil et al. 2013).

Agricultural practices like cropping system and soil management (tillage) also influence the occurrence of entomopathogenic fungi in soils. Klingen et al. (2002) found that while some fungal entomopathogens were more abundant in organic versus conventionally managed soils, there was no difference in abundance in the uncultivated "margins" adjacent to these fields. Field margins overall are found to have more entomopathogens compared to cultivated land (Goble et al. 2009).

Understanding the roles of cropping systems and management practices used in the field may improve the success of applying entomopathogenic fungi for control of insect pests. Typically, field-applied entomopathogenic fungi do not persist under the local management practices and must be reapplied regularly. This is partly because chemical fungicides, herbicides, insecticides, and nematicides used in conventional farming impact the diversity of entomopathogenic fungi in the soil, likely by differently affecting growth and sporulation of entomopathogenic fungal species

**FIGURE 15.4**  A cicada whose abdomen has been replaced with spores of *Massospora* sp., a fungus that can manipulate the insect's behavior. [Image reproduced with permission from Shutterstock: Gerry Bishop.]

(Poprawski and Majchrowicz 1995). Some entomopathogenic fungi, like *M. anisopliae*, appear to be more tolerant to pesticides than others (Mietkiewski et al. 1997; Sain et al. 2019). *M. anisopliae* also has a greater ability to persist longer in agricultural soils compared to *B. bassiana*, which is more persistent in undisturbed environments (Sanchez-Peña et al. 2011; Vänninen et al. 2000). Interestingly, native entomopathogens more readily establish in the environment than introduced species (e.g., Klingen et al. 2015). However, some introduced entomopathogens have been successfully applied to control invasive insect species. The entomopathogenic fungus *Entomophaga maimaiga* has greatly reduced the damaging effects of tussock moth caterpillars on North American forests since its introduction in the 20th century (Andreadis and Weseloh 1990; Hajek et al. 1996). In the future, careful management of agroecosystems may maintain and enhance native diversity of insect pest biocontrol fungi.

### 15.1.4  FUNGAL-INSECT INTERACTIONS ARE IMPORTANT FACTORS IN ECOSYSTEM PROCESSES

Fungal-insect interactions have important and complex impacts on agroecosystems. Insects sequester nutrients from plants, while fungi both control insect predation and can alter nutrient availability to plants, in concert with or at the expense of insects (as discussed below). At the same time, fungal-insect partnerships can also be major agents of ecological disturbance. For example, Macrotermitinae termites are often pests of cultivated tropical crops, trees and wood. Nurseries and plantations of eucalyptus, coconuts, palms, coffee, cacao, mangos, and other tropical trees are damaged by termites, in some cases with important economic consequences (Rouland-Lefèvre 2010). Similarly, leaf cutting ants impose significant herbivory pressure on agroecosystems that are located in or near wet forests, like coffee or cacao plantations. Some studies suggest that diversifying agroecosystems may reduce the economic effects of leaf cutting ants by providing the ants alternate vegetation sources and reducing their specific adaptation to coffee leaves (Varón et al. 2007).

Finally, insect-fungus interactions may have important impacts on nutrient cycling. For example, entomopathogenic fungi may increase the availability of nutrients to plants. These fungi may help restore plant nitrogen that is lost through insect herbivory by releasing insect nitrogen back to the soil. More directly, *Metarhizium* spp. appear to have the ability to form endophytic relationships with plants (Sasan and Bidochka 2012), transferring insect nitrogen to the plant (Behie et al. 2012; Behie and Bidochka 2014) while plant carbon is transferred to the fungi (Behie et al. 2017). Fungus farming termites similarly impact nutrient cycles by creating nutrient rich patches that are directly available to plants (Jouquet et al. 2005). It has further been demonstrated that attine ant nests reduce $CO_2$ levels in wet soils by venting the $CO_2$ in and around nests to the atmosphere, representing an estimated 0.2%–0.7% of the carbon footprint in the wet forests (Fernandez-Bou et al. 2019). It is possible this enrichment of atmospheric $CO_2$ results in local enhancement of photosynthesis and the associated aeration of soil from increased root respiration. In contrast, Collembola hexapod species that selectively feed on root-associated hyphae can interrupt the flow of nutrients between the plant host and mycorrhizal fungi (LeFait et al. 2019). Because mycorrhizal mutualisms are important for the exchange and availability of nitrogen, phosphorus, and carbon, heavy Collembola infestations on mycorrhizae can reduce their plant hosts' growth (Warnock et al. 1982, Endlweber and Scheu 2007) and consequently impact the availability of nutrients and carbon cycling on a larger scale. These findings suggest that insect-fungal associations have an important role in the carbon and nitrogen cycles beyond the impacts of the parties in isolation.

## 15.2 NEMATODES

Nematodes are a diverse group of translucent roundworm invertebrates, covering almost every environment in the world. Some nematodes are microscopic, but some species can reach a size that can be seen with naked eye. There are over 25,000 species known, some living as parasites, others as free-living organisms. The majority of nematode species are free-living organisms feeding on microorganisms like bacteria, fungi, algae and other nematodes. Around 15% of nematode species are important parasites of animals and humans, causing important health problems. Nematodes can also parasitize plants, including those we cultivate. Plant parasitic nematodes (around 10% of known species) are some of the most significant pests in agricultural production worldwide.

### 15.2.1 NEMATODES CONSUME FUNGI, RESULTING IN VARIED IMPACTS ON AGROECOSYSTEMS

Nematodes represent a significant portion of soil biomass, and it is estimated there are approximately $4.4 \times 10^{20}$ nematodes inhabiting the various soils of the Earth (van den Hoogen et al. 2019). Fungi, which comprise another major component of the below-ground biomass, are food for fungus-feeding nematodes. These nematodes use a stylet to puncture the fungal cell wall, then suck out the internal cell contents. Different members of the genera *Aphelenchoides*, *Tylenchus*, *Ditylenchus*, and *Aphelenchus* are common fungivores found in agricultural soils (Freckman and Caswell 1985). Fungivorous nematodes can feed on different fungal species, including pathogens, saprophytes, and also beneficial fungi and fungal crops.

Fungivorous nematodes can be important pests of mushrooms and beneficial fungi. For example, the nematodes *Ditylenchus myceliophagus* and *Aphelenchoides composticola* cause major damage to cultivated mushrooms by consuming mycelium in the substrate (Rinker 2017; Singh and Sharma 2016). The fungus-eating nematode *Aphelenchoides* sp. feasts on *Trichoderma harzianum*, a biological control agent used against the pathogen *Sclerotinia sclerotiorium*, reducing its efficacy against the disease (Bae and Knudsen 2001). Further, nematodes of the genus *Aphelenchus* prevent the colonization of pine roots by the nutrition-enhancing arbuscular mycorrhizal (AM) *Glomus* spp. (Ragozzino and D'Errico 2011). Some mushroom-forming fungi produce lectins or proteins that provide lethal protection against fungivorous nematodes (Bleuler-Martinez et al. 2011; Plaza et al. 2016). These fungus-produced compounds can be less toxic than, and may provide alternatives to, commercial nematicides (Bogner et al. 2017).

However, fungivorous nematodes can also benefit agroecosystems by suppressing plant pathogenic fungi. Fungivory by nematodes has been exploited for the control of important fungal plant diseases. For example, the nematodes *Aphelenchus avenae* and *Aphelenchoides* spp. suppress

*Rhizoctonia solani* (Basidiomycota) and *Pythium* spp. (oomycetes, not true fungi) in the soil thereby reducing incidence of the rots these organisms cause in crops (Lagerlöf et al. 2011; Li et al 2014; Rhoades and Linford 1959).

In a seeming paradox, fungivory of plant-health promoting fungi appears to sometimes benefit plants. Some fungivorous nematodes prefer to graze on mycorrhizal fungi, resulting in both negative and positive effects on plant-mycorrhiza interactions. In an experiment with potted clover (*Trifolium subterraneum* L.), the nematode population of *Aphelenchus avenae* increased in the presence of the AM fungi *Gigaspora margarita* and *Glomus coronatum*, which ultimately decreased mycorrhizal colonization of the clover roots (Bakhtiar et al. 2001). Despite decreased mycorrhization of their roots, clover plants grew better in the presence of nematodes. Hua et al. (2014) found that co-inoculation of AM fungi and nematodes increased plant growth and tolerance of soil arsenic. It is possible that nematodes help cycle soil nutrients, which are then absorbed and transferred by the fungi to the plant.

## 15.2.2 DIVERSE FUNGI ATTACK AND CONSUME NEMATODES

Fungi have many different ecological strategies and environments, and some of these niches have limited amounts of required nutrients. For example, decaying wood is poor in nitrogen relative to carbon (see Chapters 16 and 17 in this volume). **Nematophagous** fungi are species-diverse and exhibit diverse strategies for capturing and digesting nematodes, which can be an important source of nitrogen. However, nematophagous fungi are also common in nitrogen-rich substrates like animal dung. There are over 700 different nematophagous fungal species described from multiple fungal phyla, although the most charismatic of these, the nematode trap formers discussed as Type I below, are from the single ascomycete order Orbiliales (Li et al. 2015). Some of these nematophagous fungi are common edible mushrooms producing unique structures and mechanisms to infect their hosts.

Nematophagous fungi are divided into four groups depending on the mechanism they use to attack nematodes: (1) nematode trapping fungi, (2) toxin-producing fungi, (3) endoparasitic fungi, and (4) egg- and female-parasitic fungi (Li et al. 2015). In each of these strategies, the result is the nematode becoming completely digested by the fungus.

1. Nematode trapping fungi produce one of several specialized hyphal structures – traps – that immobilize the nematode and facilitate invasion of its body. Some traps are mechanical, while others have an adhesive function. Different hyphal arrangement types have been described, including constricting and non-constricting **rings**, adhesive nodules and networks, non-differentiated adhesive hyphae, and three dimensional adhesive networks (Li et al. 2015) (see Figure 15.5). The fungus *Arthrobotrys oligospora* produces three-dimensional nets that can trap and immobilize nematodes, but interestingly this fungus can take on multiple nutritional lifestyles such as mycoparasite, saprophyte, and plant root colonizer (Yang et al. 2011).

2. Other fungi use toxins to immobilize their victims prior to infection. The oyster mushroom, *Pleurotus ostreatus*, as well as other members of the *Pleurotus* genus are known to produce nematotoxic compounds (Soares et al. 2018). Over 200 compounds produced by different fungi adversely affect nematodes (Li et al. 2007; Li and Zhang 2014). While the mechanisms of action of these toxins are not all known, cytotoxic and hemolytic effects and paralysis through neuro-muscular depolarization have been suggested (Lee et al. 2020).

3. Some endoparasitic nematophagous fungi use spores (conidia or zoospores) rather than traps to infect the nematode. Spores may adhere to the cuticle (outer cell layer) or be ingested by the nematode (Dijksterhuis et al. 1990). Endoparasitic fungi are obligately nematophagous and are more host specific than the nematode trapping fungi. The endoparasites *Drechmeria coniospora* and *Harposporium* sp., commonly associated with

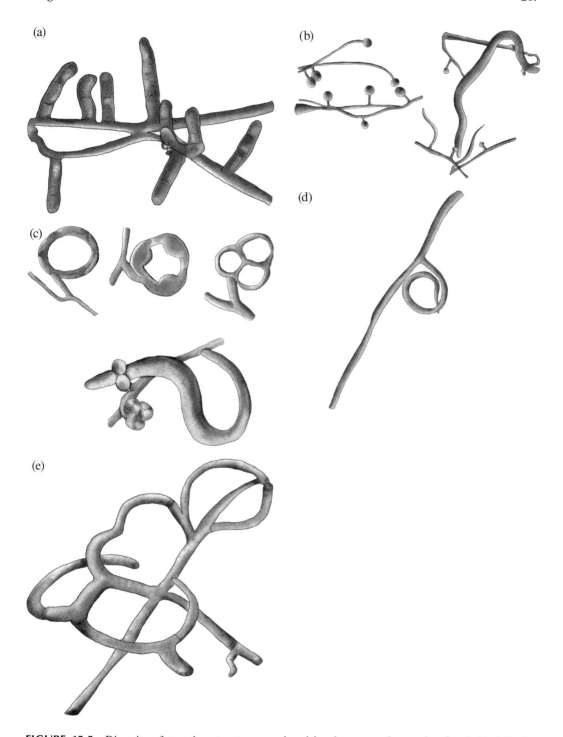

**FIGURE 15.5** Diversity of trapping structures produced by the nematode-trapping fungi. (a) Adhesive hyphae. (b) Adhesive knobs. The right of figure (b) shows a nematode being trapped by adhesive knobs. (c) Constricting rings. The bottom of figure (c) shows a nematode trapped by a constricting ring. (d) Non-constricting rings. (e) Adhesive networks. (Illustrations provided by Sarah B. Scott.)

*Caenorhabditis elegans* in the field, are some of the most common fungi used in biological nematode control (Félix and Duveau 2012).

4. Nematophagous fungi can also attack eggs and cysts made by female nematodes. These fungi tend to produce a specialized penetration structure (**appressorium**) and other penetration structures to infect the eggs (Lopez-Llorca et al. 2008). A number of different fungi, including *Lecanicillium psalliotae*, *Pochonia chlamydosporia*, *Trichoderma* sp., and others, have been studied to control the eggs of parasitic nematodes in the lab and greenhouses (Hussain et al. 2018; Manzanilla-López et al. 2013; Al-Hazmi and TariqJaveed et al. 2016).

Another group of nematophagous fungi has been proposed where the fungi produce a very interesting and unique structure like "medieval weapons" to breach the nematode cuticle and facilitate infection. The fungus *Coprinus comatus* (the shaggy mane mushroom) produces a spiny ball resembling a medieval caltrop; the fungus from the genus *Stropharia rugosoannualata* (the wine cap mushroom) produces spiked cells called acanthocytes that resemble the medieval spike ball and chain; the fungus from the genus *Hyphoderma* sp. produces a globular cell with basal spikes that resembles the medieval morning star; and the oomycete from the genus *Haptoglossa* produces a globular terminal cell called "gun cell", which resembles a cannon with a harpoon-shaped cell inside (reviewed in Soares et al. 2018) (see Figure 15.6).

### 15.2.3 Nematodes and Plant Pathogenic Fungi Can Act Synergistically to Cause Plant Disease

Plant parasitic nematodes and root pathogenic fungi are important root pests in different crop systems. Some plant parasitic nematodes and root pathogenic fungi have worked together to create synergistic interactions, increasing disease severity. The causal agent of fusarium wilt of cotton (*Fusarium oxysporum* f. sp. *vasinfectum*) and the root knot nematode (*Meloidogine* spp.) was the first complex described, in which the presence of the nematode increased the severity of fusarium wilt. Other disease complexes resulting from interactions between nematodes and plant pathogenic fungi have been described (reviewed in Back et al. 2002). In each of these, several mechanisms may be responsible for the synergistic effect and disease severity. For example, nematodes can induce physical and physiological changes in plant roots, creating a more suitable environment for fungal pathogens. Cyst nematodes are hypothesized to change the available nutrition of infected tissues and favor their attractiveness to plant pathogenic fungi (Back et al. 2002). Nematodes can also facilitate fungal invasion through entry wounds and ruptures in nematode galls and may also trigger the signal for fungi to switch to a pathogenic lifestyle.

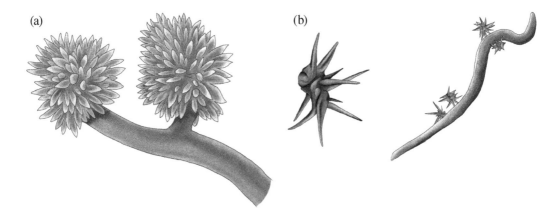

(a)                                                                                                    (b)

**FIGURE 15.6** Specialized structures used by some fungi to rupture nematode cuticles. (a) Spiny ball. (b) Spiked cells called acanthocytes. The right of figure (b) shows a nematode being pierced by acanthocytes. (Illustrations provided by Sarah B. Scott.)

## 15.2.4 COEVOLUTION OF FUNGI AND NEMATODES HAS RESULTED IN DIVERSE ASSOCIATIONS

Fungi and nematodes exhibit diverse ecological interactions, utilizing complex structures and chemistry that suggests there has been long-term coevolution of these two organismal groups. Some of these interactions, like mycophagy of plant pathogens and nematophagy of plant parasitic nematodes, provide important counterbalance to pest damage in cropping systems and can be exploited for pest management. Mycophagy of beneficial fungi can have both negative and positive net outcomes for plants due to system-level processes. However, plant parasitic nematodes and fungi can also work synergistically to increase the severity of plant disease.

## ACKNOWLEDGMENT

Drawings provided by Sarah B. Scott; her art and time are gratefully acknowledged.

## REFERENCES

Aanen, D.K., Eggleton, P., Rouland-Lefevre, C., Guldberg-Froslev, T., Rosendahl, S., and J.J. Boomsma. 2002. The evolution of fungus-growing termites and their mutualistic fungal symbionts. *Proceedings of the National Academy of Sciences USA* 99:14887–14892, doi:10.1073/pnas.222313099.

Adams, R.M.M., Mueller, U.G., Holloway, A.K., Green, A.M., and J. Narozniak. 2000. Garden sharing and garden stealing in fungus-growing ants. *Naturwissenschaften* 87:491–493, doi:10.1007/s001140050765.

Al-Hazmi A.S. and M. TariqJaveed. 2016. Effects of different inoculum densities of *Trichoderma harzianum* and *Trichoderma viride* against *Meloidogyne javanica* on tomato. *Saudi Journal of Biological Sciences* 23:288–292.

Altizer, S.M., Thrall, P.H., and J. Antonovics. 1998. Vector behavior and the transmission of anther-smut infection in *Silene alba*. *The American Midland Naturalist* 139:147–163, doi:10.1674/0003-0031(1998) 139[0147:VBATTO]2.0.CO;2.

Araújo, J.P. and D.P. Hughes. 2016. Diversity of entomopathogenic fungi: Which groups conquered the insect body? In *Advances in Genetics*, eds. B. Lovett and R.J. St. Leger (Vol. 94, pp. 1–39). London: Academic Press.

Andreadis, T.G., and R.M. Weseloh. 1990. Discovery of *Entomophaga maimaiga* in North American gypsy moth, *Lymantria dispar*. *Proceedings of the National Academy of Sciences USA* 87(7):2461–2465.

Back, M.A., Haydock, P.P.J., and P. Jenkinson. 2002. Disease complexes involving plant parasitic nematodes and soilborne pathogens: Nematodes and soilborne pathogens. *Plant Pathology* 51:683–697.

Bae, Y.-S., and G.R. Knudsen. 2001. Influence of a fungus-feeding nematode on growth and biocontrol efficacy of *Trichoderma harzianum*. *Phytopathology* 91:301–306.

Bakhtiar, Y., Miller, D., Cavagnaro, T., and S. Smith. 2001. Interactions between two arbuscular mycorrhizal fungi and fungivorous nematodes and control of the nematode with fenamifos. *Applied Soil Ecology* 17:107–117.

Behie, S.W. and M.J. Bidochka. 2014. Ubiquity of insect-derived nitrogen transfer to plants by endophytic insect-pathogenic fungi: An additional branch of the soil nitrogen cycle. *Applied Environmental Microbiology* 80:1553–1560, doi:10.1128/AEM.03338-13.

Behie, S.W., Moreira, C.C., Sementchoukova, I., Barelli, L., Zelisko, P.M., and M.J. Bidochka. 2017. Carbon translocation from a plant to an insect-pathogenic endophytic fungus. *Nature Communications* 8:14245, doi:10.1038/ncomms14245.

Behie, S.W., Zelisko, P.M., and M.J. Bidochka. 2012. Endophytic insect-parasitic fungi translocate nitrogen directly from insects to plants. *Science* 336:1576–1577, doi:10.1126/science.1222289.

Biedermann, P.H.W., and M. Taborsky. 2011. Larval helpers and age polyethism in ambrosia beetles. *Proceedings of the National Academy of Sciences* 108:17064–17069, doi:10.1073/pnas.1107758108.

Bleuler-Martínez, S., Butschi, A., Garbani, M., et al. 2011. A lectin-mediated resistance of higher fungi against predators and parasites: Fungal resistance lectins. *Molecular Ecology* 20:3056–3070.

Bogner, C.W., Kamdem, R.S.T., Sichtermann, G., et al. 2017. Bioactive secondary metabolites with multiple activities from a fungal endophyte. *Microbial Biotechnology* 10:175–188.

Boyce, G.R., Gluck-Thaler, E., Slot, J.C., et al. 2019. Psychoactive plant- and mushroom-associated alkaloids from two behavior modifying cicada pathogens. *Fungal Ecology* 41:147–164, doi:10.1016/j.funeco.2019.06.002.

Bunyard, B. 2018. Deadly amanita mushrooms as food: A survey of the feeding preferences of *Mycophagous diptera* from across North America, with notes on evolved detoxification. *Fungi Magazine* Winter 2018.

Caballero Ortiz, S., Trienens, M., and M. Rohlfs. 2013. Induced fungal resistance to insect grazing: Reciprocal fitness consequences and fungal gene expression in the Drosophila-Aspergillus model system. *PLoS One* 8:e74951, doi:10.1371/journal.pone.0074951.

Cafaro, M.J., Poulsen, M., Little, A.E.F., et al. 2011. Specificity in the symbiotic association between fungus-growing ants and protective Pseudonocardia bacteria. *Proceedings of the Royal Society B* 278:1814–1822, doi:10.1098/rspb.2010.2118.

Cremer, S., Armitage, S.A., and P. Schimid-Hempel. 2007. Social immunity. *Current Biology* 17(16), R693–R702.

de Bekker, C., Quevillon, L.E., Smith, P.B., et al. 2014. Species-specific ant brain manipulation by a specialized fungal parasite. *BMC Evolutionary Biology* 14:166, doi:10.1186/s12862-014-0166-3.

De Fine Licht, H.H., and J.J. Boomsma. 2010. Forage collection, substrate preparation, and diet composition in fungus-growing ants. *Ecological Entomology* 35:259–269, doi:10.1111/j.1365-2311.2010.01193.x.

Dijksterhuis, J., Veenhuis, M., and W. Harder. 1990. Ultrastructural study of adhesion and initial stages of infection of nematodes by conidia of *Drechmeria coniospora. Mycological Research* 94:1–8.

Döll, K., Chatterjee, S., Scheu, S., Karlovsky, P., and M. Rohlfs. 2013. Fungal metabolic plasticity and sexual development mediate induced resistance to arthropod fungivory. *Proceedings of the Royal Society B* 280:20131219. doi:10.1098/rspb.2013.1219.

Endlweber, K., and S. Scheu. 2007. Interactions between mycorrhizal fungi and Collembola: Effects on root structure of competing plant species. *Biology and Fertility of Soils* 43:741–749, doi:10.1007/s00374-006-0157-7.

Farrell, B.D., Sequeira, A.S., O'Meara, B.C., Normark, B.B., Chung, J.H., and B.H. Jordal. 2001. The evolution of agriculture in beetles (Curculionidae: Scolytinae and Platypodinae). *Evolution* 55:2011–2027, doi:10.1111/j.0014-3820.2001.tb01318.x.

Félix, M-A. and F. Duveau. 2012. Population dynamics and habitat sharing of natural populations of *Caenorhabditis elegans* and *C. briggsae. BMC Biology* 10:59. https://doi.org/10.1186/1741-7007-10-59.

Fernandez-Bou, A.S., Dierick, D., Swanson, A.C., et al. 2019. The role of the ecosystem engineer, the leaf-cutter ant *Atta cephalotes*, on soil $CO_2$ dynamics in a wet tropical rainforest. *JGR Biosciences* 124:260–273, doi:10.1029/2018JG004723.

Freckman, D.W. and E.P. Caswell. 1985. The ecology of nematodes in agroecosystems. *Annual Review of Phytopathology* 23:275–296.

Fredericksen, M.A., Zhang, Y., Hazen, M.L., et al. 2017. Three-dimensional visualization and a deep-learning model reveal complex fungal parasite networks in behaviorally manipulated ants. *Proceedings of the National Academy of Science USA* 114:12590–12595, doi:10.1073/pnas.1711673114.

Freeman, S., Sharon, M., Maymon, M., et al. 2013. *Fusarium euwallaceae* sp. nov. – A symbiotic fungus of *Euwallacea* sp., an invasive ambrosia beetle in Israel and California. *Mycologia* 105:1595–1606, doi:10.3852/13-066.

Goble, T.A., Dames, J. F., Hill, M. P., and S. D. Moore. 2010. The effects of farming system, habitat type and bait type on the isolation of entomopathogenic fungi from citrus soils in the Eastern Cape Province, South Africa. *BioControl* 55:399–412, doi:10.1007/s10526-009-9259-0.

Hajek, A.E., Elkinton, J.S., and J.J. Witcosky. 1996. Introduction and spread of the fungal pathogen *Entomophaga maimaiga* (Zygomycetes: Entomophthorales) along the leading edge of gypsy moth (Lepidoptera: Lymantriidae) spread. *Environmental Entomology* 25(5), 1235–1247.

Hsieh, H.M., Chung, M.C., Chen, P.Y., et al. 2017. A termite symbiotic mushroom maximizing sexual activity at growing tips of vegetative hyphae. *Botanical studies* 58(1), 39. https://doi.org/10.1186/s40529-017-0191-9.

Hughes, W.O.H., Eilenberg, J., and J.J. Boomsma. 2002. Trade-offs in group living: Transmission and disease resistance in leaf-cutting ants. *Proceedings of the Royal Society B* 269:1811–1819, doi:10.1098/rspb.2002.2113.

Hussain, M., Zouhar, M., and P. Ryšánek. 2018. Suppression of *Meloidogyne incognita* by the Entomopathogenic Fungus *Lecanicillium muscarium. Plant Disease* 102:977–982.

Johnson, R.A., Thomas, R.J., Wood, T.G., and M.J. Swift. 1981. The inoculation of the fungus comb in newly founded colonies of some species of the Macrotermitinae (Isoptera) from Nigeria. *Journal of Natural History* 15:751–756, doi:10.1080/00222938100770541.

Jouquet, P., Tavernier, V., Abbadie, L., and M. Lepage. 2005. Nests of subterranean fungus-growing termites (Isoptera, Macrotermitinae) as nutrient patches for grasses in savannah ecosystems. *African Journal of Ecology* 43:191–196, doi:10.1111/j.1365-2028.2005.00564.x.

Kasson, M.T., O'Donnell, K., Rooney, A.P., et al. 2013. An inordinate fondness for Fusarium: Phylogenetic diversity of fusaria cultivated by ambrosia beetles in the genus Euwallacea on avocado and other plant hosts. *Fungal Genetics and Biology* 56:147–157, doi:10.1016/j.fgb.2013.04.004.

Klingen, I., Eilenberg, J., and R. Meadow. 2002. Effects of farming system, field margins and bait insect on the occurrence of insect pathogenic fungi in soils. *Agriculture, Ecosystems & Environment* 91:191–198, doi:10.1016/S0167-8809(01)00227-4.

Klingen, I., Westrum, K., and N.V. Meyling. 2015. Effect of Norwegian entomopathogenic fungal isolates against *Otiorhynchus sulcatus* larvae at low temperatures and persistence in strawberry rhizospheres. *Biological Control* 81, 1–7.

Koné, N.A., Yéo, K., Konaté, S., and K.E. Linsenmair. 2013. Socio-economical aspects of the exploitation of Termitomyces fruit bodies in central and southern Côte d'Ivoire: Raising awareness for their sustainable use. *Journal of Applied Biosciences* 70:5580–90.

Korb, J., and D.K. Aanen. 2003. The evolution of uniparental transmission of fungal symbionts in fungus-growing termites (Macrotermitinae). *Behavioral Ecology and Sociobiology* 53:65–71, doi:10.1007/s00265-002-0559-y.

Kostovcik, M., Bateman, C.C., Kolarik, M., Stelinski, L.L., Jordal, B.H., and J. Hulcr. 2015. The ambrosia symbiosis is specific in some species and promiscuous in others: Evidence from community pyrosequencing. *ISME Journal* 9:126–138, doi:10.1038/ismej.2014.115.

Lagerlöf, J., Insunza, V., Lundegårdh, B., and B. Rämert. 2011. Interaction between a fungal plant disease, fungivorous nematodes and compost suppressiveness. *Acta Agriculturae Scandinavica, Section B – Soil & Plant Science* 61:372–377.

Lee, C.H., Chang, H.W., Yang, C.T., Wali, N., Shie, J.J., and Y.P. Hsueh. 2020. Sensory cilia as the Achilles heel of nematodes when attacked by carnivorous mushrooms. *Proceedings of the National Academy of Sciences* 117(11):6014–22.

LeFait, A., Gailey, J., and G. Kernaghan. 2019. Fungal species selection during ectomycorrhizal grazing by Collembola. *Symbiosis.* 78:87–95, doi:10.1007/s13199-018-00596-x.

Lewis, T. 1975. Colony size, density and distribution of the leaf-cutting ant, *Acromyrmex octospinosus* (Reich) in cultivated fields. *Transactions of the Royal Entomological Society of London* 127(1), 51–64.

Li, G., Zhang, K., Xu, J., Dong, J., and Y. Liu. 2007. Nematicidal substances from fungi. *Recent Patents on Biotechnology* 1:212–233.

Li, G-H., and K-Q. Zhang. 2014. Nematode-toxic fungi and their nematicidal metabolites. In *Nematode-Trapping Fungi*, eds. K.-Q. Zhang and K.D. Hyde, pp. 313–375. Dordrecht: Springer Netherlands.

Li, J., Zou, C., Xu, J., et al. 2015. Molecular mechanisms of nematode-nematophagous microbe interactions: Basis for biological control of plant-parasitic nematodes. *Annual Review of Phytopathology* 53:67–95.

Li, Y.T., Tsay, T.T., and P. Chen. 2014. Using three fungivorous nematodes to control lettuce damping-off disease caused by *Rhizoctonia solani* (AG4). *Plant Pathology Bulletin* 23(1), pp. 43–53.

Lopez-Llorca, L.V., Maciá-Vicente, J.G., and H-B. Jansson. 2008. Mode of action and interactions of nematophagous fungi. In *Integrated Management and Biocontrol of Vegetable and Grain Crops Nematodes*, eds. A. Ciancio, and K.G. Mukerji, pp. 51–76. Dordrecht: Springer Netherlands.

Loreto, R.G., Araújo, J.P.M., Kepler, R.M., Fleming, K.R., Moreau, C.S., and D.P. Hughes. 2018. Evidence for convergent evolution of host parasitic manipulation in response to environmental conditions. *Evolution.* 72:2144–2155, doi:10.1111/evo.13489.

Lovett, B., Macias, A., Stajich, J.E., et al. 2020. Behavioral betrayal: How select fungal parasites enlist living insects to do their bidding. *PLoS Pathogens* 16(6):e1008598. https://doi.org/10.1371/journal.ppat.1008598.

Mayers, C.G., McNew, D.L., Harrington, T.C., et al. 2015. Three genera in the ceratocystidaceae are the respective symbionts of three independent lineages of ambrosia beetles with large, complex mycangia. *Fungal Biology* 119:1075–1092, doi:10.1016/j.funbio.2015.08.002.

Manzanilla-López, R.H., Esteves, I., Finetti-Sialer, M.M., et al. 2013. *Pochonia chlamydosporia*: Advances and challenges to improve its performance as a biological control agent of sedentary endo-parasitic nematodes. *Journal of Nematology* 45:1–7.

Mietkiewski, R.T., Pell, J.K., and S.J. Clark. 1997. Influence of pesticide use on the natural occurrence of entomopathogenic fungi in arable soils in the UK: Field and laboratory comparisons. *Biocontrol Science and Technology* 7:565–576, doi:10.1080/09583159730622.

Mueller, U.G., Schultz, T.R., Currie, C.R., and D. Malloch. 2001. The origin of the attine ant-fungus mutualism. *Quarterly Review of Biology* 76:169–197, doi:10.1086/393867.

Pinto-Tomas, A.A., Anderson, M.A., Suen, G., et al. 2009. Symbiotic nitrogen fixation in the fungus gardens of leaf-cutter ants. *Science* 326:1120–1123, doi:10.1126/science.1173036.

Plaza, D.F., Schmieder, S.S., Lipzen, A., Lindquist, E., and M. Künzler. 2016. Identification of a novel nematotoxic protein by challenging the model mushroom *Coprinopsis cinerea* with a fungivorous nematode. *G3 Genes|Genomes|Genetics* 6:87–98.

Poprawski, T.J., and I. Majchrowicz. 1995. Effects of herbicides on in vitro vegetative growth and sporulation of entomopathogenic fungi. *Crop Protection* 14(1), 81–87.

Poulsen, M., Cafaro, M.J., Erhardt, D.P., et al. 2010. Variation in Pseudonocardia antibiotic defence helps govern parasite-induced morbidity in Acromyrmex leaf-cutting ants. *Environmental Microbiology Reports* 2(4):534–40.

Quesada-Moraga, E., Navas-Cortés, J.A., Maranhao, E.A.A., Ortiz-Urquiza, A., and C. Santiago-Álvarez. 2007. Factors affecting the occurrence and distribution of entomopathogenic fungi in natural and cultivated soils. *Mycological Research* 111:947–966, doi:10.1016/j.mycres.2007.06.006.

Ragozzino, A., and G. d'Errico. 2012. Interactions between nematodes and fungi: A concise review. *Redia* 19(94):123–5.

Rhoades, H. L., and M.B. Linford. 1959. Control of Pythium root rot by the nematode *Aphelenchus avenae*. *Plant Disease Reporter* 43:323–328.

Rinker, D.L. 2017. Insect, mite, and nematode pests of commercial mushroom production. In *Edible and Medicinal Mushrooms*, eds. C.Z. Diego and A. Pardo-Giménez, pp. 221–237. Chichester: John Wiley & Sons, Ltd.

Rouland-Lefèvre, C. 2010. Termites as pests of agriculture. In *Biology of Termites: A Modern Synthesis*, eds. D. Bignell, Y. Roisin, and N. Lo, pp. 499–517. Dordrecht: Springer.

Sain, S.K., Monga, D., Kumar, R., Nagrale, D.T., Hiremani, N.S., and S. Kranthi. 2019. Compatibility of entomopathogenic fungi with insecticides and their efficacy for IPM of *Bemisia tabaci* in cotton. *Journal of Pesticide Science* 44:97–105, doi:10.1584/jpestics.D18-067.

Sánchez-Peña, S.R., Lara, J.S.J., and R.F. Medina. 2011. Occurrence of entomopathogenic fungi from agricultural and natural ecosystems in Saltillo, Mexico, and their virulence towards thrips and whiteflies. *Journal of Insect Science* 11(1), 1. https://doi.org/10.1673/031.011.0101.

Sasan, R.K., and M.J. Bidochka. 2012. The insect-pathogenic fungus *Metarhizium robertsii* (Clavicipitaceae) is also an endophyte that stimulates plant root development. *American Journal of Botany* 99:101–107, doi:10.3732/ajb.1100136.

Shykoff, J.A. and E. Bucheli. 1995. Pollinator visitation patterns, floral rewards and the probability of transmission of *Microbotryum violaceum*, a veneral disease of plants. *Journal of Ecology* 189–198.

Slot, J.C. and M.T. Kasson. 2021. Ecology: Fungal mimics dupe animals by transforming plants. *Current Biology* 31:R250–R252.

Snyder, J.R. 2014. *Ecological Implications of Laurel Wilt Infestation on Everglades Tree Islands, Southern Florida*. US Department of the Interior, US Geological Survey.

Soares, F.E. F., de Sufiate, B.L., and J.H. de Queiroz. 2018. Nematophagous fungi: Far beyond the endoparasite, predator and ovicidal groups. *Agriculture and Natural Resources* 52:1–8.

Tkaczuk, C., Król, A., Majchrowska-Safaryan, A., and Ł. Nicewicz. 2014. The occurrence of entomopathogenic fungi in soils from fields cultivated in a conventional and organic system. *Journal of Ecological Engineering* 15:137–44.

Tragust, S., Mitteregger, B., Barone, V., Konrad, M., Ugelvig, L.V., and S. Cremer. 2013. Ants disinfect fungus-exposed brood by oral uptake and spread of their poison. *Current Biology*. 23:76–82, doi:10.1016/j.cub.2012.11.034.

Singh, A. U., and K. Sharma. 2016. Pests of mushroom. *Advances in Crop Science and Technology* 04.

Vänninen, I., Tyni-Juslin, J., and H. Hokkanen. 2000. Persistence of augmented *Metarhizium anisopliae* and *Beauveria bassiana* in Finnish agricultural soils. *BioControl* 45(2):201–22.

Varón, E.H., Eigenbrode, S.D., Bosque-Pérez, N.A., and L. Hilje. 2007. Effect of farm diversity on harvesting of coffee leaves by the leaf-cutting ant Atta cephalotes. *Agricultural and Forest Entomology* 9:47–55, doi:10.1111/j.1461-9563.2006.00320.x.

Vega, F.E., Meyling, N.V., Luangsa-ard, J.J., and M. Blackwell. 2012. Fungal entomopathogens. In *Insect Pathology*, eds. F.E. Vega and H.K. Kaya, pp. 171–220. Amsterdam: Elsevier.

Wakil, W., Ghazanfar, M.U., Riasat, T., Kwon, Y.J., Qayyum, M.A., and M. Yasin. 2013. Occurrence and diversity of entomopathogenic fungi in cultivated and uncultivated soils in Pakistan. *Entomological Research* 43(1):70–78.

Warnock, A.J., Fitter, A.H., and M.B. Usher. 1982. The influence of a springtail *Folsomia candida* (Insecta, Collembola) on the mycorrhizal association of leek *Allium porrum* and the vesicular-arbuscular mycorrhizal endophyte *Glomus fasciculatus*. *New Phytologist* 90:285–292, doi:10.1111/j.1469-8137.1982.tb03260.x.

Will, I., Das, B., Trinh, T., Brachmann, A., Ohm, R. A., and C. de Bekker. 2020. Genetic Underpinnings of Host Manipulation by *Ophiocordyceps* as Revealed by Comparative Transcriptomics. *G3Genes-Genomes-Genetics* 10(7), 2275–2296. https://doi.org/10.1534/g3.120.401290.

Yanagawa, A., and S. Shimizu. 2007. Resistance of the termite, *Coptotermes formosanus* Shiraki to *Metarhizium anisopliae* due to grooming. *Biocontrol* 52:75–85, doi:10.1007/s10526-006-9020-x.

Yang, J., Wang, L., Ji, X., et al. 2011. Genomic and proteomic analyses of the fungus *Arthrobotrys oligospora* provide insights into nematode-trap formation. *PLoS Pathogens* 7:e1002179.

# Section III

## Products

# 16 Principles of Modern Fungal Cultivation

*Elizabeth "Izzie" Gall*

## CONTENTS

## 16.1 INTRODUCTION

Mushrooms are a vitamin-rich, non-animal-based protein source low in saturated fatty acids and cholesterol (Chang and Miles 2004). As complete proteins, most cultivated mushrooms are good sources for all the amino acids required in a healthy adult diet. They also come in a variety of textures and colors that are appealing to people from many different cultures. As such, mushrooms are an important, reliable, and palatable protein source and it is no wonder that their cultivation continues to expand.

As with any agricultural product, achieving a high-quality mushroom crop requires careful management and a thorough understanding of the crop's biology, needs, and lifestyle. This chapter briefly considers limiting factors and common practices in modern mushroom cultivation.

## 16.2 PROTEIN VALUE OF MUSHROOMS

It can be difficult to generalize the nutritional value of "mushrooms" since the category includes so many species and different methods of cultivation. For example, the maximum nitrogen content of a mushroom depends strongly on the C/N ratio of its substrate (see Chapter 17 in this volume) and the first mushroom harvest may take up more of the minerals in the substrate than the subsequent harvests. However, studies that investigate the average content of mushrooms from different species have demonstrated that mushrooms are highly nutritious, containing many B-type and other vitamins; potassium, phosphorus, calcium, magnesium, and other minerals; a good amount of fiber; and very little fat (Chang and Miles 2004).

Due to their high moisture content, it can be difficult to compare the amount of total usable protein in mushrooms with the protein content of other sources, like animal meats. Table 16.1 includes

DOI: 10.1201/9780429320415-19

**TABLE 16.1**

**Crude Protein Content of Several Mushrooms and Other Protein Sources When Fresh and Dried**

| Protein Source | Moisture Content (% Fresh Weight) | Fresh Protein Content (% Fresh Weight) | Dry Protein Content (% Dry Weight) |
|---|---|---|---|
| *Agaricus bisporus* | 78.3–90.5 | 2.3–7.6* | 23.9–34.8 |
| *Agaricus campestris* | 89.7 | 3.4* | 33.2 |
| *Auricularia* sp. (Philippine var.) | 89.1 | 0.5* | 4.2 |
| *Boletus edulis* | 87.3 | 3.8* | 29.7 |
| *Flammulina velutipes* | 89.2 | 1.9* | 17.6 |
| *Lentinula edodes* | 90.0–91.8 | 1.1–1.8* | 13.4–17.5 |
| *Pleurotus eous* | 92.2 | 2.0* | 25 |
| *Pleurotus florida* | 91.5 | 2.3* | 27 |
| *Pleurotus ostreatus* | 73.7–90.8 | 1.0–8.0* | 10.5–30.4 |
| *Pleurotus sajor-caju* | 90.2 | 2.6* | 26.6 |
| *Volvariella displasia* | 90.4 | 2.7* | 28.5 |
| *Volvariella volvacea* | 89.1 | 2.8* | 25.9 |
| Beef, lean | 73.1 | 23.2 | 86.2* |
| Veal, lean | 74.8 | 24.8 | 33.2* |
| Lamb, lean | 72.9 | 21.9 | 30.0* |
| Mutton, lean | 73.2 | 21.5 | 29.4* |

*Source:* Chang and Miles (2004), Williams (2007).

Values marked with an asterisk (*) are calculated based on data available in the sources.

the moisture content of several fresh protein sources, including many mushrooms and a few types of animal meat. Note that while the meats are 70–75% water when fresh, the mushrooms range from 73% up to 90% water by fresh weight. A higher moisture content means proportionally less protein, so the fresh protein content of mushrooms appears to be very low compared with that of the meats (Fresh Protein Content in Table 16.2). However, we can account for the high water content by instead comparing protein sources on a dry weight basis. Note in Table 16.1 that many of the mushrooms, including some *Pleurotus* and *Volvariella* species, have a dry protein content rivaling that of lean[1] animal meats like veal, lamb, and mutton.

Sheer amount of protein is not the whole picture, however. Proteins are created with biological "building blocks" called **amino acids**. Of the 20 "standard" amino acids used to create human proteins, adult humans can create 11 of them. The remaining nine must be sourced from the diet and are known as the **essential amino acids**: histidine (abbreviated His), isoleucine (Ile), leucine (Leu), lysine (Lys), methionine (Met), phenylalanine (Phe), threonine (Thr), tryptophan (Trp), and valine (Val). There are additional amino acids that children and infants must have in the diet, but these nine are needed for humans of any age. Any protein source that supplies all nine of these essential amino acids is considered a **complete protein**. Animal proteins are complete proteins; many plant-based proteins are **incomplete**, lacking at least one of the essential amino acids. This is why vegans are advised to mix legumes (which are typically low in methionine but high in lysine) with cereal grains (which are high in methionine, but low in lysine).

Because all nine essential amino acids are needed for protein construction, the essential amino acid that is present in the lowest relative amount will limit the amount of protein that can be constructed. Therefore, the amino acid that is least available in a protein source relative to how much

---

[1] Lean meats have had the external fat trimmed but may have some marbled fat in with the muscle.

**TABLE 16.2**

**Abundance of the Nine Essential Amino Acids in Several Edible Fungi and Several Animal Proteins**

| Protein Source | Essential Amino Acid Analysis (Grams of Amino Acid per 100 g of Protein) | | | | | | | | | |
|---|---|---|---|---|---|---|---|---|---|---|
| | His | Ile | Leu | Lys | Met | Phe | Thr | Trp | Val | Total |
| *A. bisporus* | 2.7 | 4.5 | 7.5 | 9.1 | 0.9 | 4.2 | 5.5 | 2.0 | 2.5 | 38.9 |
| *Lentinula edodes* | 1.9 | 4.9 | 7.9 | 3.9 | 1.9 | 5.9 | 5.9 | NR | 3.7 | 36.0 |
| *Pleurotus florida* | 2.8 | 5.2 | 7.5 | 9.9 | 3.0 | 3.5 | 6.1 | 1.1 | 6.9 | 46.0 |
| *Pleurotus ostreatus* | 1.7 | 4.2 | 6.8 | 4.5 | 1.5 | 3.7 | 4.6 | 1.3 | 5.1 | 33.4 |
| *Pleurotus sajor-caju* | 2.2 | 4.4 | 7.0 | 5.7 | 1.8 | 5.0 | 5.0 | 1.2 | 5.3 | 37.6 |
| *Volvariella diplasia* | 4.2 | 7.8 | 5.0 | 6.1 | 1.2 | 7.0 | 6.0 | 1.5 | 9.7 | 48.5 |
| *V. volvacea* | 3.8 | 3.4 | 4.5 | 7.1 | 1.1 | 2.6 | 3.5 | 1.5 | 5.4 | 32.9 |
| Hen's egg | 2.4 | 6.6 | 8.8 | 6.4 | 3.1 | 5.8 | 5.1 | 1.6 | 7.3 | 47.1 |
| Beef[1] | 2.8 | 5.0 | 8.5 | 8.2 | 2.2 | 4.1 | 4.2 | 1.3 | 5.6 | 41.9 |
| Lamb[1] | 2.9 | 4.7 | 7.2 | 7.5 | 2.4 | 3.8 | 4.8 | 1.2 | 5.1 | 39.6 |
| Pork[1] | 3.1 | 4.8 | 7.6 | 7.9 | 2.6 | 4.3 | 5.2 | 1.5 | 5.2 | 42.2 |

Unmarked data are from Chang and Miles (2004).

[1] Ahmad et al. (2018).

The least abundant amino acid is highlighted for each protein source. For *L. edodes,* the two lowest (equal) amino acids are highlighted.

NR, not reported.

is needed by the human body is known as the **limiting amino acid**. To identify the limiting amino acid, we first find a food's **amino acid analysis** or **amino acid score**. This analysis identifies the quantity of each amino acid in the protein source and reports it as a percent of the total **crude** (isolated) protein. Table 16.2 contains the essential amino acid analysis for many mushrooms and several animal proteins (including the vegetarian option of a chicken egg) for comparison.

Note that tryptophan (Trp) is the least abundant amino acid for most of the protein options listed. This does *not* make it the limiting amino acid for those proteins. Of the essential amino acids, tryptophan is needed in the smallest amount. Adults only need about 3.5 to 6.0 mg of tryptophan per kilogram of body weight per day (Richard et al. 2009). Therefore, a 68 kg (150 pound) person needs about 0.2–0.4 grams of tryptophan daily, an amount supplied by 65 grams (2.3 ounces) of fresh beef or by 63 grams (2.2 ounces) of dried mushroom (*Pleurotus sajor-caju*).

Instead, the limiting amino acid must be found by comparing the amino acid score of each protein source with the score of a **reference protein**, a complete protein known to have the essential amino acids in the right proportion for the human body's needs. To find the limiting amino acid, the scores for all the amino acids from one protein source are divided by the scores for the reference protein. For example, the protein in *A. bisporus* is 7.5% leucine, while the ovalbumin protein in a hen's egg (the reference protein) is 8.8% leucine (Table 16.2). Dividing the leucine score of *A. bisporus* over the leucine score of a hen's egg, we find that *A. bisporus* includes about 85% of the amount of leucine found in a hen's egg. We conduct this comparison with all the essential amino acids' scores; the amino acid with the lowest score relative to the hen's egg is the limiting amino acid for the protein. Table 16.3 highlights the limiting amino acid values for each protein source relative to the values in a hen's egg.

## 16.3   LIMITING FACTORS IN MUSHROOM CULTIVATION

Fungi are heterotrophs; they cannot generate their own energy and biomolecules from scratch, so they must be supplied with a ready source of nutrients. As absorptive feeders, fungi secrete enzymes

**TABLE 16.3**

**Amino Acid Scores for the Nine Essential Amino Acids in Several Edible Fungi and Several Animal Proteins, Relative to the Amino Acid Scores for a Reference Protein (Hen's Egg)**

| Protein Source | Amino Acid Scores Relative to Hen's Egg Reference Protein (Percentage) | | | | | | | | |
|---|---|---|---|---|---|---|---|---|---|
| | His | Ile | Leu | Lys | Met | Phe | Thr | Trp | Val |
| *A. bisporus* | 113 | 68 | 85 | 142 | 29 | 72 | 108 | 125 | 34 |
| *Lentinula edodes* | 79 | 74 | 90 | 61 | 61 | 102 | 116 | No data | 51 |
| *Pleurotus florida* | 117 | 79 | 85 | 155 | 97 | 60 | 120 | 69 | 95 |
| *Pleurotus ostreatus* | 71 | 64 | 77 | 70 | 48 | 64 | 90 | 81 | 70 |
| *Pleurotus sajor-caju* | 92 | 67 | 80 | 89 | 58 | 86 | 98 | 75 | 73 |
| *Volvariella diplasia* | 175 | 118 | 57 | 95 | 39 | 121 | 118 | 94 | 133 |
| *V. volvacea* | 158 | 52 | 51 | 111 | 35 | 45 | 69 | 94 | 74 |
| Hen's egg | 100 | 100 | 100 | 100 | 100 | 100 | 100 | 100 | 100 |
| Beef | 117 | 76 | 97 | 128 | 71 | 71 | 82 | 81 | 77 |
| Lamb | 121 | 71 | 82 | 117 | 77 | 66 | 94 | 75 | 70 |
| Pork | 129 | 73 | 86 | 123 | 84 | 74 | 102 | 94 | 71 |

The limiting amino acid is highlighted for each protein source. For beef, the two equally limiting amino acids are highlighted. This table includes calculations from the values in Table 16.2.

into the environment to digest large molecules, then bring the smaller "building blocks" into their cells to use as energy or to construct the molecules required for their growth and metabolism.

Most macrofungi currently grown at an industrial scale are basidiomycetes (see Chapter 4 in this volume). Saprotrophic varieties are typically grown on a **compost**, a pile of dead and decaying vegetable matter sometimes combined with animal waste. Depending on the available materials, compost piles for mushroom cultivation can include corn cobs and stalks, cocoa bean husks, and other agricultural and industrial waste products (see Chapter 17 in this volume). Wood-rotting fungi like shiitake mushrooms are typically cultivated on logs, wood chips, sawdust, and other non-fermented materials.

### 16.3.1 CARBON DIOXIDE CONCENTRATIONS

Like humans, fungi need access to oxygen gas in order to conduct metabolism, and that metabolism releases carbon dioxide gas ($CO_2$). The balance between these two gases has an important impact on mushroom fruiting. The ideal concentrations of oxygen and $CO_2$ for growth vary between mushroom species, but generally, a relatively high $CO_2$ concentration of 1–2% encourages vegetative growth while a decrease to about 0.1% $CO_2$ initiates fruiting (Senseair 2020). Carbon dioxide concentrations higher than 0.1% during fruiting can cause changes in the growth and appearance of mushrooms, such as a decrease in cap size and a longer stipe (stem). The popular enoki mushroom *Flammulina velutipes* is often grown in high-$CO_2$ environments to encourage the growth of long stipes, which are considered desirable. However, mushroom formation is completely inhibited at $CO_2$ concentrations of 0.5% (Senseair 2020). For comparison, in 2019 the average $CO_2$ level in the global atmosphere was 0.04%, the highest atmospheric carbon dioxide level in 800,000 years (Lindsey 2020). The ideal $CO_2$ concentrations for fungal growth and fruiting are much higher than this average, so they must be achieved locally through metabolism, environmental features, or human control. This is why mushrooms grow well in areas like caves, forest understories, and mushroom houses, where $CO_2$ can build up. The lower $CO_2$ levels that favor fruiting may be reached with air circulation, either natural (wind) or human-directed (ventilation systems).

### 16.3.2 CARBON AND NITROGEN

The **carbon** in a substrate provides both the materials for structural growth and the energy for metabolism. As discussed in Chapters 4 and 17, fungi can easily source their carbon from structural plant molecules like starch, cellulose, and lignin. Fungi can also use protein backbones in the substrate as a carbon source, if needed (Jennings and Lysek 1999). The ideal concentration of carbon can range from 2% to 20% of the substrate mass, depending on the fungal species (Chang and Miles 2004).

**Nitrogen** is required for the construction of proteins, nucleic acids, and some complex structural sugars (Chang and Miles 2004). If ammonium ($NH_4^+$) is present in low concentrations, it is the preferred nitrogen source for fungi. However, mushroom species such as *Agaricus bisporus* are highly sensitive to ammonia, a form of nitrogen that can form easily from ammonium, and mushroom growth can actually be inhibited by high amounts of it. Industrial composting for cultivation of saprobic mushrooms (see below) continues until most of the ammonium has been removed from the substrate and turned into proteins. When ammonium is not available, fungi will harvest nitrogen from proteins and nucleic acids in their substrate (Jennings and Lysek 1999). Some fungi can also utilize nitrate, $NO_3$, as a nitrogen source. Nitrate is highly water-soluble and is a common form of synthetic fertilizer.

The amount of nitrogen required in the substrate depends on the mushroom being cultivated. The relative amount of carbon and nitrogen in the substrate is often referenced as the **C/N ratio**, usually a whole number. For example, the species *Agaricus bisporus* requires 17 units of carbon for every unit of nitrogen in the substrate, a C/N ratio of 17 (Chang and Miles 2004). A very high C/N ratio, such as the wastepaper C/N ratio of 400 (Osunde et al. 2019), indicates a low relative nitrogen content. The ideal C/N ratio is different for each fungal species and the C/N ratio of different substrates varies widely (see Chapter 17 in this volume). The closer a substrate's C/N ratio is to the ideal growth C/N ratio for a fungal species, the better that species will perform on that substrate, whether in terms of growth rate or mushroom productivity. The substrate C/N ratio can be decreased by adding nitrogen-rich compounds, such as grain bran or chicken manure.

### 16.3.3 OTHER MACRONUTRIENTS

Besides carbon and nitrogen, there are certain other compounds that fungi require in relatively high amounts. One such macronutrient is **phosphorus** (elemental symbol P), which is an important part of metabolism due to its presence in the biological energy molecule adenosine triphosphate. Phosphorus is also needed for creating the backbones of nucleic acids and for making cellular membranes. In nature, fungi source phosphorus from the nucleic acids, membranes, and other phosphorus-containing compounds inside the organisms they digest (Jennings and Lysek 1999).

Another important macronutrient is **potassium** (elemental symbol K), which is important for maintaining electrical signaling potential and managing the flow of water within and between cells (Jennings and Lysek 1999). In cultivation media, potassium and phosphorus are supplied together as potassium phosphate salt (Chang and Miles 2004).

**Sulfur** (S) is needed for the creation of certain amino acids (e.g., methionine and cysteine), vitamins, and secondary metabolites. For example, sulfur is an important part of the structure of penicillin, an important antibacterial compound produced by some fungi (Chang and Miles 2004). In nature, fungi use **inorganic** (non-biologically derived) sulfur compounds from the environment. In mushroom cultivation, sulfur is supplied along with **magnesium** in magnesium sulfate. Magnesium (Mg) is essential to all fungi for its ability to activate enzymes (Chang and Miles 2004).

### 16.3.4 WATER

During the last few days of growth, an *Agaricus* mushroom doubles in size every 24 hours (Beyer 2017). The sudden, massive growth of most cultivated mushrooms is mostly due to the movement of water into the fruit, contributing 73% to 91% of the fresh mushroom's weight (Chang and Miles

**TABLE 16.4**

**Substrate Water and Humidity Requirements for Some Edible and Medicinal Mushrooms**

| Mushroom | Substrate | Substrate Water Content | Humidity |
|---|---|---|---|
| *Lentinula edodes* (shiitake) | Hardwood log | 40%–45% | NR |
| *Volvariella volvacea* (straw mushroom) | Wheat straw compost | 60%–65% | NR |
| *Flammulina velutipes* (enoki) | Sawdust media | 60%–65% | To stimulate fruiting: 75%–85% |
| *Ganoderma lucidum* (Reishi) | Unspecified log | 60%–65% | Vegetative growth: 60%–70% To stimulate fruiting: 85%–90% Fruit growth: 70%–85% |
| *Agaricus blazei* (relative of the button mushroom) | Sugar cane bagasse compost | Spawning: 55%–60% Casing: 60%–65% | All phases: 75%–85% |
| *Grifola frondosa* (maitake) | Hardwood sawdust | 60%–63% | Vegetative growth: 60%–65% Fruit growth: 85%–95% |
| *Hericium erinaceus* (Lion's mane) | Sawdust and rice bran | 65% | Fruit growth: 85%–90% |

*Source:* Chang and Miles (2004).
NR, not reported.

2004). Fungi also require a considerable amount of water during vegetative growth. However, too much water in the substrate can encourage the growth of molds, mushroom diseases, and other non-target organisms that can affect the mass of the crop harvest. For mushrooms grown on compost, this issue can be resolved by using a thoroughly wetted, pasteurized compost; mushrooms can then gain almost half of their water from what is already in the substrate and additional watering can be limited without affecting harvest (SureHarvest 2017). Logs used for mushroom cultivation are typically dried for about one month after they are cut, which brings their internal moisture level from about 50% to 40–45%, the moisture level that usually encourages healthy mycelial growth (Chang and Miles 2004). After the trophic growth phase is complete, the logs may be sprayed with water or briefly soaked to provide enough moisture to support fruiting.

Table 16.4 shows how wet the substrate needs to be for ideal growth of vegetative mycelia for several edible mushroom varieties. Note that in addition to the substantial amounts of water sourced from the substrate, mushrooms require very humid environments to avoid water loss during fruiting.

### 16.3.5 Light

Because they do not photosynthesize, fungi do not depend on sunlight for growth. Many mushrooms can grow and fruit regardless of light or darkness levels in their surroundings. Some agriculturally important fungal species do require some light to encourage growth, to act as a signal to begin fruiting, or to encourage fruits to mature (Chang and Miles 2004). For example, to begin fruiting, the Reishi mushroom *Ganoderma lucidum* requires 500–1000 lux of light, the level of illumination in an average office. As the fruits mature, Reishi require a higher light level of 750–1500 lux, a bit brighter than an overcast day (Chang and Miles 2004; National Optical Astronomy Observatory 2015). This amount of light can be easily supplied by commercially available "daylight" fluorescent or LED lights (McCoy 2016).

### 16.3.6 Micronutrients

Micronutrients are elements that are essential for growth but are only required in very low concentrations. Many of the micronutrients needed for fungal growth overlap with those needed by plants, but fungi often have more effective mechanisms than plants for taking up these elements from their surroundings. Fungal micronutrients include calcium, iron, copper, zinc, molybdenum, and manganese, the last of which is especially important for digesting lignin (Jennings and Lysek 1999).

## 16.4 MODERN MUSHROOM CULTIVATION

Here we outline the very general process of modern mushroom cultivation. This chapter does not provide nearly enough information to begin cultivating fungi! Several of the processes outlined here, like composting, require special equipment and specialized management. Furthermore, each mushroom variety has specific needs such as preferred growth substrate, temperature range, and humidity levels. To thoroughly explore the cultivation methods for a wide variety of mushrooms, we highly recommend Chang and Miles' textbook *Mushrooms: Cultivation, Nutritional Value, Medicinal Effect, and Environmental Impact* (2004).

### 16.4.1 STARTING SUBSTRATE

The cultivation of saprobic fungi like the button mushroom *Agaricus bisporus* begins with a compost heap. Ideally the compost contains the waste products of locally farmed crops and locally managed livestock. Using crop waste in this way helps close the farm system into a self-sustaining "loop", as opposed to a linear system where inputs must be brought in from elsewhere and waste must be sent out for processing or disposal (Figure 16.1).

The composting process we outline here has been codified for a mixture of wheat straw and horse manure, the traditional substrate of button mushrooms. As the most popular saprobic mushroom on the global market, the button mushroom has been heavily studied and the composting process for its substrate is very well understood (Chang and Miles 2004). However, the general execution will be similar for most composted materials.

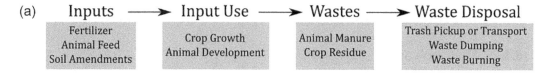

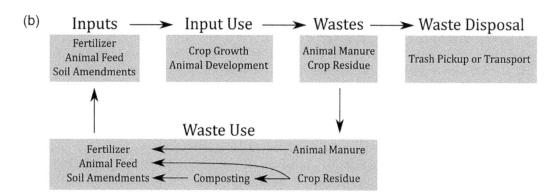

FIGURE 16.1 (a) In a linear waste stream, agricultural byproducts are disposed of via burning, dumping, etc. (b) In a closed-loop system, useful agricultural byproducts are put back into the system as inputs, reducing the amount of waste that is burned, dumped, or otherwise unproductively disposed of.

Composting begins very simply by heaping agricultural waste in a pile in a designated area, wetting it down, and turning the pile regularly for 3–15 days (Beyer 2017). There are already many fungal species present in a compost heap: plant endophytes, epiphytes, and pathogens; excreted gut fungi in animal feces; and soil symbionts and saprobes. In addition to these fungi and their spores, there may be spores of foreign species brought to the compost heap via wind, rain, or the clothing of farm workers. Other saprobes like bacteria and insects are also present, and they may be accompanied by their predators like nematodes and larger insects. Together, this complex ecosystem works to break down the dead plant material in the compost heap, using up the simple sugars that appeal to mushroom competitors like molds and making biological compounds more easily available to the mushroom species that will be cultivated on the compost (Chang and Miles 2004). Large mushroom farms use specialized machines that turn compost piles into rectangular strips for easy processing, then stir and wet them regularly as the materials inside decompose naturally (Beyer 2017). During composting, the material becomes very warm due to the metabolic action of all the organisms in the pile; the $CO_2$ concentration in the center of the pile can also get prohibitively high and kill off organisms if the pile is not stirred. Therefore, it is often necessary to use forced air to keep the interior of the pile oxygenated. Compost amendments like gypsum and animal fertilizer can be added to the piles as needed to bring the C/N ratio into the ideal range for the target mushrooms. This first round of composting is known as **Phase I**.

After the agricultural waste material has been broken down sufficiently and the compost is high in ammonia and simple sugars, Phase I is considered complete. **Phase II** composting begins with the **pasteurization** (partial sterilization) of the compost in preparation for inoculation with the target mushroom species (Beyer 2017). It is important that the Phase I composting agents be killed off so they do not compete with, eat, or cause disease to the cultivated mushrooms. However, the pasteurization temperatures are not sufficient to kill off the **thermophilic** ("heat-loving") microbes in the pile, which continue the composting process. As the compost pile is slowly cooled down over several days, the thermophilic microbes use up most of the ammonia created during Phase I, turning it from a mushroom toxin into proteins that saprobic fungi can readily use (Beyer 2017). After the pile has been fully cooled to room temperature, Phase II is complete.

### 16.4.2 Mushroom Spawn

**Spawn** is a mixture of grain and mycelium that is used to "seed" the substrate with the mushroom species of choice. Clean samples of the desired fungus' mycelium are stored in vials or Petri dishes at low temperature (usually −4°C/24.8°F). When a new batch of mushrooms needs to be started, sterile tools are used to remove small samples of the mycelium. The mycelium is then mixed into a container of dry grains (often rice or barley) and allowed to grow at optimal temperature for several days. Once the grains are thickly inoculated with mycelium, they can be mixed with larger batches of grain to increase the volume of spawn.

When enough spawn has been produced, it is mixed into the compost, creating an even distribution of the mushroom mycelium in the nutrient-rich substrate. Greater spawn density in the compost leads to quicker colonization of the substrate, leaving less time for pests or competitors to establish in the event of contamination (Beyer 2017). The mixture may then be loaded into raised growing beds, polyethylene (special plastic) bags, or glass jars, depending on the mushroom variety and the process preferred by the farmer. The humidity, temperature, and other factors in the growing chamber should be closely monitored to ensure the growth of healthy mycelium throughout the substrate. For log-grown mushrooms, small sections of spawn are tamped gently into holes drilled into the logs, then covered with wood plugs that prevent contamination and moisture loss. The period of vegetative growth is known as the **spawn run** or **mycelial running**. The spawn run usually takes two to three weeks for button mushrooms (Beyer 2017) but can take up to a year for log-grown varieties (Chang and Miles 2004).

When the vegetative mycelium reaches the end of its substrate, the hyphae at the substrate edge send signals that lead to the development of aerial hyphae and the formation of the fruiting body. Fruiting in saprobic mushrooms can also be triggered with the addition of a **casing**, a layer of soil or pH-adjusted peat. For log-grown mushrooms, fruiting can be encouraged with watering or by moving logs to a more humid, shaded environment (Chang and Miles 2004). When mushrooms first begin to emerge, they appear as tiny spheres on the substrate surface which are known as **pins** or **primordia**. These primordia bloom into fully fledged mushrooms over several days. Mature fruiting bodies are removed and can be sold fresh, canned, or dried, depending on the local mushroom market's scope and demands. Once the first crop of mushrooms is harvested, the mycelium can often give rise to additional harvests, though the first harvest will supply the greatest mass. After the final harvest, used compost, now known as **spent mushroom substrate**, can be discarded from the mushroom cultivation process, weathered, and used as a soil amendment (see Chapter 17 in this volume). Solid growth containers can be sterilized and filled with the next batch of spawn and compost. Logs can generate several years' worth of harvests before being discarded.

## 16.5   GENETIC ASPECTS OF MUSHROOM CULTIVATION

It is important to emphasize that fungi are generally "planted" clonally, using a culture of existing mycelium rather than planting genetically distinct spores. The initial sample of mycelium is usually isolated from a sterile (interior) portion of a mushroom that has a desirable taste, smell, and appearance. The mycelium can be cultured on multiple test plates to ensure that it has a good potential to continue fruiting under controlled conditions. **Clonal propagation**, the expansion of growth material from a single genetic source, ensures that all the growing mushrooms have consistent needs and are of similar quality. However, it also means that each mushroom in the population is susceptible to the same threats (see Chapter 2 in this volume). Therefore, mushroom growing spaces need to be carefully maintained and cleaned regularly to limit the outbreak of pests and diseases.

As in any biological system, mutations may occur in fruiting mycelial stock. If any differences in mycelial color or texture appear, the mycelial stock with typical appearance should be re-plated numerous times before being used for fruiting again (Beyer 2017). Unfortunately, some mutations are invisible in the vegetative mycelia and will only be revealed upon fruiting – or failure to fruit. Some varieties of gourmet mushroom have been lost from the global market because the only available mycelial strains stopped fruiting (Gall 2016). Therefore, it is useful to keep domesticating new wild strains of desirable mushrooms; not only will they increase the number of mushroom varieties available to the global market, but having multiple options will help insure mushroom farmers against the loss of one fruiting variety.

## REFERENCES

Ahmad, R. S., A. Imran, and M.B. Hussain. 2018. Nutritional Composition of Meat. Essay. In *Meat Science and Nutrition*, ed. M. S. Arshad. IntechOpen. DOI:10.5772/intechopen.77045.

Beyer, D. M. 2017. Basic Procedures for Agaricus Mushroom Growing. Penn State Extension. https://extension.psu.edu/basic-procedures-for-agaricus-mushroom-growing (accessed April 29, 2020).

Chang, S. T., and P. G. Miles. 2004. *Mushrooms: Cultivation, Nutritional Value, Medicinal Effect and Environmental Impact*, 2nd ed. Boca Raton, Florida: CRC Press.

Finnigan, T., L. Needham, and C. Abbott. 2017. Mycoprotein: A Healthy New Protein with a Low Environmental Impact. Essay. In *Sustainable Protein Sources,* ed. S. R. Nadathur, P.D. Wanasundara, and L. Scanlin 305–325. London: Academic Press.

Gall, I. 2016. (*Taming of the Shroom*). *Morel Dilemma*. Podcast audio. Aug 15, 2016. https://www.podbean.com/eu/pb-ekgi4-678a43 (accessed April 28, 2020).

Jennings, D.H., and G. Lysek. 1999. *Fungal Biology: Understanding the Fungal Lifestyle*, 2nd ed. Oxford, UK: BIOS Scientific Publishers Ltd.

Lindsey, R.. 2020. Climate Change: Atmospheric Carbon Dioxide. Climate.gov. National Oceanic and Atmospheric Administration, August 14, 2020. https://www.climate.gov/news-features/understanding-climate/climate-change-atmospheric-carbon-dioxide (accessed April 26, 2020).

McCoy, P. 2016. Mushroom Cultivation and Application Course. Brooklyn; NY.

National Optical Astronomy Observatory. 2015. Recommended Light Levels. Mauna Kea, Hawaii.

Osunde, M. O., A. Olayinka, C. D. Fashina, and N. Torimiro. 2019. Effect of Carbon-Nitrogen Ratios of Lignocellulosic Substrates on the Yield of Mushroom (*Pleurotus pulmonarius*). *Open Access Library Journal* 6:e5777. https://doi.org/10.4236/oalib.1105777

Richard, D. M., M. A. Dawes, C. W. Mathias, A. Acheson, N., Hill-Kapturczak, and D. M. Dougherty. 2009. L-Tryptophan: Basic Metabolic Functions, Behavioral Research and Therapeutic Indications. *International Journal of Tryptophan* Research 2009(2):45–60. DOI: 10.4137/ijtr.s2129

Senseair, 2020. Mushroom Farming. Senseair. Asahi Kasei Group. https://senseair.com/knowledge/application-notes/life-science-safety/agriculture/mushroom-farming/ (accessed April 28, 2020).

SureHarvest. 2017. The Mushroom Sustainability Story: Water, Energy, and Climate Environmental Metrics, 2017 Report (March). The Mushroom Council. https://www.mushroomcouncil.com/wp-content/uploads/2017/12/Mushroom-Sustainability-Story-2017.pdf (accessed April 30, 2020).

Williams, P.G. 2007. Nutritional composition of red meat. *Nutrition & Dietetics* 64(Suppl 4):S113–S119.

# 17 Integrating Fungi into Existing Farms

*Elizabeth "Izzie" Gall*

## CONTENTS

## 17.1 INTRODUCTION

Fungal mycelium is a multi-use material. In the solid state, it can be made into leather substitutes, biodegradable packaging, bricks for temporary housing, and animal-free "meat" products (Marchant 2020). In the liquid state, fungal fermentation can create biofuels and generate enzymes used in the textile, detergent, and food industries (Kumla et al. 2020; see Chapter 18 in this volume). However, these are all applications that require specialized equipment and devoted buildings or warehouses. Fortunately, mushrooms also have the potential to generate usable products at low startup cost to farmers, making their integration into agricultural systems relatively simple. The easiest step is to incorporate fungi into waste recycling procedures, as fungi are unparalleled at degrading agro-industrial waste. If wastes can be properly separated, the different streams can be used as specialized growth medium for mushrooms, which represent a highly nutritious and valuable additional product for a farm. Farmers with enough agricultural waste of the right types could even be pioneers of new gourmet mushroom varieties by partnering with government or university laboratories to domesticate local species. Even the "spent" waste of mushroom cultivation can be a useful material or sellable product after the mushrooms have been harvested, so incorporating mushrooms into agroecological systems is both an ecologically and economically sound practice.

DOI: 10.1201/9780429320415-20

## 17.2  MUSHROOM CULTIVATION RECYCLES AGRO-INDUSTRIAL WASTE

**Agro-industrial waste** is any waste generated during agricultural activities or the processing of animal or crop products. This category includes everything from crop residue (seed husks, stalks, and roots left in the field after harvest) to the pulp and bagasse left over after extracting juice from fruit or sugar from sugarcane (Kumla et al. 2020). The plant-based component of agro-industrial waste is highly **lignocellulosic**, which means that it is mainly composed of the molecules cellulose, hemicellulose, and lignin. The proportion of these molecules in the waste depends on the source plant species, the tissue the waste is composed of, and the developmental stage of the source plant (Kumla et al. 2020).

**Cellulose** is a highly regular molecule made up of simple sugars linked together in a repetitive pattern. It is produced by all plants and is the most common biological molecule on Earth. Because of its regular structure and prevalence, many microbes can break down cellulose with a single class of enzymes.

**Hemicellulose** is similar in structure to cellulose but is made up of multiple types of simple sugars, which makes it harder to break down. However, it is still a regularly structured molecule that many organisms can break down.

By contrast, **lignin** is a rigid, irregular molecule with a complex structure that does not repeat. A single plant may lay down different configurations of lignin depending on environmental conditions (Kumla et al. 2020). Therefore, it takes the action of many different, specialized enzymes working in concert to fully break down lignin. Lignin degradation often releases cellulose and hemicellulose, so efficiently breaking down lignocellulosic waste requires the breakdown of lignin first (Kumla et al. 2020). Fungi are important degraders in terrestrial ecosystems because unlike many organisms, they can digest all three molecular components of lignocellulosic waste.

The chemical breakdown of lignocellulosic waste by fungi (or other organisms) is known as **fermentation**. Filamentous and yeast fungi can perform fermentation in liquid culture, turning lignocellulosic waste into highly desired molecules. This is how many enzymes, some preservatives, and other compounds are produced on an industrial scale. Liquid fermentation by fungi is also the method used to generate many biofuels, such as ethanol, from agro-industrial wastes like wheat and rice straw, corn residue, and sugarcane bagasse (Chomnunti and Fulds 2019). Filamentous and mushroom-forming fungi can perform **solid-state fermentation**, turning clean, wetted lignocellulosic waste into mycelium and mushrooms. Wood-inhabiting and saprotrophic fungi, especially white-rotting basidiomycetes, are the top degraders of lignin on Earth, which makes them an excellent resource for recycling lignocellulosic wastes. Introducing mushrooms to agricultural systems, especially in low-income areas, has the potential to close a waste loop and reduce the costs of waste disposal while generating a nutritious product with a high average protein content (see Chapter 16 in this volume).

The substrate's carbon to nitrogen ratio, abbreviated as the **C/N ratio**, has a large impact on the protein content of the mushroom product. This ratio represents the number of carbon atoms available per nitrogen atom in the substrate. Because nitrogen is necessary for the production of proteins and nucleic acids like those that build DNA, a higher C/N ratio makes it harder for a mushroom to concentrate nitrogen and produce high-nitrogen compounds. The final mineral and protein composition of the mushroom also depends on fungal species and the environmental conditions during growth and fruiting (Kumla et al. 2020). Recall from Chapter 6 that plant crop efficiency is measured with **harvest index**, the dry mass of harvestable product a plant has produced relative to the plant's overall dry biomass. Because fungi grow throughout their substrate as fine hyphae before fruiting, it can be very difficult to measure their dry biomass. Mushrooms also have a much higher water content than commodity crops like soybeans – up to 90% of a fresh mushroom is water (Chang and Miles 2004)! Therefore, mushroom productivity is evaluated as **biological efficiency**, the *fresh* weight of harvested mushrooms relative to the *dry* weight of the cultivation substrate (Kumla et al. 2020). This productivity measure accounts for both the high moisture content of the

**TABLE 17.1**

**Lignin Content and C/N Ratios of Selected Lignocellulosic Wastes**

| Lignocellulosic Waste Substrate | Lignin Content (%) | C/N Ratio |
|---|---|---|
| Banana leaves | 25 | 38 |
| Barley straw | 14–19 | 82–120 |
| Corn cob | 4[1]–15 | 50–123 |
| Corn straw | 8 | 50 |
| Cotton stalk | 22 | 70–78 |
| Cotton waste | 10.2[2] | 39.5[2] |
| Cottonseed hull | 18 | 59–67 |
| Oat straw | 17–18 | 48–83 |
| Oil palm empty fruit bunch | 12–15 | 77 |
| Rice straw | 18–36 | 35–72 |
| Soya stalk | 20 | 20–40 |
| Sugarcane bagasse | 11–23 | 50 |
| Sugarcane straw | 16–26 | 70–120 |
| Sunflower stalk | 13 | 97 |
| Wheat straw | 6.65[2]–21 | 50–80 |

*Source:* Unless otherwise marked, data are from Kumla et al. (2020). Marked data are from:
[1] Irshad and Asgher (2011); [2] Philippoussis et al. (2001).
Lignin content is reported as percent of substrate dry weight. The C/N ratio is reported as units of carbon per unit of nitrogen.

product and the fact that, as heterotrophs, mushrooms' molecular content depends on the elemental makeup of the substrate.

Profitability of mushroom cultivation requires a biological efficiency of at least 50% (Kumla et al. 2020), but most agro-industrial waste is low in nitrogen, which can limit mushroom productivity. Substrates can be supplemented with amendments that are high in nitrogen, like manure, cereal bran, or soybean meal (Kumla et al. 2020; Thongklang 2019). It is also possible to lower the C/N ratio favorably by mixing different types of lignocellulosic waste together. Farmers and scientists often conduct tests, growing a mushroom variety on multiple substrates and substrate mixtures to find the best C/N ratio for the mushrooms before scaling up to larger production. The C/N ratios of various lignocellulosic wastes are given in Table 17.1 and the preferred substrate C/N ratios of some edible mushrooms are given in Table 17.2. Table 17.3 shows the biological efficiency of the same edible mushrooms on various types of lignocellulosic wastes. In Table 17.3, highlighted boxes indicate profitable biological efficiency ranges and mushrooms with a dry protein content comparable to, or higher than, that of milk. Rows in Table 17.3 that have both columns highlighted represent strain-substrate combinations that are both profitable and provide consumers with a strong protein source.

## 17.3 ABUNDANCE OF LIGNOCELLULOSIC WASTE

In 2016, the world generated more than nine billion metric tonnes of agricultural waste, more than 32 billion tonnes of industrial waste, and about one billion tonnes of compostable municipal (household) waste (Kaza et al. 2018). That is enough lignocellulosic waste to rebuild the Great Wall of China 798 times![1] As urbanization and population growth continue to increase, the amount of waste humans generate is also increasing (Kaza et al. 2018).

---

[1] Calculations conducted in 2020 are based on Frohlich and Harrington 2018.

**TABLE 17.2**

**Ideal and Maximum Substrate C/N Ratios for the Growth of Selected Edible Mushrooms**

| Edible Mushroom Species | Ideal C/N Ratio Range | Maximum Reported C/N Ratio |
|---|---|---|
| *Agaricus bisporus* | 19 | 22 |
| *Agaricus subrufescens* | 27 | 33 |
| *Agrocybe aegerita* | 72–81[5] | 96[5] |
| *Agrocybe cylindracea* | NR | NR |
| *Corprinus comatus* | 5–10[2] | 10 |
| *Ganoderma lucidum* | 70–80 | 80 |
| *Hericium erinaceus* | 23–38[3] | 38[3] |
| *Lentinula edodes* | 30–35 | 55 |
| *Lentinus sajor-caju* | 45–55 | 90 |
| *Pleurotus citrinopileatus* | 40–48[1] | 48 |
| *Pleurotus columbinus* | NR | NR |
| *Pleurotus cystidiosus* | NR | NR |
| *Pleurotus djamor* | 40–48[2] | 48 |
| *Pleurotus eous* | NR | NR |
| *Pleurotus eryngii* | 45–55 | 70 |
| *Pleurotus florida* | 45–60 | 150 |
| *Pleurotus ostreatus* | 45–60 | 90 |
| *Pleurotus pulmonarius* | NR | 400[7] |
| *Pleurotus sapidus* | 7.5[6] | 26.4[6] |
| *Schizophyllum commune* | 20[4] | 30[4] |
| *Volvariella volvacea* | 40–60 | 60 |

*Source:* Unless otherwise marked, data are from Kumla et al. (2020). Marked data are sourced from: [1] Atila (2017); [2] Dijkstra (1976); [3] Feng et al. (2007); [4] Irshad and Asgher (2011); [5] Isikhuemhen et al. (2008); [6] Mshandete (2011); [7] Osunde et al. (2019).

NR, not reported in the literature.

Currently, most agro-industrial waste is burned, left to decompose, or put into landfills, despite its high energy content and potential for sustainable re-use (Chomnunti and Faulds 2019; Kumla et al. 2020). Thirty-three percent of global waste is openly dumped, leading to inefficient decomposition, if any; almost 40% goes to landfills, many of which are not capped properly and therefore release unfiltered greenhouse gases (Kaza et al. 2018). The inefficient decomposition of crop residues in the field is a major contributor to the greenhouse gas effect. Between 2010 and 2016, almost 47% of greenhouse gas emissions in Asia came from the inefficient disposal of crop residues. Burning the waste also generates many harmful or even poisonous airborne substances that increase air pollution and can impact human health. Despite the harm, abandonment and burning remain common methods for disposing of agro-industrial waste because of their low costs (Chomnunti and Faulds 2019). Therefore, alternative methods of treating lignocellulosic waste need to be inexpensive to have maximum appeal and adoption.

Treatment of lignocellulosic waste should also be flexible enough not to rely on municipal waste collection. Waste collection is not standardized in low-income countries or rural areas, so households usually manage their own waste, usually through dumping or burning (Kaza et al. 2018). Improving waste management in such areas has the potential to significantly improve human health and reduce pollution. Using lignocellulosic waste to cultivate a nutritional, high-protein food item such as mushrooms would not only help reduce waste but could also improve the nutrition of

**TABLE 17.3**

**Biological Efficiency and Crude Protein Content of Selected Edible Mushrooms Grown on Various Types of Lignocellulosic Waste Substrates**

| Edible Mushroom Species | Lignocellulosic Waste Substrate | Mushroom Biological Efficiency (%) | Mushroom Crude Protein Content (% of Dry Weight) |
|---|---|---|---|
| *Agaricus bisporus* | Oat straw | 47.2–52.9 | 26.8–36.2 |
| | Wheat straw | 47.2–51.1 | 21.0–27.0 |
| *Agaricus subrufescens* | Wheat straw | 53.7 | 28.4 |
| *Agrocybe aegerita* | Cotton waste | 30–34.9[3] | NR |
| | Wheat straw | 47.5–49.8[3] | 27.1[1] |
| *Agrocybe cylindracea* | Corn cob | 33.5 | 14.8[2] |
| | Wheat straw | 61.4 | 1.5–(15.9)[2] |
| *Corprinus comatus* | Rice straw | 18 | 10.9 |
| *Ganoderma lucidum* | Oat straw | 2.3 | 9.9 |
| *Hericium erinaceus* | Rice straw | 33.9 | 24.1 |
| | Wheat straw | 39.4–43.5 | 26.8 |
| *Lentinula edodes* | Barley straw | 64.1–88.6 | 15.1–16.8 |
| | Rice straw | 48.7 | 16.2 |
| | Sugarcane bagasse | 130.0–133.0 | 13.1–13.8 |
| | Sugarcane straw | 83.0–98.0 | 14.4 |
| | Wheat straw | 66.0–93.1 | 15.2–15.4 |
| *Lentinus sajor-caju* | Rice straw | 78.3 | 23.4 |
| | Soya stalk | 83 | 25.8 |
| | Sunflower stalk | 63.1 | 21 |
| | Wheat straw | 74.9 | 22.9 |
| *Pleurotus citrinopileatus* | Rice straw | 76.5–89.2 | 22.8 |
| | Wheat straw | 98.3–105.6 | 25.3 |
| *Pleurotus columbinus* | Corn cob | 79.1 | 1.9 |
| | Rice straw | 71.4 | 4.8 |
| | Wheat straw | 69.2 | 2.9 |
| *Pleurotus cystidiosus* | Corn cob | 50.1 | 24.5 |
| | Sugarcane bagasse | 49.5 | 22.1 |
| *Pleurotus djamor* | Rice straw | 82.7 | 24.8 |
| | Sugarcane bagasse | 101.7 | 25.1 |
| *Pleurotus eous* | Rice straw | 79.8 | 29.3 |
| | Soya stalk | 82.3 | 30.5 |
| | Sunflower stalk | 61.5 | 27.4 |
| | Wheat straw | 75.1 | 19.5 |
| *Pleurotus eryngii* | Corn cob | 51.8 | 23.8 |
| | Cotton waste | 28.6–38.47[3] | NR |
| | Rice straw | 45.9 | 21.8 |
| | Sugarcane bagasse | 41.3 | 20.5 |
| | Wheat straw | 48.2–(110.2)[3] | 21.5 |
| *Pleurotus florida* | Corn cob | 55 | 29.1 |
| | Corn straw | 31.6 | 26.3 |
| | Cotton stalk | 25.1 | 29.8 |
| | Cottonseed hull | 13.6 | 20 |
| | Soya stalk | 87.6 | 23.5 |
| | Sugarcane bagasse | 75.6 | 8.7 |
| | Wheat straw | 66.4 | 27.9 |

*(Continued)*

**TABLE 17.3** (*Continued*)

**Biological Efficiency and Crude Protein Content of Selected Edible Mushrooms Grown on Various Types of Lignocellulosic Waste Substrates**

| Edible Mushroom Species | Lignocellulosic Waste Substrate | Mushroom Biological Efficiency (%) | Mushroom Crude Protein Content (% of Dry Weight) |
|---|---|---|---|
| *Pleurotus ostreatus* | Banana leaves | ND | 15 |
| | Barley straw | 21.3 | 12.8 |
| | Corn cob | 31.7–66.1 | 15.4–29.7 |
| | Cotton stalk | 44.3 | 30.1 |
| | Cottonseed hull | 8.9 | 17.5 |
| | Rice straw | 25.6–84.6 | 12.5–23.4 |
| | Soya stalk | 85.2 | 24.7 |
| | Sugarcane bagasse | 65.7 | 27.1 |
| | Wheat straw | 22.6–52.6 | 11.6–14.6 |
| *Pleurotus pulmonarius* | Banana leaves | 17.9 | 16.9–23.5 |
| | Cotton stalk | 42.3 | 29.3 |
| | Cotton waste | 92.9–97.1[3] | NR |
| | Rice straw | 23.5 | 21.1 |
| | Wheat straw | 81.4–123.1[3] | NR |
| *Pleurotus sapidus* | Rice straw | 64.7 | 23.4 |
| | Soya stalk | 72.7 | 26.8 |
| | Sunflower stalk | 45.9 | 20.1 |
| | Wheat straw | 62.2 | 14.9 |
| *Schizophyllum commune* | Oil palm empty fruit bunch | 3.7 | 6.1 |
| *Volvariella volvacea* | Banana leaves | 15.2 | 23.9 |
| | Corn straw | ND | 23 |
| | Cotton waste | 11.3–20.2[3] | NR |
| | Oil palm empty fruit bunch | 3.6–6.5 | 33.5–41.0 |
| | Rice straw | 10.2–15.0 | 36.9–38.1 |
| | Wheat straw | 3.4–8.5[3] | NR |

*Source:* Unless otherwise marked, data are from Kumla et al. (2020). Marked data are sourced from:
[1] Isikhuemhen et al. (2008); [2] Koutrotsios et al. (2014); [3] Philippoussis et al. (2001).
Biological efficiency is measured as the wet weight of mushroom divided by the dry weight of substrate. Crude protein content is measured as a percentage of mushroom dry weight. Highlighted boxes in the Biological Efficiency column are higher than 50%, indicating profitability. Highlighted boxes in the Crude Protein Content column have a dry crude protein content equal to or higher than that of milk (Chang and Miles 2004).
NR, not reported in the literature.

low-income and rural families. Utilizing waste that is already present in those areas also reduces the costs of both waste disposal and obtaining substrates for mushroom cultivation (Thongklang 2019).

Some agro-industrial waste is present in most countries, like waste from staple grains. For example, in Europe and Brazil, the main composting materials used for *Agaricus bisporus* cultivation are wheat straw, hay, and corn cobs, and in Asia where wheat is not a major crop, chopped rice straw is the main mushroom substrate. However, many areas have locally unique waste, such as the waste generated by a farm's non-staple products. Because most agro-industrial waste is lignocellulosic, almost any variety can be used for mushroom cultivation. Depending on the C/N ratio of the available substrates, some nitrogen enrichment might be necessary; again, the nature of the supplement can vary based on what is locally available. In Europe and the Americas, cereal bran may be used; in Asia, substrates can be enriched with rice bran instead (Thongklang 2019). The C/N ratios and lignin contents of many types of lignocellulosic waste are listed in Table 17.1.

## 17.4  SELECTING SUBSTRATES AND FUNGAL STRAINS

Since mushrooms are heterotrophs, their final protein content depends on the nitrogen availability of their substrate, as well as other factors (Kumla et al. 2020; Osunde et al. 2019). Therefore, the highest yields and most nutritious mushrooms come from nutrient-rich or nutrient-enriched substrates with a low C/N ratio. For example, *Pleurotus pulmonaris* (oyster) mushrooms grown on corn cobs (C/N ratio of 120) have significantly greater cap size, weight, protein content, and vitamin content than those grown on unenriched wastepaper (C/N ratio of 400) (Osunde et al. 2019).

Selection of the cultivation substrate also depends on whether the mushrooms are primary decomposers (wood rots), which can be grown on any lignocellulosic material (Karunarathna 2019a), or secondary decomposers (saprotrophs), which must be grown on Phase II compost (see Chapter 16 in this volume). Saprotrophic fungi like *Agaricus bisporus* (the button mushroom) grow on compost that has already been partially digested by other fungi and bacteria, which requires stirring and aeration equipment and large pasteurization chambers to prepare. By contrast, mushrooms cultivated directly on lignocellulosic wastes only require the waste and a suitable container, such as polyethylene bags of sawdust, open planting boxes, beds of wood chips, or cut logs. Wood rotting mushrooms have greater potential for use in rural and low-income areas due to the high availability of the waste and the lack of significant startup materials or equipment.

The genus *Pleurotus* is a good example of a wood rotting genus with many edible members that can be cultivated on multiple types of locally available lignocellulosic waste. For example, one study found that *P. ostreatus* grew best on brizantha grass (palisade grass), which is native to tropical and southern Africa (USDA 2021), over sugarcane- or wheat-derived substrates (Thongklang 2019). In terms of sawdust, *P. ostreatus* can be cultivated on softwood including coconut, cashew, rubber, and mango tree dust (Karunarathna 2019a) as well as on many varieties of hardwood. While biological efficiency of the mushroom should be a priority to generate the most nutritious or profitable product, the ultimate factor determining the substrate should be whatever waste type is locally available (Karunarathna 2019a).

Variety or species selection is another important aspect of integrating mushrooms into agroecological systems. While there are more than 50 internationally cultivated species of mushroom (Kumla et al. 2020), there is also considerable opportunity to bring local mushroom varieties into cultivation. Of the 7,000 mushroom species known to be edible, more than 3,000 are considered desirable (both edible and delicious) and about 700 are known for medicinal uses (either biochemically discovered or traditional; Xu 2019). Therefore, several hundred desirable, edible species are available for domestication. With as many as 92% of fungi yet undescribed in the scientific literature (Karunarathna 2019b), the number of desirable edible species may be even larger. Several studies in recent years have reported edible tropical species that are new to science. A 2018 study found that 93% of the *Agaricus* species identified in Thailand within the last few years were unknown to science, and many of them were edible, such as *Agaricus flocculosipes* (first reported in 2012) and *A. subtilipes* (first reported in 2016) (Thongklang 2019). Most of those species are saprobic, growing on compost, but there are likely many wood-rotting fungi that have yet to be discovered and domesticated.

Isolating a mushroom species is a simple procedure, but it does require several steps. A clean (ideally sterile) environment is necessary to obtain a genetically pure sample of fungal mycelium. Fortunately, once such an area is available, only a single fruiting body must be broken open for sample collection. The mushroom selected for this job should be mature and free from insects. The cap can be broken open (not cut) and a sample of the inner tissue placed on **agar** or another standardized growing medium, which can be prepared with a pressure cooker and glass jars. Additional agar and jars will be needed if the initial culture is not genetically pure or is contaminated, until a pure, genetically stable strain is isolated. Some type of refrigerated storage system is then needed to store the mycelium so reliable samples can be taken for cultivation.

When a local edible mushroom has been identified, farmers can run small-scale tests to find the best growing conditions for it. One benefit of identifying local mushrooms is that the fungi are likely already adapted to local lignocellulosic substrates, humidity, and temperature conditions (Karunarathna 2019b). Cooler temperature mushrooms are useful for adding a revenue stream during the winter, while heat-tolerant mushrooms can be good candidates for field co-cropping during the summer. Mushrooms that can grow at lower humidity can be cultivated alongside crop plants with low leaf cover, while those that demand more humid environments are useful for adding revenue to managed forests or fields of leafy vegetables (see Section 17.5.3).

In addition to fruiting tests in the environment, it is vitally important that the local mushroom's edibility be confirmed (Thongklang 2019). The confirmation should come via scientific methods that do not require eating the mushrooms, like DNA identification and mineral analysis. University or government laboratories in agricultural areas will probably be the best source of this type of research and confirmation. Nutritionists and mycologists will likely be interested in a new species and willing to conduct such tests for publication credit as well as to help the local farm economy. More advanced tests could also determine the protein content and other nutritional qualities of the mushroom, which could considerably increase demand (see Chapters 8 and 16 in this volume). If the edibility of the mushroom cannot be confirmed, it should not be used for consumption, merely as a means of helping recycle agro-industrial wastes.

Developing a local strain can present several hurdles to small-scale farms: identifying a promising mushroom strain with market desirability, isolating and maintaining a genetically pure sample, determining the best fruiting conditions and substrate, and finding a researcher willing to confirm edibility and protein content of the mushroom. Ideally, such work should be supported by government programs, but government funding is often hard to come by even for valuable projects. Currently, few local strains are grown on the mushroom market, which is instead dominated by established international strains like the button mushrooms *Agaricus bisporus*. However, establishing stable, high-quality strains of local mushrooms will bolster the local mushroom market and provide more nutritional, high-value protein products to locals (Thongklang 2019). And of course, there is no problem with locally cultivating mushrooms that are currently on the international market as long as the proper waste substrate is available and there is local demand for the mushrooms (Karunarathna 2019b).

## 17.5   MUSHROOM CULTIVATION LOCATIONS

When a farmer considers closing waste loops with mushrooms, it is important to consider where the mushrooms will be cultivated. Constructing devoted mushroom houses represents a large upfront cost that is prohibitive for many small-scale farmers (Xu 2019). Fortunately, it is possible to cultivate mushrooms in existing buildings, outdoors in wooded areas, or outdoors in fields either with or without plant co-crops.

### 17.5.1   MUSHROOM CULTIVATION IN BUILDINGS

Mushrooms can be cultivated in just about any enclosed space, including abandoned outbuildings, high tunnel greenhouses, and empty storage facilities or basements, making the cost of startup very low. Compared with outdoor cultivation, indoor cultivation of mushrooms has the benefit of more consistent temperature and greater protection of the crop from infection and pests. This makes it easier to predict and adjust production indoors. If the systems are already in place for outbuildings, farmers can also automatically manage the humidity level, air flow, and lighting to maximize fruiting potential (Cornell Small Farms Program 2021a). However, maintaining the indoor spaces with water and electricity can become expensive and may contribute greenhouse gases to the atmosphere, making mushroom cultivation less economically or ecologically sustainable (Xu 2019).

### 17.5.2 MUSHROOM CULTIVATION IN WOODS

If farmers have sustainably managed forests on their property, it will be easy to source logs for mushroom cultivation. Most of the labor associated with log-grown mushrooms is the upfront cost of felling trees and inoculating the logs with mycelium – in fact, that setup represents about half of the effort required in cultivating shiitake mushrooms (Cornell Small Farms Program 2021a). Since forest management is often a winter activity and logs can be inoculated at any time, it is possible to front-load the effort for mushroom cultivation during the plant off-season rather than adding pressure to the already busy growing season. Even if farmers do not have the correct substrate trees on their property, a log needs only to be felled or purchased once, and a single section of log can produce shiitake mushrooms for more than four seasons in some cases. Therefore, the upfront cost of log cultivation is relatively low given each log's productive lifetime (Cornell Small Farms Program 2021a). By beginning with 100 logs and rotating in 100 more each season, farmers can slowly but steadily build up a large shiitake farm that provides profits in the second year, with small (500-log) farms making as much as U.S. $9,000 over the first five years (Cornell Small Farms Program 2021a).

Shiitake are the most reliable outdoor cultivated mushroom, but there are other wood-grown varieties that do well in managed forests. Oyster mushrooms can be cultivated on stumps or upright logs called "totems", fruiting up to three times per growing season. Lion's mane mushrooms can also be grown on logs or totems, though they typically only fruit once per year, in the fall. Nameko mushrooms are another variety that fruits in the fall, though they are more substrate specific than the other varieties discussed here, preferring black cherry or sugar maple wood. Logs inoculated with any mushroom variety can be stacked in woods, where the humidity and temperature levels will be ideal, or can be surrounded by woodchips in a woodchip bed to help maintain humidity (Cornell Small Farms Program 2021b). The average maintenance labor cost is just one hour per log through the log's entire productive lifetime, which includes checking for pests and harvesting mushrooms (Cornell Small Farms Program 2021a). Therefore, even less reliably fruiting varieties have low labor costs per harvest and can be worthwhile to grow.

While cultivation on logs is cost-effective for smallholders with forests on their property, there are drawbacks. Without the protection of a building, mushrooms are exposed to temperature fluctuations and pests. All outdoor mushroom cultivation needs to be carefully managed so that the harvested fruits can be high quality and not be disturbed or eaten by insects or other small animals. Sustainably managed forests can ease this pest pressure considerably with the presence of local birds, who can feed on insects that might otherwise reduce the quality of mushroom crops.

### 17.5.3 MUSHROOM CULTIVATION IN FIELDS

In farms using crop rotation, mushrooms can be rotated in with low start-up costs. Mushrooms can be cultivated alongside plants or during plant off-seasons, representing a cost-effective way to sustainably intensify lands already under cultivation. For example, in tropical and subtropical regions, the "wine cap" *Stropharia rugosoannulata* and "shaggy mane" *Coprinus comatus* can be grown at 22°C (71°F) in autumn and winter, then followed by heat-tolerant mushrooms like *Volvariella volvacea* and *Phlebopus portentosus* in the summer when temperatures rise to around 35°C (95°F). In Cambodia, some farmers spread *Volvariella* spawn over rice straw in paddies during the off season to create saleable product when the rice is not growing. Field cultivation of mushrooms is not restricted by soil type; there are mushroom varieties that thrive on lignocellulosic wastes placed atop clay (e.g., *Volvariella volvacea*), sandy soils (e.g., *Coprinus comatus*), and loam (e.g., *Dictyophora indusiata*, "bamboo mushroom") (Mortimer and Karunarathna 2019; see Table 17.4).

**TABLE 17.4**

**Examples of Mushroom Species Suitable for Field Cultivation, Including Soil Characteristics, Substrates, and Climatic Requirements for the Listed Species**

| Species | Soil Type | Soil pH | Substrate | Temperature (°C) | Humidity (%) |
|---|---|---|---|---|---|
| *Coprinus comatus* | Sandy | 7 | Cottonseed meal | 16–22 | 85–95 |
| *Dictyophora indusiate* | Loam | 6.5–7 | Bamboo litter, straw, sawdust | 22–30 | 60–65 |
| *Ganoderma* sp. | Sandy | 4.2–5.3 | Sawdust | 23–34 | 80–90 |
| *Lentinula edodes* | Clay loam | 4.5–6 | Sawdust, wheat-rice bran | 5–20 | 80–90 |
| *Morchella* sp. | Sandy loam | 7–7.5 | Sawdust, wheat bran, humus | < 20 | 50–70 |
| *Oudemansiella radicata* | Clay loam | 6.5–7.2 | Wood chips/corn cobs | 20–30 | 60–80 |
| *Phlebopus portentosus* | Clay loam, humus | 4–6 | Sawdust | 37–37 | 55–80 |
| *Polyporus umbellatus* | Sandy loam, humus | 5.5–7 | Leaves, humus | 18–24 | 60–80 |
| 90 | Humus rich, sandy | 5–6 | Wood chips/rice straw | 22–28 | 70–75 |
| *Volvariella volvacea* | Clay | 7.5–8 | Rice straw | 22–40 | |

*Source:* Reproduced from Mortimer and Karunarathna (2019) under the Creative Commons Attribution 4.0 International License (http://creativecommons.org/licenses/by/4.0/). Internal sources have been removed from the table for this reproduction.

*Stropharia* mushrooms have a reputation for being foolproof, as they can be grown in any area supplied with woodchips or straw; trays, pots, and outdoor woodchip beds are all suitable locations. They can be inoculated at any point during their growing season, produce for multiple seasons, and can produce fruits as soon as two months after inoculation. Cultivation in wood chip beds also makes them excellent candidates for co-cropping with vegetables, berries, or fruit trees (Cornell Small Farms Program 2021b), so even a subsistence garden can supply wine cap mushrooms.

Some outdoor mushrooms require more care than the wine cap. The "almond agaricus", *Agaricus subrufescens*, is a summer-fruiting mushroom that can be cultivated either in greenhouses or in any outdoor area where humidity can be monitored and increased if necessary. On a small scale, the almond mushroom can be cultivated in window boxes or with large potted plants. On a larger scale, the humidity can be provided naturally by co-cropping *A. subrufescens* in shaded woody areas or with big, leafy plants that provide adequate shade (Cornell Small Farms Program 2021b).

In addition to producing an edible or commercial product, mushroom cultivation in fields increases nutrient cycling and deposition of organic matter into the soil, helping it retain moisture and increasing plant crops' productivity even when non-mycorrhizal mushrooms are used. However, mycorrhizal plant crops can receive unparalleled benefits from mushroom co-crops.

Mycorrhizal fungi deliver benefits to plants via an intimate symbiosis (see Chapter 11 in this volume). When establishing an ectomycorrhizal co-crop in an existing plant field or orchard, it is important to prioritize benefits to the crop plants as well as mushroom flavor and yield (Kumla and Lumyong 2019). The best approach is to scout edible ectomycorrhizal fungi (ECM) in the laboratory or greenhouse for strong colonization of seedlings, then test their mushroom production after out-planting (see Chapter 12 in this volume). The highest potential for co-crop ECM fungi is in tropical areas, where soils are low in phosphate and so there is selection pressure for plants to partner with mycorrhizae (Chaiyasen and Lumyong 2019). Ectomycorrhizal fungi that produce spores internally, known as puffballs or as the artificial grouping Gasteromycetes (from the Greek *gaster* = "stomach"), are especially useful for cultivation because their spores are easy to source in large quantities for use as inoculum. For example, Shoro mushrooms (*Rhizopogon rubescens*) are edible Gasteromycetes with established success in inoculation studies with pine seedlings (Kumla and Lumyong 2019).

## 17.5.4 CO-CROPPING WITH ECTOMYCORRHIZAL FUNGI

Truffles are economically important ectomycorrhizal fungi that develop spores internally, making them good candidates for co-cropping with trees. Cultivation of truffles is often done in monoculture or polyculture of tree varieties specifically selected for their ability to host truffles and utilize marginal land (e.g., land with a steep incline that prohibits the use of farm machinery), rather than for their co-cropping potential (Benucci et al. 2011). For example, while hazelnut trees are commonly used in truffle plantations, the hazelnuts are not generally harvested.

There is enormous potential for truffles to be co-cropped with pecan trees for a cultivation system that improves productivity for both crops (Benucci et al. 2011). The trees can produce both pecans and truffles for their productive lifetimes, then their wood can offer an additional product for use or sale. Pecans are cultivated in Australia and Italy as well as across South America and the southern United States. Pecan orchards across the United States are known to produce edible truffles of various species (Benucci et al. 2011), especially the American native *Tuber lyonii* (Brenneman 2018). The cultivation of truffles in such plantations will begin with the sterile culture of pecan seedlings, followed by the inoculation of pecan roots with a spore slurry, and finally with outplanting (Benucci et al. 2011; see Chapter 12 in this volume for more details). This method has been used to successfully inoculate pecan seedlings with *Tuber lyonii*, producing substantial truffle harvests to serve local areas in Georgia, United States (Brenneman 2018). American pecan seedlings have also been successfully inoculated with ectomycorrhizae of the more familiar European truffle varieties *Tuber aestivum* and *T. borchii*, though the viability of those relationships after outplanting has not been established (Benucci et al. 2011).

*Tuber lyonii* truffles can sell for as much as $400 a pound, with the price going higher for large and perfectly ripe specimens. Demand currently exceeds supply, so there is plenty of room for growth in the pecan truffle market even using native American species of *Tuber* (Brenneman 2018).

Other edible ectomycorrhizal mushrooms should be investigated for cultivation in managed forests and lumberyards, with the potential to produce sellable products well in advance of mature lumber. Inoculation with both early and late successional fungi can provide a variety of sellable products as such forests age. Other mycorrhizal fungi can also be investigated for growth with other productive trees; for example, the porcini mushroom *Boletus edulis* regularly occurs with spruce trees, fruiting alongside the inedible but gorgeous *Amanita muscaria* ("fly agaric") in Christmas tree plantations and wild woods across Austria, Sweden, England, and Italy, as well as in the United States and New Zealand (Hall et al. 2003). While they are not currently domesticated, porcinis seem fully capable of colonizing tree plantations and could be harvested when they do occur to bring in supplemental income. Even the Amanitas could provide extra income if eco-tourists are allowed in to witness their beautiful, brightly colored fruits in the stark winter.

As always when co-cropping with fungi, it is important to use plant pest and disease management techniques that will not threaten the desirable and beneficial fungi. The use of herbicides, pesticides, and (of course) fungicides will decrease the chance of a harvest and can affect the edibility of mushrooms that do occur. Physical pest prevention methods, like fences or netting, should be used in place of chemicals whenever possible. Those methods will also help protect the mushroom crop from small animals, like rodents, who might be happy to nibble on the fruits. Emphasis should be placed on the primary plant crop as a reliable source of income, so whatever the plant requires should be considered foremost, but mushrooms should be treated as ecologically responsible bonus products that can bolster smallholder income when they fruit in properly managed forests and fields.

## 17.6   OTHER POTENTIAL USES OF MUSHROOMS

### 17.6.1   MUSHROOM GROWING KITS

Once a mushroom growing operation has been established, it represents a steady supply of mushroom mycelium. Farmers interested in additional business and who have achieved reliable harvests may be able to package and sell mushroom growing kits to locals interested in the hobby. Such kits usually involve a container filled with lignocellulosic waste which has been pre-inoculated with mycelium. During the COVID-19 pandemic of 2020–2021, mushroom kit sales soared up to 400% over previous years. The increase is largely attributed to people's desire to see the satisfying process of mushroom growth when they were unable to go about their normal activities, but the pandemic also spurred an interest in many people to become more self-sufficient (Matei 2021). Even outside of crises, mushroom kits hold appeal for their dramatic transformation, with the fruits doubling in size every 24 hours for the last few days before harvest. Mushrooms in unusual colors, such as pink and blue oyster mushrooms, are particularly desirable to consumers.

### 17.6.2   SPENT MUSHROOM SUBSTRATE

Once mushrooms have been cultivated on lignocellulosic waste, the altered waste is known as **spent mushroom substrate** (**SMS**). Spent mushroom substrate is high in organic matter and is highly structured thanks to the mycelium that has grown throughout it. Due to these traits, SMS improves water retention of soil, making it an excellent soil amendment. It is also useful for protecting newly seeded vegetables and lawns because it locks in moisture while blocking birds' access to the germinating seeds (Beyer 2011). These uses make SMS a great product for local gardeners or for use in a farmer's own fields. The final composition of SMS depends on the starting lignocellulosic material and on whether the substrate was composted prior to mushroom cultivation, so each load will have a different ratio of elements and minerals. The one drawback is that SMS tends to have a high salt content, so it must be weathered for six or more months before it is suitable for most uses. After 16 months of weathering, SMS has one-fifth the sodium content, a higher phosphorus content, and a more neutral pH than fresh SMS (Beyer 2011). While there are some plant nutrients in SMS, such as phosphorus, potassium, calcium, and organic nitrogen, the main draw of SMS over other soil amendments is its improved water retention (Beyer 2011).

### 17.6.3   MYCOREMEDIATION OF DAMAGED LANDSCAPES

Many mushrooms naturally concentrate salts, heavy metals, or other undesirable compounds from the soil. This trait can be used to **remediate** polluted or unsuitable soils, making them more suitable for cultivation. For example, cultivating salt-tolerant fungi in salty soils can help concentrate the salt, removing it from the soil and enabling the growth of more vegetation (Harischandra and Yan 2019). Most known salt-tolerant (halotolerant) fungi are filamentous, so they can be established with the same methods used to establish arbuscular mycorrhizal fungi in a field (see Chapter 12 in this volume).

Metal-concentrating fungi can absorb heavy metals like iron, zinc, and lead, removing them from damaged soils (Harischandra and Yan 2019). This is especially useful to places like old apple orchards, where heavy metals were previously used as pesticides and may now contaminate otherwise edible products like morels.

Because mushrooms are absorptive heterotrophs, they can break down non-metallic pollutants without ingesting their harmful forms. This means they are faster at degrading pollutants than ingestive organisms, which would have to ingest and tolerate a certain amount of the pollutant before breaking it down. Furthermore, many pollutants have structures similar to that of lignin, including herbicides, insecticides like dichloro-diphenyl-trichloroethane (DDT) and lindane, and banned industrial chemicals like polychlorinated biphenyls (PCBs) (Harischandra and Yan 2019). Therefore, white-rot basidiomycete fungi can degrade these and other **recalcitrant** (difficult to break down) materials, restoring soil and waterways that have previously been polluted. Ectomycorrhizal associations can also improve the survival rate of seedlings being grown in contaminated lands, such as areas near mine tailings, by concentrating the metals in the rhizosphere before they can reach the seedling (Kumla and Lumyong 2019).

The mushrooms and fungi used in **mycoremediation** should not be consumed by humans until and unless analyzed in laboratory to make sure that the contaminants have been fully digested or are not concentrated enough to cause health problems. If the mushrooms are inedible, they still provide the benefit of removing dangerous compounds or metals from the environment. Fungi used to concentrate and remove heavy metals may need to be incinerated, which is the fate of plants used for heavy metal remediation as well. Such incineration needs to be carefully controlled and the ash (containing the heavy metals) disposed of in properly managed hazardous waste landfills. In this case, the major benefit of mycoremediation (and the plant version, phytoremediation) is that the metals are drawn out of the contaminated soil and into a harvestable mass that can be removed and managed separately. Myco- and phytoremediation substantially reduce the mass of hazardous waste.

### 17.6.4  BIOPROSPECTING: "PESTS" WITH AGRICULTURAL UTILITY

There is considerable potential for species traditionally regarded as pests to be utilized in the processing of lignocellulosic waste. For example, a pathogenic fungus that attacks lavender plants also breaks down spent lavender stalks very efficiently (Chomnunti and Faulds 2019). If properly controlled, this pathogen could represent the most efficient means of recycling lavender agricultural waste. Biotrophic pathogens are highly host-specific, so it may be possible to find other highly efficient degraders for currently problematic waste streams by **bioprospecting**. In its simplest form, bioprospecting involves putting a plant tissue out in the open and observing the microbial species that colonize it as well as observing the molecules and enzymes those microbes produce. Bioprospecting often leads to the discovery of industrially useful enzymes. Bioprospecting on various lignocellulosic materials could lead to the efficient degradation of certain materials that have been difficult to dispose of, such as olive waste and citrus pulp, which are currently largely burned.

## 17.7  CONCLUSION

Fungi are efficient degraders of lignocellulosic materials like agro-industrial waste, which is abundant throughout the world. Currently, the dumping or inefficient disposal of agro-industrial waste presents issues including pollution and negative effects on human health. Waste disposal is particularly an issue in low-income countries and rural areas, where municipal garbage collection is uncommon and having garbage hauled away for efficient disposal is expensive. Fungi represent an efficient, ecologically friendly, and inexpensive means to dispose of lignocellulosic waste on small farms.

Many mushroom-producing fungi grow and fruit on a variety of lignocellulosic wastes, meaning that locally available wastes can be recycled into a protein-rich product that can supplement smallholder nutrition if the mushrooms are consumed by the farm manager or laborers. If there is market demand for mushrooms, cultivating fungi on lignocellulosic waste can transform a disposal issue into supplemental income for small farmers. There is also great potential to identify and domesticate new species of fungi specific to local climates and waste streams, increasing the number of options for waste disposal and potentially expanding the local mushroom market if edibility of the mushrooms can be confirmed scientifically.

## REFERENCES

Benucci, G.M.N., Bonito, G., Falini, L.B., and M. Bencivenga. 2011. Mycorrhization of pecan trees (*Carya illinoinensis*) with commercial truffle species: *Tuber aestivum* Vittad. and *Tuber borchii* Vittad. *Mycorrhiza* 22:383–92. DOI:10.1007/s00572-011-0413-z.

Beyer, D. 2011. *Spent Mushroom Substrate*. Penn State Extension, Pennsylvania State University. May 3, 2011. https://extension.psu.edu/spent-mushroom-substrate (accessed: May 8, 2021).

Chaiyasen, A., and S. Lumyong. 2019. Arbuscular mycorrhizae as biofertilizers. In *The Amazing Potential of Fungi: 50 Ways We Can Exploit Fungi Industrially*, eds. K.D. Hyde, J. Xu, S. Rapior, R. Jeewon, S. Lumyong, and A.T. Niego. *Fungal Diversity*, Vol. 97, 1–136. DOI:10.1007/s13225-019-00430-9.

Chang, S-T., and P.G. Miles. 2004. *Mushrooms: Cultivation, Nutritional Value, Medicinal Effect and Environmental Impact*, 2nd ed. Boca Raton, FL: CRC Press.

Chomnunti, P., and C. Faulds. 2019. Agricultural waste disposal. In *The Amazing Potential of Fungi: 50 Ways We Can Exploit Fungi Industrially*, eds. K.D. Hyde, J. Xu, S. Rapior, R. Jeewon, S. Lumyong, and A.T. Niego. *Fungal Diversity*, Vol. 97, 1–136. DOI:10.1007/s13225-019-00430-9.

Cornell Small Farms Program. 2021a. *Economics and Markets*. Cornell College of Agriculture and Life Sciences. Cornell University. https://smallfarms.cornell.edu/projects/mushrooms/economics-and-markets/ (accessed May 7, 2021).

Cornell Small Farms Program. 2021b. *Outdoor Production*. Cornell College of Agriculture and Life Sciences. Cornell University. Retrieved May 07, 2021, from https://smallfarms.cornell.edu/projects/mushrooms/outdoor-production/ (accessed May 8, 2021).

Dijkstra, F. 1976. *Submerged Cultures of Mushroom Mycelium as Sources of Protein and Flavor Compounds: 2. The Nutritional Requirements of Agaricus bisporus and Coprinus comatus*. PhD dissertation, Delft University of Technology. Retrieved from http://www.fransdijkstra.eu/diss/chapter2.htm (accessed May 8, 2021).

Feng, G., Li, S., Li M., Tian J., Wang J, and W. Li. 2007. Effect of composts with difference C/N on the development of *Hericium erinaceus* (Bull. Pers.) Mycelial and fruitbody. *Acta Agriculturae Boreali-Sinica*, 22(S2): 131–135. DOI:10.7668/hbnxb.2007.S2.033 [in Chinese].

Frohlich, T. and J. Harrington. 2018. *The Heaviest Objects in the World*. 24/7 Wall St. https://247wallst.com/special-report/2018/05/11/the-heaviest-objects-in-the-world/6/ (accessed January 11, 2020).

Hall, I.R., Yun, W., and A. Amicucci. 2003. Cultivation of edible ectomycorrhizal mushrooms. *Trends in Biotechnology* 21(10): 433–438.

Harishchandra, D., and J. Yan. 2019. Mycoremediation: Fungi to the rescue. In *The Amazing Potential of Fungi: 50 Ways We Can Exploit Fungi Industrially*, eds. K.D. Hyde, J. Xu, S. Rapior, R. Jeewon, S. Lumyong, and A.T. Niego. *Fungal Diversity*, Vol. 97, 1–136. DOI:10.1007/s13225-019-00430-9.

Irshad, M., and M. Asgher. 2011. Production and optimization of ligninolytic enzymes by white rot fungus Schizophyllum commune IBL-06 in solid state medium banana stalks. *African Journal of Biotechnology* 10(79): 18234–18242.

Isikhuemhen, O. S., Mikiashvili, N. A., and V. Kelkar. 2008. Application of solid waste from anaerobic digestion of poultry litter in *Agrocybe aegerita* cultivation: Mushroom production, lignocellulolytic enzymes activity and substrate utilization. *Biodegradation* 20(3), 351–361. DOI:10.1007/s10532-008-9226-y.

Karunarathna, S. 2019a. Growing mushrooms in bags. In *The Amazing Potential of Fungi: 50 Ways We Can Exploit Fungi Industrially*, eds. K.D. Hyde, J. Xu, S. Rapior, R. Jeewon, S. Lumyong, and A.T. Niego. *Fungal Diversity*, Vol. 97, 1–136. DOI:10.1007/s13225-019-00430-9.

Karunarathna, S. 2019b. New edible mushrooms. In *The Amazing Potential of Fungi: 50 Ways We Can Exploit Fungi Industrially*, eds. K.D. Hyde, J. Xu, S. Rapior, R. Jeewon, S. Lumyong, and A.T. Niego. *Fungal Diversity*, Vol. 97, 1–136. DOI:10.1007/s13225-019-00430-9.

Kaza, S., Yao, L.C., Bhada-Tata, P., and F. van Woerden. 2018. *What a Waste 2.0: A Global Snapshot of Solid Waste Management to 2050*. Urban Development; Washington, DC: World Bank. © World Bank. https://openknowledge.worldbank.org/handle/10986/30317 License: CC BY 3.0 IGO.

Koutrotsios, G., Mountzouris, K. C., Chatzipavlidis, I., and G.I. Zervakis. 2014. Bioconversion of lignocellulosic residues by *Agrocybe cylindracea* and *Pleurotus ostreatus* mushroom fungi – Assessment of their effect on the final product and spent substrate properties. *Food Chemistry* 161: 127–135. DOI:10.1016/j.foodchem.2014.03.121.

Kumla, J., and S. Lumyong. 2019. Application of ectomycorrhizal fungi in forestry. In *The Amazing Potential of Fungi: 50 Ways We Can Exploit Fungi Industrially*, eds. K.D. Hyde, J. Xu, S. Rapior, R. Jeewon, S. Lumyong, and A.T. Niego. *Fungal Diversity*, Vol. 97, 1–136. DOI:10.1007/s13225-019-00430-9.

Kumla, J., Suwannarach, N., Sujarit, K., et al. 2020. Cultivation of mushrooms and their lignocellulolytic enzyme production through the utilization of agro-industrial waste. *Molecules*, 25(12): 2811. DOI:10.3390/molecules25122811.

Marchant, N. 2020. Why this mushroom could be your next house, handbag or 'hamburger'. *World Economic Forum*. https://www.weforum.org/agenda/2020/12/mycelium-mushroom-sustainable-packaging-fashion-meat/ (accessed May 25, 2021).

Matei, A. 2021. Why Growing Mushrooms at Home Is Everyone's New Pandemic Hobby. https://www.the-guardian.com/lifeandstyle/2021/mar/17/mushrooms-as-houseplant (accessed May 24, 2021).

Mortimer, P.E., and S.C. Karunarathna. 2019. Growing mushrooms in the field. In *The Amazing Potential of Fungi: 50 Ways We Can Exploit Fungi Industrially*, eds. K.D. Hyde, J. Xu, S. Rapior, R. Jeewon, S. Lumyong, and A.T. Niego. *Fungal Diversity,* Vol. 97, 1–136. DOI:10.1007/s13225-019-00430-9.

Mshandete, A. 2011. Cultivation of *Pleurotus* HK-37 and *Pleurotus sapidus* (oyster mushrooms) on cattail weed (*Typha domingesis*) substrate in Tanzania. *International Journal of Research in Biological Sciences* 1(3): 35–44.

Osunde, M. O., Olayinka, A., Fashina, C.D., and N. Torimiro. 2019. Effect of carbon-nitrogen ratios of ligno-cellulosic substrates on the yield of mushroom (*Pleurotus pulmonarius*). *Open Access Library Journal* 6: e5777. DOI:10.4236/oalib.1105777.

Philippoussis, A., Zervakis, G., and P. Diamantopoulou. 2001. Bioconversion of agricultural lignocel-lulosic wastes through the cultivation of the edible mushrooms *Agrocybe aegerita, Volvariella volvacea* and *Pleurotus* spp. *World Journal of Microbiology and Biotechnology* 17(2), 191–200. DOI:10.1023/a:1016685530312.

Thongklang, N. 2019. Growing mushrooms in compost. In *The Amazing Potential of Fungi: 50 Ways We Can Exploit Fungi Industrially*, eds. K.D. Hyde, J. Xu, S. Rapior, R. Jeewon, S. Lumyong, and A.T. Niego. *Fungal Diversity,* Vol. 97, 1–136. DOI:10.1007/s13225-019-00430-9.

USDA. 2021. *Taxon: Urochloa brizantha* (Hochst. ex A. Rich.). R. D. Webster. GRIN-Global, United States Department of Agriculture. https://npgsweb.ars-grin.gov/gringlobal/taxon/taxonomydetail?id=401367 (accessed May 20, 2021).

# 18 Fungi in Food Processing

*Noureddine Benkeblia*

## CONTENTS

## 18.1 INTRODUCTION

While today we may imagine food processing to be an industrial or modern process, in truth humans have been preserving or altering food by chemical means for thousands of years, and many of these methods rely on fungi. Many ancient Egyptian murals and tomb ornaments depict bread and wine making. Judging from archaeological finds, fungi, especially yeasts, have been used for both food and medicinal purposes since before recorded history. Filamentous fungi have been used in the ripening of cheeses, while yeasts have been used in the fermentation of fruits to produce wines and of cereals to produce beers and breads (Campbell-Platt and Cook 1989; Moore and Chiu 2001).

Historically, Western (e.g., European) cuisine relied less on the addition or utilization of microorganisms than Eastern (e.g., Asian or African) food. However, as the global population increased and travel became more common, Western cultures realized the health benefits of the Eastern fermented foods, so that nowadays fermented foods and beverages play a prominent role in the diets of billions of people of different regions of the world (Beuchatl 1983; Campbell-Platt and Cook 1989).

With the development of molecular biology and biotechnology, the potential of fungi has been greatly explored; currently, they are employed to produce numerous traditional foods and make major contributions to the development of novel food products, the large-scale production of previously rare ingredients, and conversion of food by-products into usable forms (Chan et al. 2018; Copetti 2019).

## 18.2 FERMENTED AND AGED FOOD PRODUCTS

Fermented foods are foods produced or altered through controlled microbial growth, whereby microbial enzymes convert food components to other biochemical or chemical compounds (Dimidi

DOI: 10.1201/9780429320415-21

et al. 2019), some of which prevent the colonization of the food product by other microbes, thus preserving it. Fermentation is thought to be one of the oldest and most cost-effective methods for producing and preserving food, having been used for thousands of years (Jeyaram 2009; Ross et al. 2002). Throughout the world, large quantities of diverse fermented products are produced, and this modern diversity depends on the cultural practices and history of specific countries or even smaller regions (Batra and Millner 1974).

Lactic acid bacteria (LAB) are primarily responsible for the biochemical transformations that take place in the most common fermented food products. In tested fermented foods, the LAB community may be composed of many different genera of bacteria such as *Enterococcus, Lactoctoccus, Lactobacillus, Leuconostoc, Pediococcus*, and *Streptococcus*. Although the preservation action is attributed to lactic acid, which is the major end product of the bacterial fermentative action, other metabolites possessing anti-microbial activities are produced, such as propionic acid and proteinaceous inhibitors known as bacteriocins (Caplice and Fitzgerald 1999; de Vuyst and Vandamme 1994; Perez et al. 2014).

### 18.2.1 Fermented Foods Are Diverse and Numerous

The scope of global food fermentation is enormous and highly detailed. The wide variety of substrates, fermenting microorganisms, fermentation methods and periods, final products, and useful side products could fill a textbook (or several). This chapter is not meant to provide an exhaustive list or description of fermented foods around the world, but only to supply some examples of the incredible density and diversity of traditionally fermented products (see Table 18.1).

As one illustration of this variety, numerous regional fungal fermented foods are produced in Asia, and different products are obtained by fermentation using a broad variety of molds and yeasts (Maheshwari et al. 2020; Nout and Aidoo 2011; Samson 1993). These products can be classified based on the type of products obtained, such as condiments or flavorings (e.g., Japanese soy sauce); protein-rich meat substitutes (e.g., Indonesian tempeh); and bread- or cake-like products (e.g., Chinese mantou) (Lim 1991). In their reviews, Ko (1986) and Nout et al. (2007) have also categorized various food products based on the functional molds and yeasts that conduct the fermentation.

In India, many traditional fermented foods are a fundamental part of ethnic heritage (Rawat et al. 2018). For example, in India, Nepal, Pakistan, Sikkim, Tibet, and neighboring countries, numerous fermented products are known, including the following (Batra and Millner 1974; Sekar and Mariappan 2007):

- Idli: Lentil and rice cakes fermented by *Torulopsis candida* and *Trichosporon pullulans*
- Jalebis: Chickpea and wheat flour batter fermented by *Saccharomyces bayanus*
- Kanji: Rice porridge fermented by *Hansenula anomala* var. *anomala*
- Kinema: Soybeans fermented by *Candida parapsilosis* and *Geotrichum candidum*
- Murcha: A starter culture for many beverages, which can be composed of a variety of cereal flours and is fermented by many fungi, including *Hansenula anomala* var. *schneggii*, *Mucor rouxianus*, and *Rhizopus arrhizus*
- Panjabi wadies and papadams: Lentil dumplings and crackers made from a variety of flours, respectively, both fermented by *Candida* sp. and *Saccharomyces cerevisiae*
- Toddy: A palm wine fermented by *S. cerevisiae*

In Indonesia, many different filamentous and yeast species are used to ferment dehulled and boiled soybeans and/or cereals to produce tempeh (Hachmeister and Fung 1993; Nowak 1992). In South Korea fungi and yeasts are used to ferment numerous vegetables and products, including Meju, a fermented soybean cake used to produce many condiments (Shin and Jeong 2015). These are just a few examples of the incredible variety of fermented foods available in Asia and elsewhere.

**TABLE 18.1**

**Examples of Some Fungi Used in the Production of Various Types of Fermented Beverages and Food Products**

| Food Category | Substrate(s) | Product | Fungi and Yeasts Used | References |
|---|---|---|---|---|
| Alcoholic beverages | Fruits | Wine | *Saccharomyces cerevisiae* | Wang et al. (2015) |
| | Malt (germinated barley) | Beer | *Saccharomyces cerevisiae* | Holt et al. (2018) |
| | Rice | Sake | *Aspergillus oryzae* | Bokulich et al. (2014) |
| | Apple juice | Cider | *Candida sake, Pichia fermentans, Saccharomyces cerevisiae* | Lorenzini et al. (2018) |
| | Coffee | Fermented coffee | *Saccharomyces marscianus* | Silva et al. (2008) |
| | Rice, sweet potato, barley | Kōji | *Aspergillus flavus* var. *oryzae, Aspersillus sojae* | Jeong et al. (2003) |
| | Milk | Koumiss (Kumis or airag) | *Candida pararugosa, Dekkera anomala, Geotrichum sp., Issatchenkia orientalis, Kazachstania unispora, Kazachstania marxianus, Kluyveromyces marxianus, Meyerozyma caribbica, Pichia deserticola, Pichia fermentans, Pichia manshurica, Pichia membranaefaciens Saccharomyces cerevisiae, Saccharomyces unisporus, Torulaspora delbrueckii* | Guo et al. (2019) Mu et al. (2012) Tang et al. (2020) |
| Non-alcoholic beverages | Polished amber rice | Komesu (rice vinegar) | *Aspergillus oryzae, Saccharomyces cerevisiae* | Wang et al. (2016) |
| | Unpolished brown rice | Kurosu (black rice vinegar) | | Wu et al. (2012) |
| | Water | Water kefir | *Candida albicans, Candida friedricchi, Candida holmii, Candida kefir;* | Cai et al. (2020) |
| | Milk | Milk kefir | *Issatchenkia orientalis, Kluyveromyces lactis, Kluyveromyces marxianus, Pichia fermentas, Saccharomyces cerevisiae, Saccharomyces delbruecki, Saccharomyces exiguous, Saccharomyces unisporus, Saccharomyces humaticus, Saccharomyces turicensis, Torulopsis holmii, Torulospora delbrueckii* | Dertli and Çon (2017) |
| | Ethanol, cider, or wine | Vinegar | *Candida lactis-condensi, Candida stellata, Hanseniaspora valbyensis, Hanseniaspora osmophila, Saccharomyces cerevisiae, Zygosaccharomyces bailii, Zygosaccharomyces bisporus, Zygosaccharomyc espseudorouxii, Zygosaccharomyces lentus, Saccharomycodes ludwigii, Zygosaccharomyces rouxii; Zygosaccharomyces mellis* | Li et al. (2015) Rainieri and Zambonelli (2009); Solieri and Giudici (2009) |
| | Leaves of *Camellia sinensis* (tea) | Kombucha | *Hanseniaspora valbyensisi, Hanseniaspora vineae, Lachancea fermentati, Torulaspora delbrueckii, Zygosaccharomyces bailii, Zygosaccharomyces kombuchaensis* | Burinia et al. (2021) |

*(Continued)*

**TABLE 18.1 (Continued)**
**Examples of Some Fungi Used in the Production of Various Types of Fermented Beverages and Food Products**

| Food Category | Substrate(s) | Product | Fungi and Yeasts Used | References |
|---|---|---|---|---|
| Condiments | Soybeans, wheat | Soy sauce | *Aspergillus oryzae, Aspergillus sojae, Candida versatilis, Zygosaccharomyces rouxii* | Devanthi and Gkatzionis (2019); Wei et al. (2013) |
| | Soybeans, rice, barley, seaweed | Miso | *Aspergillus oryzae, Aspergillus sojae, Zygosaccharomyces rouxii* | Allwood et al. (2021); Kusumoto et al. (2021) |
| Protein sources | Soybeans | Tempeh | *Rhizopus microspores* var. *oligosporus* | Rizal et al. 2020 |
| | Soybeans | Sufu | *Actinomucor* spp., *Mucor* spp., *Rhizopus* spp. | Han et al. (2004) |
| | Carbohydrate mixture (industrial) | Quorn (fungal mycelium) | *Fusarium venenatum* | Whittaker et al. (2020) |
| | Meat | Meat and sausage | *Debaryomyces hansenii* | Núñez et al. (2015); Ramos-Moreno et al. (2021) |
| | Milk (various animals) | Cheese | *Penicillium camemberti, Penicillium roqueforti* | Ropars et al. (2012) |
| | Soy, peanut, cassava, or coconut | Oncom | *Neurospora intermedia, Neurospora sitophila, Rhizopus microsporus* var. *oligosporus* | Hedger (1978) |
| Bread | Teffgrass flour | Injera | *Candida humilis, Rhodotorula mucilaginosa* var. *mucilaginosa, Kluyveromyces marxianus, Debaryomyces hansenii* | Stewart and Getachew (1962) |
| | Wheat, barley, oat, or rye flour | Sourdough | *Kazachstania exigua, Candida humilis, Saccharomyces cerevisiae* | Carbonetto et al. (2020) |
| | Cassava flour | Gari | *Geotricum candida* (with bacterium *Corynebacterium manilaot*) | Akinrele (1964) |
| Dessert and candy | Cacao beans | Cacao | *Kluyveromyces marxianus, Hanseniaspora guilliermondii, Hanseniaspora opuntiae, Debaryomyces hansenii, Kodamaea ohmeri, Pichia kluyverii* | Pereira et al. (2016); Schwan et al. (2015) |
| Food colorants | Rice | Angkak | *Monascus purpureus, Monascus ruber, Saccharomyces marscianus* | Panda et al. (2010); Singh et al. (2021) |

## 18.2.2  BREADS

Sourdough bread, one of the most ancient fermented foods known, is leavened using a starter dough fermented by local airborne LAB and yeasts. This natural sourdough is obtained using a mixture of flour and water left open at room temperature until acidification occurs. In ancient times, leavened bread was exclusively produced using sourdough, whereas most modern bread is produced using a starting culture of baker's yeast, *Saccharomyces cerevisiae* (Catzeddu 2019). In the leavening process, yeasts have the primary role, while the LAB produce important flavor components (Liu et al. 2020).

In 2021, Johansson et al. surveyed the fermenting yeasts present in sourdoughs of various cereals, including oat, rye, barley, and wheat. The doughs, prepared under sterile lab conditions, showed somewhat varying fungal communities among different flour types, indicating that some of the leavening yeasts in sourdoughs arise from the cereal grains themselves rather than the air. The species isolated from these sourdoughs included *Cyberlindnera fabianii, Hanseniaspora uvarum, Hyphopichia burtonii, Kazachstania servazzii, Kluyveromyces marxianus, Pichia fermentans* and *P. kudriavzevii, Torulaspora delbrueckii, Wickerhamomyces anomalus*, and *W. ciferrii* (Johansson et al. 2021). *Saccharomyces cerevisiae* was notably absent from the sourdoughs in the study, demonstrating that the baker's yeast that dominates most breads does not come from the grain flour itself but from the environment (Johansson et al. 2021). Indeed, the locally available LAB and airborne yeasts are known to confer specific flavor characteristics to bread, so a sourdough mixture will taste markedly different depending on where it is prepared (Chavan and Chavan 2011; Suo et al. 2021; ur-Rehman et al. 2006).

## 18.2.3  FERMENTED MEATS

Molds and yeasts are frequently used in the fermentation of meat and meat products, particularly sausages. Fermented sausages are a blend of ground meat, salt, and curing agents stuffed into casings and subjected to a fermentation by "spontaneous" flora (i.e., airborne bacteria and fungi rather than a specific starter culture). Most of these fermented sausages are dried, making them easy and safe to store at room temperature (Lücke 1994). Although LAB are the predominant microflora found in fermented sausages, the yeast *Debaryomyces hensenii* (Leistner and Bern 1970; Comi and Cantoni 1980) and the mold *Penicillium nalgiovense* (Leistner 1986) have also been found in the spontaneous fermentation flora.

Fish are often fermented for safe storage and the addition of desirable flavor. As with sausage meats, fish are often fermented spontaneously and LAB are more common than yeasts in the microbial community. Traditional fermented fish dishes in Africa include koobi and momone in Ghana, lanhouin in Benin and Togo, and dagaa in Uganda (Agyei et al. 2020). An investigation of the yeasts in the traditional fermented fish dish adjuevan, from the Ivory Coast, revealed eight yeast genera: *Candida, Debaryomyces, Hanseniaspora, Hansenula, Pichia, Rhodotorula*, and *Saccharomyces*, though the latter is rarely encountered in fermented fish due to high salt concentrations. Most of these yeasts can also be found on the skin or intestines of fresh fish (Clementine et al. 2012).

## 18.2.4  CHEESES

Mold-ripened cheeses are believed to be among the earliest manufactured foods and were consumed in large quantities long before fundamental dairy microbiology acquired any importance. Indeed, cheese processing is thought to have occurred approximately 4,000 years ago, and according to some legends, Europeans were adding molds and other microorganisms to cheese for flavor approximately 2,000 years ago (Babel 1953; Copley et al. 2005a,b). In an interesting review, Irlinger et al. (2015) reviewed the microbial genera detected on the surface of soft cheeses, reporting that cheese rinds host a specific microbiota of eukaryotes consisting of a total of 35 fungal genera, with 17 filamentous molds and 18 yeasts. Indeed, the visible fungal mycelium on the product is considered a fundamental part of many cheeses; the two most familiar examples are Camembert, made by fermenting cow's milk with *Penicillium camemberti*, and Roquefort or "blue cheese", made by fermenting milk with *Penicillium roqueforti* (Figure 18.1). Unlike its wild relatives in the *Penicillium* genus, the fungus *P. camemberti* is a domesticated species producing few conidia (so the cheese surface does not appear or taste powdery or "fuzzy"), and it does not produce detectable mycotoxins or many other volatiles during ripening (Jollivet et al. 1993; Pitt et al. 1986).

The microbiota of cheese is a hybrid of wild species from the environment and domesticated species added as starter cultures. For example, *Penicillium* molds have changed during their adaptation to anthropogenically supplied dairy substrates, producing domesticated phenotypes with industrial properties which are now used to make soft cheeses (Bodinaku et al. 2019). During ripening and aging, *Penicillium* species colonizing the surfaces of cheeses are of both industrial origin (starters and inoculum) (Nielsen et al. 1998) and spontaneous origin (species from the local, natural population) (Kure et al. 2004; Lund et al. 1995). The microflora of cheese might also be dominated by other filamentous fungi and yeasts which do not ripen the cheeses, but contribute to cheese flavor. For example, the yeast-like fungus *Geotrichum candidum* is also largely used in cheese ripening, colonizing already fermented surfaces (Marcellino and Benson 2013); upon analysis of 25 different Tilsit cheeses (semi-hard yellow cheeses), the yeast-like but non-fermenting fungus *Debaryomyces hansenii* was found to be the predominant species during both aging and ripening (Bockelmann et al. 1997).

## 18.2.5  YOGURTS AND OTHER FERMENTED DAIRY PRODUCTS

Yogurt and many other fermented dairy products (e.g., sour cream, butter, and buttermilk) can be created through the fermentation of milk from a variety of animals, including cows, goats, buffalo, camels, and sheep (Agyei et al. 2020; Aryana and Olson 2017). In spontaneous fermentations, yeasts and LAB may be sourced from the air, from fermentation vessels or machinery, from the hands and clothes of workers (in nonsterile conditions), or from the raw milk. For example, in a sampling of traditional fermented milk beverages from Cameroon, 13 species of yeast were discovered, four of which are also found in raw milk and five of which were derived from the environment. Only three yeast species were present in all sampled products, both artisanal and traditional, regardless of sampling location: *Galactomyces candidum, Torulaspora delbrueckii,* and *Saccharomyces cerevisiae/paradoxus* (Maïworé et al. 2019).

Many fermented dairy products are made with starters, small amounts of desirable bacteria and yeasts that colonize the substrate and outcompete spoilage organisms. Some starters arise from previous spontaneous fermentation, such as in "back-slopping", when a small amount of a finished fermented product is used to start fermentation of a new batch (Agyei et al. 2019). The final microflora composition of a spontaneously fermented or "back-slopped" dairy product depends heavily on the yeasts present in the starting culture and the geographic location of fermentation. For example, yeasts from indigenous African fermented dairy products vary widely (Agyei et al. 2020):

- Amabere (from Kenya): *Candida famata* and *C. albicans, Saccharomyces cerevisiae, Trichosporon mucoides*
- Graiss (from Sudan): *Kluyveromyces marxianus, Issatchenkia orientalis*
- Lben (from Algeria): *Kluyveromyces lactis, Saccharomyces cerevisiae*
- Nunu (from Ghana): *Candida parapsilosis, rugosa,* and *tropicalis; Galactomyces geotrichum, Pichia kudriavzevii, Saccharomyces cerevisiae*
- Suusac (from Kenya and Somalia): *Candida famata, inconspicua, krusei,* and *lusitaniae; Cryptococcus laurentii, Geotrichum penicillatum, Rhodotorula mucilaginosa, Saccharomyces cerevisiae, Trichosporon cutaneum* and *mucoides*

(a)

(b)

(c)

**FIGURE 18.1** Examples of visible molds on cheese: (a) Camembert; (b) Brie; (c) Roquefort (blue cheese). [Reproduced with permission from Shutterstock: (a) vitals; (b) gresei; (c) Supida Khemawan.]

Industrial yogurt preparations are made with starters using only the LAB *Streptococcus thermophilus* and *Lactobacillus bulgaricus*, but many yeasts and other LAB are commonly found in them. One such yeast genus, *Torulopsis*, contributes to the long-term survival of the starter LAB (Aryana and Olson 2017).

## 18.3  FERMENTED BEVERAGES

It is uncertain when alcoholic beverages were first made; however, it is presumed that they originated from a happy accident that occurred thousands of years ago. Based on archaeological analyses of beer jugs, beer was probably first made about 10,000 B.C.E., during the Neolithic period (Patrick 1952); wine is at least 6,000 years old, having appeared in Egyptian pictographs dated to around 4,000 B.C.E. (Braidwood et al. 1953; Lucia 1963).

### 18.3.1  Alcoholic Fruit-Based Beverages (Wines)

Wine is obtained by the alcoholic fermentation of whole grapes or grape juice using mainly *Saccharomyces* spp. Hundreds of strains of *Saccharomyces cerevisiae,* considered the "conventional" brewing yeast, have been developed to satisfy different requirements in brewing of wine (Molinet and Cubillos 2020). However, more than a dozen fungi from different families have been found in wines, such as *Brettanomyces custerianus, B. nanus,* and *B. naardenensis; Candida colliculosa* and *C. stellata; Dekkera anomala* and *D. bruxellensis; Hanseniaspora uvarum; Kloeckera apiculata; Metschnikowia pulcherrima; Torulaspora delbrueckii; Pichia anomala* and *membranifaciens;* and *Zygosaccharomyces bailii* (Connell et al. 2002; Ivey and Phister 2011; Perry-O'Keefe et al. 2001; Röder et al. 2007; Xufre et al. 2006). There is interest in sequential or mixed inoculation for wines, in which non-conventional yeasts are used as part of the inoculation community with *S. cerevisiae.* This can have unpredictable effects on the flavor profile and requires further study and refinement, but already *Pichia kluyverii* is known to give Sauvignon blanc wines a sweet, tropical aroma (Holt et al. 2018). Wild strains of yeast, including varieties of *Saccharomyces cerevisiae*, are being investigated for their utility and flavor potential in the production of wines with unique flavor profiles. For example, wine produced with wild yeasts isolated from oak tree bark has an earthy, citrusy, and floral flavor (Molinet and Cubillos 2020). Indeed, wine fermentation is a complex process and many species have been found to produce alcohols, esters and terpenols that enhance the beverage's flavor; unfortunately, other species produce undesirable biochemical compounds which might negatively affect the organoleptic and sensorial qualities of wine (Comitini et al. 2011, Medina et al. 2013; Sun et al. 2014).

Besides grape wine, many other fruit-based alcoholic beverages are produced by fermentation (see Table 18.1). Wines can be made from date palms (Awe and Nnadoze 2015; Broshi 2007), berries of various species (Dubinina et al. 2020), bananas (Kundu et al. 1976; Stover and Simmonds, 1987), figs or pineapples (Joshi 1997), pears (Joshi et al. 2011), plums (Joshi et al. 2000), apricots (Joshi et al. 1990), pumpkins (Thakur et al. 2014), peaches (Joshi et al. 2005), and strawberries (Joshi et al. 2006), among other fruits. However, the most popular such beverage is cider, obtained by the fermentation of fresh apple juice and varying from 1.2% to 8.5% of alcohol by volume (ABV) (AICV 2021; Guiné et al. 2021). The most predominant yeasts found in ciders of different origins are *Candida sake, Pichia fermentans,* and *Saccharomyces cerevisiae* (Coton et al. 2006; Cousin et al. 2017; Morrissey et al. 2004), but other species have been reported such as *Candida oleophila, C. stellate,* and *C. tropicalis; Dekkera anomala; Hanseniaspora uvarum* and *H. valbyensis; Kluyveromyces marxianus; Lachancea cidri; Metschnikowia pulcherrima; Pichia delftensis, P. misumaiensis* and *P. nakasei;* and *Saccharomyces bayanus* (Beech 1993; Coton et al. 2006; Morrissey et al. 2004; Pando et al. 2010; Valles et al. 2007).

## 18.3.2 Alcoholic Grain-Based Beverages (Beers)

Beer is the most-consumed fermented alcoholic beverage product in the world; in 2018, beer production was six times greater than wine production by volume (Pontes Eliodório et al. 2019). Beer is produced through the alcoholic fermentation of sprouted and cooked cereal grains known as **malt**. Like wine, beer production is dominated by "conventional" *Saccharomyces* yeasts, *S. cerevisiae* for ale and *S. pastorianus* for lager, but involves the participation of many other fungi from various taxonomic groups (Pontes Eliodório et al. 2019). The most common "non-conventional" beer brewing yeasts are *Brettanomyces* species, which impart a complex, sour flavor desirable in many spontaneously fermented beverages including the Belgian gueuze and lambic beers. *Brettanomyces* yeasts are of interest because they are known spoilage yeasts in both wine and beer, making the final product take medicinal or leathery; however, when controlled properly, certain strains produce highly desirable, "exotic" or tropical flavors like pineapple, mango, and pear (Colomer et al. 2019). Brewing yeasts also include species from the genera *Candida, Clavispora, Hanseniaspora, Kluyveromyces, Naumovia, Rhodotorula,* and *Pichia* (Burinia et al. 2021; Couluibaly et al. 2021).

In some cases of "spontaneous" beer brewing, the chief brewing yeasts arise from the environment; for example, traditional sorghum beers in West Africa are dominated by *S. cerevisiae* but then by *Candida tropicalis* in Ghana and Cote d'Ivoire, by *Clavispora lusitaniae* in Benin, and by *Candida inconspicua* in Rwanda (Couluibaly et al. 2021). However, in Belgian lambic beer production, industrial and craft brewing show the same community succession, demonstrating that the brewing yeasts originate from the wood in barrels or from residue from previously brewed batches (Pontes Eliodório et al. 2019). If the fermentation process is stopped when the beer reaches 6%–7% ABV, the yeast can be used for five to ten additional fermentations (Bamforth 2017).

Several candidate yeasts for sequential fermentation of malt with *S. cerevisiae* are under investigation (Holt et al. 2018). For example, beers can also be given a "clove-like aroma" when fermented with *Torulaspora delbrueckii* or "spicy" notes from certain *Brettanomyces* species (Holt et al. 2018). Many other yeasts are known to produce specific desirable or undesirable flavors in beer, such as (Burinia et al. 2021):

- *Hanseniaspora uvarum*: Solvent (undesirable in high amounts) and banana
- *Lachancea fermentati*: Citrusy, winey
- *Pichia kluyveri*: Banana, apple
- *S. cerevisiae* var. *boulardii*: Fruity
- *S. eubayanus*: Spicy, smoky
- *Torulaspora delbrueckii* strains: Banana, pineapple, floral, or strong spicy notes
- *Wickerhamomyces anomalus*: Pear, apple, peach

Unfortunately, as in wine, there are also yeasts which have undesirable effects on beer flavor, or mixed effects. For example, *Pichia kluyverii* and *Hanseniaspora uvarum* increase both desirable banana flavors and "solvent-like", undesirable flavors (Burinia et al. 2021; Holt et al. 2018).

Many cereals can be used in the production of beer, including malted and non-malted barley, wheat, maize, finger millet, sorghum, teff grass, linseed (flax), and flours of many types. Beers can also have numerous flavor additives; most commonly hops are used, but other flavoring ingredients include stems of the gesho shrub (traditional in Ethiopia), taro root, vegetables, processed sugar, cherries, and chili peppers (Colomer et al. 2019; Fentie 2020).

## 18.3.3 Other Alcoholic Beverages

The nomads of the Mongolian steppes, of Turkic and Mongol origin, are known for a naturally fermented mare's milk called koumiss (or kumis; *Airag* in Mongolian), an alcoholic beverage with a sharp acidic and alcoholic flavor (Akuzawa and Surono 2011; Zhang 2012). Koumiss is a typical yeast-lactic acid fermented product made with the yeasts *Dekkera anomala*; *Kazachstania marxianus* and *K. unispora*; *Meyerozyma caribbica*; and *Saccharomyces cerevisiae* and *S. unisporus* as well as certain strains of LAB (Montanari et al. 1996; Mu et al. 2012; Tang et al. 2020). However, many other wild yeast species have been detected in koumiss, including, for example, *Candida pararugosa*; *Geotrichum* sp.; *Issatchenkia orientalis*; *Kluyveromyces marxianus*; *Pichia deserticola, P. fermentans, P. manshurica*, and *P. membranaefaciens*; and *Torulaspora delbrueckii* (Mu et al. 2012).

Mead is an alcoholic beverage made by fermenting a diluted honey solution and was probably first brewed in Persia and India 8,000 years ago (Diniz Felipe et al. 2019). The fermentation of mead can take months and can have unreliable results due to the osmotic (water) stress of the yeasts growing in the thick honey mixture. Nevertheless, *Saccharomyces cerevisiae* remains the powerhouse for mead production in most of the world, as even stressed, it outcompetes the few yeasts naturally present in honey (Iglesias et al. 2014). There are many variations of mead that include flavoring or fermenting ingredients like fruit juices, citric acid, and local pollen. For example, metheglin is mead flavored with herbs and spices; melomel is mead produced with fruit juices; and hippocras is a grape melomel with additional herbs and spices (Iglesias et al. 2014). The traditional Ethiopian wines tej, ogol, and booka are made with the fermentation of honey with other ingredients, like gesho twigs and bark from the native tree *Blighia unijungata* L. (Fentie 2020).

### 18.3.4 Non-Alcoholic Fermented Beverages

A large number of non-alcoholic fermented beverages are produced throughout the world using various yeasts and filamentous fungi and versatile raw materials (Blandino et al. 2003; Marsh et al. 2014). Among these different and various non-alcoholic fermented beverages, kefir and vinegar are the most popular worldwide.

Kefir is produced by adding fermenting "kefir grains" (small starter cultures of LAB, acetic acid bacteria, and yeast) to either milk (for milk kefir) or sugar water (for water kefir). These grains can be recovered after fermentation to make additional batches (Guzel-Seydim et al. 2011, 2021; Lynch et al. 2021). Beside the numerous acetic acid bacteria and LAB detected in kefir, more than a dozen yeast species have also been reported, including *Candida albicans, C. friedricchi, C. holmii*, and *C. kefir*; *Issatchenkia orientalis*; *Kluyveromyces lactis* and *K. marxianus*; *Pichia fermentas*; *Saccharomyces cerevisiae, S. delbruecki, S. exiguous, S. humaticus, S. turicensis*, and *S. unisporus*; *Torulopsis holmii*; and *Torulospora delbrueckii* (Angulo et al. 1993; Gulitz et al. 2011; Guzel-Seydim et al. 2011; Latorre-Garcia et al. 2007; Simova et al. 2002; Wang et al. 2008; Witthuhn et al. 2005). Though it is considered a non-alcoholic beverage, kefir ethanol content actually ranges from 0.02% to 2.0% (Laureys and De Vuyst, 2014; Martínez-Torres et al., 2017). The level of ethanol content which is produced during the late stage of fermentation depends on the production method and the fermentation duration; however, under controlled conditions allowing the growth of *Lactobacteria* only and excluding other microorganisms that form much higher amounts of ethanol, the level does not exceed 0.02 g/L, and alcohol content might be as low 0.002%–0.005% by volume (Farnworth 2008; Gorgus et al. 2016). This is far below the threshold of 0.5% alcohol by volume that is used as the threshold to mark a beverage as alcoholic in many countries (Bellut and Arendt 2019; Wszolek et al. 2001).

Vinegar, also known as white, distilled, or spirit vinegar, has been known worldwide for thousands of years. White vinegar was previously produced from the fermentation of different foods like sugar beets, potatoes, molasses, or milk whey, and generally contained 4%–7% acetic acid. Currently, most white vinegar is produced from the fermentation of grain alcohol (ethanol), and since alcohol does not contain nutrients, other ingredients such as yeast or phosphates are added to start the fermentation process. However, other sources can also be used to produce vinegar such as wine (Ciani 1998), balsam (Solieri et al. 2006), brown rice (Lee et al. 2010), onion (Horiuchi et al. 2000), apple juice or cider (Joshi and Sharma 2009), and olive oil mill wastewaters (De Leonardis et al. 2018) among other minor sources.

Although acetate bacteria play the major role in vinegar fermentation, the metabolic activity of yeasts and molds is also pivotal in the production of some types of vinegar (Rainieri and Zambonelli 2009). Different studies have reported more than 45 different yeast species involved in the fermentation process of vinegar production, such as *Candida lactis-condensi* and *C. stellata*; *Hanseniaspora osmophila* and *H. valbyensis*; *Saccharomyces cerevisiae* and *S. ludwigii*; and *Zygosaccharomyces bailii*, *Z. bisporus*, *Z. lentus*, *Z. mellis*, *Z. pseudorouxii,* and *Z. rouxii* (Solieri and Giudici 2008; Wu et al. 2012).

In Japan, two traditional rice vinegars, komesu (produced from a polished amber rice) and kurosu (produced from an unpolished brown rice), are both fermented by *Aspergillus oryzae* and *Saccharomyces cerevisiae* (Murooka and Yamshita 2008). In Chinese Zhenjiang aromatic vinegar, a black vinegar produced from rice, species of the fungal genus *Saccharomyces* are predominant (Xu et al. 2011).

Low- and non-alcoholic beers are gaining interest in international markets due to the growing evidence of health risks of alcohol consumption, desire to drink familiar beverages in areas where alcohol is prohibited, and other factors (Johannson et al. 2021). These are produced using non-conventional yeasts which can ferment simple sugars into flavor compounds, but do not engage in alcohol fermentation of maltose, the most common sugar in wort (malt extract used for brewing) and the one *S. cerevisiae* so efficiently turns into alcohol (Johansson et al. 2021; Holt et al. 2018). Several such yeast species have been identified, including species from the genera *Brettanomyces, Candida, Cyberlindnera, Kluyveromyces, Hanseniaspora, Lachancea, Pichia, Torulaspora,* and *Zygosaccharomyces* (Burinia et al. 2021; Holt et al. 2018; Johansson et al. 2021). For example, *Pichia fermentans* shows high potential for creating low- or non-alcoholic Belgian wit and German weizen beers, which rely on the clove-like and phenolic flavors that this yeast species produces in fermentation (Johansson et al. 2021).

## 18.4 FOOD ADDITIVES AND INGREDIENTS

From ancient times, various ingredients have been added to foods to improve their flavor, coloration, and even nutritional value, and modern advances in food technology, chemistry, and biotechnology have made such additives more numerous and affordable. In the modern day, food additives may be added to maintain or improve safety, freshness, and/or nutritional value, and/or to improve organoleptic and sensorial qualities such as taste, texture, and appearance (see Table 18.2).

These additives are often now produced by cell cultures, large vats of microbes which produce large volumes of potential food ingredients. To give a few examples, sweeteners like thaumatin and monellin; flavors like citrus, strawberry, and vanilla; essential oils like mint oil, jasmine oil, and anis-seed oil; and coloring agents like carotenoids, betacyanins, and anthocyanins are all currently produced by cell cultures (see Figure 18.2) (Archer 2000; Dufossé 2018; Copetti 2019; Jacobson and Wasileski 1994; Punt et al. 2002). Filamentous and yeast-form fungi play a pivotal role in industrial fermentation and the production of many modern food ingredients, including organic acids, vitamins, pigments, colorants, and fatty acids.

The fermentation process consists of converting either liquid or solid substrates into various desirable compounds through the metabolic action of microbes (Table 18.3). Fermentation should be conducted under sterile and carefully controlled conditions to produce high yields of the targeted metabolites, contaminated with few undesirable metabolites, at low cost (Christi 2014; Keshavarz 2014). Recent advances in molecular and genetic engineering, combined with increasingly stringent nutritional requirements, have led to the selection of non-toxic fungal species which produce high volumes of food-grade fermented products that also comply with national regulations (Singh et al. 2020). Moreover, newly developed or engineered fungi are also bringing new classes of ingredients to industrial scale, such as azaphilone pigments and antioxidants (Mapari et al. 2010) and mycoprotein (Whittaker et al. 2020). Mycoprotein, generated by vats of filamentous fungi, is of particular note because it has excellent protein content, low fat content, and no cholesterol (Ghorai et al. 2009;

**TABLE 18.2**

**Some Types of Common Food Additives, Their Uses, and Examples of the Names of Food Products Where They Are Found**

| Types of Ingredients | What They Do | Examples of Uses | Names Found on Product Labels |
|---|---|---|---|
| Preservatives | Prevent food spoilage from bacteria, molds, fungi, or yeast (antimicrobials); slow or prevent changes in color, flavor, or texture and delay rancidity (antioxidants); maintain freshness | Fruit sauces and jellies, beverages, baked goods, cured meats, oils and margarines, cereals, dressings, snack foods, fruits and vegetables | Ascorbic acid, citric acid, sodium benzoate, calcium propionate, sodium erythorbate, sodium nitrite, calcium sorbate, potassium sorbate, BHA, BHT, EDTA, tocopherols (vitamin E) |
| Sweeteners | Add sweetness with or without the extra calories | Beverages, baked goods, confections, table-top sugar, substitutes, many processed foods | Sucrose (sugar), glucose, fructose, sorbitol, mannitol, corn syrup, high fructose corn syrup, saccharin, aspartame, sucralose, acesulfame potassium (acesulfame-K), neotame |
| Color additives | Offset color loss due to exposure to light, air, temperature extremes, moisture and storage conditions; correct natural variations in color; enhance colors that occur naturally; provide color to colorless and "fun" foods | Many processed foods, (candies, snack foods margarine, cheese, soft drinks, jams/jellies, gelatins, pudding and pie fillings) | FD&C Blue Nos. 1 and 2, FD&C Green No. 3, FD&C Red Nos. 3 and 40, FD&C Yellow Nos. 5 and 6, Orange B, Citrus Red No. 2, annatto extract, beta-carotene, grape skin extract, cochineal extract or carmine, paprika oleoresin, caramel color, fruit and vegetable juices, saffron (Note: Exempt color additives are not required to be declared by name on labels but may be declared simply as colorings or color added) |
| Flavors and spices | Add specific flavors (natural and synthetic) | Pudding and pie fillings, gelatin dessert mixes, cake mixes, salad dressings, candies, soft drinks, ice cream, BBQ sauce | Natural flavoring, artificial flavor, and spices |
| Flavor enhancers | Enhance flavors already present in foods (without providing their own separate flavor) | Many processed foods | Monosodium glutamate (MSG), hydrolyzed soy protein, autolyzed yeast extract, disodium guanylate or inosinate |
| Fat replacers (and components of formulations used to replace fats) | Provide expected texture and a creamy "mouth-feel" in reduced-fat foods | Baked goods, dressings, frozen desserts, confections, cake and dessert mixes, dairy products | Olestra, cellulose gel, carrageenan, polydextrose, modified food starch, microparticulated egg white protein, guar gum, xanthan gum, whey protein concentrate |
| Nutrients | Replace vitamins and minerals lost in processing (enrichment), add nutrients that may be lacking in the diet (fortification) | Flour, breads, cereals, rice, macaroni, margarine, salt, milk, fruit beverages, energy bars, instant breakfast drinks | Thiamine hydrochloride, riboflavin (vitamin $B_2$), niacin, niacinamide, folate or folic acid, beta carotene, potassium iodide, iron or ferrous sulfate, alpha tocopherols, ascorbic acid, Vitamin D, amino acids (L-tryptophan, L-lysine, L-leucine, L-methionine) |

*(Continued)*

**TABLE 18.2 (*Continued*)**

| Types of Ingredients | What They Do | Examples of Uses | Names Found on Product Labels |
|---|---|---|---|
| Emulsifiers | Allow smooth mixing of ingredients, prevent separation. Keep emulsified products stable, reduce stickiness, control crystallization, keep ingredients dispersed, and to help products dissolve more easily | Salad dressings, peanut butter, chocolate, margarine, frozen desserts | Soy lecithin, mono- and diglycerides, egg yolks, polysorbates, sorbitan monostearate |
| Stabilizers and thickeners, binders, texturizers | Produce uniform texture, improve "mouth-feel" | Frozen desserts, dairy products, cakes, pudding and gelatin mixes, dressings, jams and jellies, sauces | Gelatin, pectin, guar gum, carrageenan, xanthan gum, whey |
| pH Control Agents and acidulants | Control acidity and alkalinity, prevent spoilage | Beverages, frozen desserts, chocolate, low acid canned foods, baking powder | Lactic acid, citric acid, ammonium hydroxide, sodium carbonate |
| Leavening agents | Promote rising of baked goods | Breads and other baked goods | Baking soda, monocalcium phosphate, calcium carbonate |
| Anti-caking agents | Keep powdered foods free-flowing, prevent moisture absorption | Salt, baking powder, confectioner's sugar | Calcium silicate, iron ammonium citrate, silicon dioxide |
| Humectants | Retain moisture | Shredded coconut, marshmallows, soft candies, confections | Glycerin, sorbitol |
| Yeast nutrients | Promote growth of yeast | Breads and other baked goods | Calcium sulfate, ammonium phosphate |
| Dough strengtheners and conditioners | Produce more stable dough | Breads and other baked goods | Ammonium sulfate, azodicarbonamide, L-cysteine |
| Firming agents | Maintain crispness and firmness | Processed fruits and vegetables | Calcium chloride, calcium lactate |
| Enzyme preparations | Modify proteins, polysaccharides and fats | Cheese, dairy products, meat | Enzymes, lactase, papain, rennet, chymosin |
| Gases | Serve as propellant, aerate, or create carbonation | Oil cooking spray, whipped cream, carbonated beverages | Carbon dioxide, nitrous oxide |

*Source:* Reproduced from International Food Information Council (IFIC) and U.S. Food and Drug Administration (FDA). 2010. *Overview of Food Ingredients, Additives & Colors.* U.S. Food and Drug Administration, which is in the public domain.

see Chapter 16 in this volume). The first major mycoprotein brand, Quorn, appeared in the 1980s in the United Kingdom, and the company has made considerable innovations in processing that make mycoprotein a more desirable meat substitute. For example, Quorn pioneered the assembly of fiber-gel composites that exploit the filamentous nature of the mycelium and create authentic meat-like textures (Lonchamp et al. 2019; Whittaker et al. 2020).

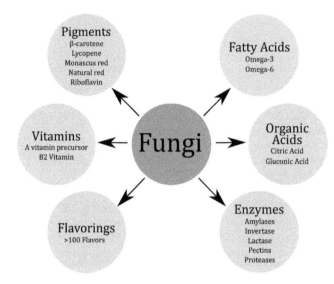

**FIGURE 18.2** Food ingredients produced by fungi at industrial scale. (Modified from Copetti, M. V. 2019. Fungi as industrial producers of food ingredients. *Current Opinion in Food Science* 25:52–6, with permission from Elsevier.)

---

**TABLE 18.3**

**Some Fungal Metabolites Obtained from Industrial Fermentation and Used as Ingredients in the Food Industry**

| Ingredient | Compound | Producing Fungus | Application in Food Industries |
|---|---|---|---|
| Enzymes | Amylase | *Aspergillus niger*<br>*Aspergillus oryzae* | Production of glucose syrup, bread improvement, etc. |
| | Invertase | *Saccharomyces cerevisiae*<br>*Saccharomyces uvarum* | Soft-centered candies, artificial honeys, confectioneries, liqueurs, etc. |
| | Galactosidase | *Mortierella vinaceae* | Beet sugar refining |
| | Lactase | *Aspergillus niger*<br>*Aspergillus oryzae*<br>*Kluyveromyces marxianus*<br>*Kluyveromyces fragilis* | Production of lactose-free milk and dairy products, upgrading cheese whey |
| | Pectics | *Aspergillus niger*<br>*Aspergillus* spp. | Juice clarification, improvement of grape juice yield, removing coffee mucilage, etc. |
| | Proteases | *Aspergillus oryzae*<br>*Aspergillus* spp. | Bread improvement, chill proofing of beer, milk coagulation, meat tenderization, etc. |
| Fatty acids | Ω-3 and Ω-6 | *Mortierella alpine*<br>*Saccharomyces cerevisiae*<br>*Candida lipolytica* | Addition of polyunsaturated fat acids (bioactive compounds) to the composition of food and food products |
| Flavoring | Blue cheese flavor | *Penicillium roqueforti* | Impress blue cheese flavor in food products |
| | Bitter almond flavor | *Ischnoderma* spp. | Impress almond flavor in food products |
| | Roselike odor | *Saccharomyces* spp.<br>*Kluyveromyces* spp. | General food flavoring |
| | Fruity, nutty, and fatty odor | *Candida lipolytica*<br>*Pichia ohmeri* | General food flavoring |
| Organic acids | Citric acid | *Aspergillus niger*<br>*Candida lipolytica* | Soft drinks, jams, jellies, candies, frozen fruits, dairy products, wine, etc. |

*(Continued)*

**TABLE 18.3** (*Continued*)

| Ingredient | Compound | Producing Fungus | Application in Food Industries |
|---|---|---|---|
| | Gluconic Acid | *Aspergillus niger* | Cleansing milk, beer and soft drinks bottles, baking products, etc. |
| Pigments & vitamins | β-carotene | *Blakeslea trispora* | Orange–red food colorants, vitamin A precursor and antioxidant |
| | Lycopene | *Blakeslea trispora* | Red food colorant and bioactive compound |
| | Monascus pigments | *Monascus* spp. | Spice and yellow, orange and red food colorant in Asia, meat preservative |
| | Natural red | *Penicillium oxalicum* | Red food colorant |
| | Riboflavin | *Ashbya gossypii* | Yellow colorant and $B_2$ vitamin |

*Source:* Reprinted with permission from Elsevier: Copetti, M. V. 2019. Fungi as industrial producers of food ingredients. *Current Opinion in Food Science* 25:52–6.

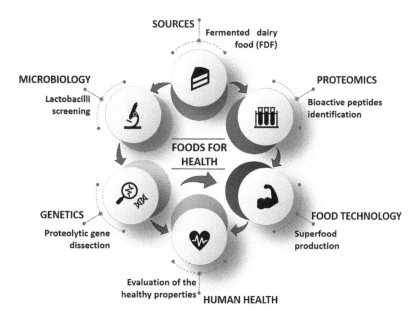

**FIGURE 18.3** Fermented dairy foods and pro-health molecules. (Graphical abstract from Tagliazucchi, D., M. Martini, and L. Solieri. 2019. Bioprospecting for bioactive peptide production by lactic acid bacteria Isolated from fermented dairy food. *Fermentation* 5:96; open access article distributed under the Creative Commons attribution license.)

The production, chemical stability, and safety of any new metabolite classes need to be thoroughly investigated and research efforts should be intensified before they can be given GRAS (generally recognized as safe) status in the United States and elsewhere (Dufossé 2018; Mapari et al. 2009, 2010).

## 18.5 BIOPROCESSING AND NOVEL FOOD PRODUCTION BY FUNGI

"Bioprocessing" or "bioprospecting" is the noninvasive study of naturally occurring organisms, metabolites, and genes with the purpose of discovering biochemical processes that will be useful for food, medicine, or other industries (Mateo et al. 2001). In this regard, fermented dairy foods and their associated LAB are being considered as good sources of health-promoting molecules, such as the bioactive peptides produced by microbial protein digestion. Beyond their anti-microbial activities, these bioactive peptides have shown interesting biofunctions such as anti-hypertensive, antioxidant, and immuno-modulatory effects (see Figure 18.3) (Tagliazucchi et al. 2014).

During the last few decades, with the development of modern microbiology and molecular engineering, bioprocessing food production using fungi has expanded and found numerous applications. Laccase, a lignin-digesting enzyme commonly produced by fungi, has potential applications in food processing, including by making bread dough less sticky and more machinable. Laccase can also be useful in reducing the waste of the food industry while simultaneously increasing the efficiency of biofuel production (Mayolo-Deloisa et al. 2020). Fungal phytases are several enzymes that break down phytates, the biological storage form of phosphate, and are widely used to improve phosphate bioavailability in both human food and animal feed (Jatuwong et al. 2020). Inulinases are another type of hydrolyzing enzyme produced by fungi and are important for the industrial production of high fructose syrup and fructooligosaccharides (FOS: alternative sweeteners) (Singh and Kanika 2018). A better functional understanding of further fungal enzymes will likely provide other applications for the food industry in the future.

## REFERENCES

Abee, T., L. Krockel, and C. Hill. 1995. Bacteriocins: Modes of action and potentials in food preservation and control of food poisoning. *International Journal of Food Microbiology* 28:169–85.

Agyei, D., J. Owusu-Kwarteng, F. Akabanda, and S. Akomea-Frempong. 2020. Indigenous African fermented dairy products: Processing technology, microbiology and health benefits. *Critical Reviews in Food Science and Nutrition* 60:991–1006.

AICV. 2021. *Richness of European Ciders*. Brussels: Association des Industries des Cidres et Vins de Fruits. https://aicv.org/en/publications [accessed: August 2, 2021].

Akuzawa, R., and I. S. Surono. 2011. Fermented milks of Asian. In *Encyclopaedia of Dairy Science*, vol. 2, eds. H. Roginski, J. W. Fuquay, and P. F. Fox, pp. 507–511. San Diego: Academic Press.

Angulo, L., E. Lopez, and C. Lema. 1993. Microflora present in kefir grains of the Galician region (North-west of Spain). *Journal of Dairy Research* 60:263–7.

Akinrele, I. A. 1964. Fermentation of cassava. *Journal of the Science of Food and Agriculture* 15:589–94.

Allwood, J. G., L. T. Wakeling, and D. C. Bean. 2021. Fermentation and the microbial community of Japanese koji and miso: A review. *Journal of Food Science* 87:2194–207.

Archer, D. B. 2000. Filamentous fungi as microbial cell factories for food use. *Current Opinion in Biotechnology* 11:478–83.

Aryana, K.J., and D. W. Olson. 2017. 100-Year review: Yogurt and other cultured dairy products. *Journal of Dairy Science* 100:9987–10013.

Awe, S., and S. N. Nnadoze. 2015. Production and microbiological assessment of date palm (*Phoenix dactylifera* L.) fruit wine. *Microbiology Research Journal International* 8:480–8.

Babel, F. J. 1953. The role of fungi in cheese ripening. *Economic Botany* 7:27–42.

Bamforth, C. W. 2017. Progress in brewing science and beer production. *Annual Review of Chemical and Biomolecular Engineering* 8:161–76.

Batra, L. R., and P. D. Millner. 1974. Some Asian fermented foods and beverages, and associated fungi. *Mycologia* 66:942–50.

Beech, F. W. 1993. Yeasts in cider-making. In *The Yeasts*, eds. A. H. Rose, and J. S. Harrison, 169–213. San Diego: Academic Press.

Bellut, K., and E. K. Arendt. 2019. Chance and challenge: Non-saccharomyces yeasts in nonalcoholic and low alcohol beer brewing – A review. *The Science of Beer* 77:77–91.

Beuchatl, R. 1983 Indigenous fermented foods. In *Biotechnology*, eds. H. J. Rehm, and G. Reed, 477–528. Weinheim: Verlag Chemie.

Blandino, A., M. E. Al-Aseeri, S. S. Pandiella, D. Cantero, and C. Webb. 2003. Cereal-based fermented foods and beverages. *Food Research International* 36:527–43.

Bockelmann, W., U. Krusch, G. Engel, N. Klijn, G. Smit, and K. J. Heller. 1997. The microflora of Tilsit cheese. Part 1. Variability of the smear flora. *Food/Nahrung* 41:208–12.

Bodinakua, I., J. Shaffer, A. B. Connors, et al. 2019. Rapid phenotypic and metabolomic domestication of wild *Penicillium* molds on cheese. *Applied and Environmental Science* 10:e02445–19. https://doi.org/10.1128/mBio.02445-19.

Bokulich, N. A., M. Ohta, M. Lee, and D. A. Mills. 2014. Indigenous bacteria and fungi drive traditional gimoto sake fermentations. *Applied and Environmental Microbiology* 80:5522–9.

Braidwood, R. J., J.D. Sauer, H. Helbaek, et al. 1953. Did man once live by beer alone? *American Anthropologist* 55:515–26.

Broshi. M. 2007. Date beer and date wine in antiquity. *Palestine Exploration Quarterly* 139:55–9.

Burinia, J.A., J.I. Eizaguirre, C. Loviso, and D. Libkind. 2021. Levaduras no convencionales como herramientas de innovación y diferenciación en la producción de cerveza. *Revista Argentina de Microbiología* 53: 359–77.

Cai, Y., A. Sounderrajan, and L. Serventi. 2020. Water Kefir: A review of its microbiological profile, antioxidant potential and sensory quality. *Acta Scientific Nutritional Health* 4:10–7.

Campbell-Platt, C., and P. E. Cook. 1989. Fungi in the production of foods and food ingredients. *Journal of Applied Bacteriology* 67:117S–31S.

Caplice, E., and G. F. Fitzgerald. 1999. Food fermentations: Role of microorganisms in food production and preservation. *International Journal of Food Microbiology* 50:131–49.

Carbonetto, B., T. Nidelet, S. Guezenec, M. Perez, D. Segond, and D. Sicard. 2020. Interactions between *Kazachstania humilis* yeast species and lactic acid bacteria in sourdough. *Microorganisms* 8:240. https://doi.org/10.3390/microorganisms8020240.

Catzeddu, P. 2019. Sourdough breads. In *Flour and Breads and Their Fortification in Health and Disease Prevention*, eds. V. R. Preedy, and R. R. Watson, 177–188. London: Academic Press.

Chan, L. G., J. L. Cohen, and J. M. L. N. de Moura Bell. 2018. Conversion of agricultural streams and food-processing by-products to aalue-added compounds using filamentous fungi. *Annual Review of Food Science and Technology* 9:503–23.

Chavan, R. S. and S. R. Chavan. 2011. Sourdough technology – A traditional way for wholesome foods: A review. *Comprehensive Reviews in Food Science and Food Safety* 10:169–80.

Ciani, M. 1998. Wine vinegar production using base wines made with different yeast species. *Journal of the Science of Food and Agriculture* 78:290–4.

Clementine, K. A.; C. Mohamed; K. Epiphane; B. kouakou David; M. Koffi Dje and D. Montet. 2012. Identification of yeasts associated with the fermented fish, adjuevan, of Ivory Coast by using the molecular technique of PCR-denaturing gradient gel electrophoresis (DGGE). *African Journal of Microbiology Research* 6: 4138–45.

Colomer, M.S., Funch, B., Forster, J. 2019. The raise of Brettanomyces yeast species for beer production. *Current Opinion in Biotechnology* 56:30–5.

Comi, G., and C. Cantoni. 1980. I lieviti in insaccati crudi stagionati. *Industrie Alimentari* 19:857–60.

Comitini, F., M. Gobbi, P. Domizio, et al. 2011. Selected non-saccharomyces wine yeasts in controlled multi-starter fermentations with *Saccharomyces cerevisiae*. *Food Microbiology* 28:873–82.

Connell, L., H. Stender, C. G. Edwards. 2002. Rapid detection and identification of Brettanomyces from winery air samples based on peptide nucleic acid analysis. *American Journal of Enology and Viticulture* 53:322–4.

Copetti, M. V. 2019. Fungi as industrial producers of food ingredients. *Current Opinion in Food Science* 25:52–6.

Copley, M. S., R. Berstan, A. J. Mukherjee, et al. 2005b. Dairying in antiquity. III. Evidence from absorbed lipid residues dating to the British Neolithic. *Journal of Archaeological Science* 32:523–46.

Copley, M. S., R. Berstan, S. N. Dudd, V. Straker, S. Payne, and R. P. Evershed. 2005a. Dairying in antiquity. I. Evidence from absorbed lipid residues dating to the British iron age. *Journal of Archaeological Science* 32:485–503.

Coton, E., M. Coton, D. Levert, S. Casaregola, and D. Sohier. 2006. Yeast ecology in French cider and black olive natural fermentations. *International Journal of Food Microbiology* 1108:130–5.

Coulibaly, W. H., K. M. J-P. Bouatenin, Z. B. I. A. Boli, et al. 2021. Volatile compounds of traditional sorghum beer (tchapalo) produced in Cote d'Ivoire: Comparison between wild yeasts and pure culture of *Saccharomyces cerevisiae*. *World Journal of Microbiology and Biotechnology* 37:75. https://doi.org/10.1007/s11274-021-03026-1.

Cousin, F. J., R. Le Guellec, M. Schlusselhuber, M. Dalmasso, J. M. Laplace, and M. Cretenet. 2017. Microorganisms in fermented apple beverages: Current knowledge and future directions. *Microorganisms* 5:39. https://doi.org/10.3390/microorganisms5030039.

Chisti, Y. 2014. Fermentation (industrial) basic considerations. In *Encyclopedia of Food Microbiology*, eds. C. A. Batt, and M. L. Tortorello, 751–761. Amsterdam: Academic Press.

De Leonardis, A., V. Macciola, M. Iorizzo, S. J. Lombardi, F. Lopez, and E. Marconi. 2018. Effective assay for olive vinegar production from olive oil mill wastewaters. *Food Chemistry* 240:437–40.

Dertli, E., and A. H. Çon. 2017. Microbial diversity of traditional kefir grains and their role on kefir aroma. *LWT – Food Science and Technology* 85:151–7.

Devanthi, P. V. P., and K. Gkatzionis. 2019. Soy sauce fermentation: Microorganisms, aroma formation, and process modification. *Food Research International* 120:364–74.

de Vuyst, L., and E. J. Vandamme. 1994. *Bacteriocins of Lactic Acid Bacteria.* Glasgow: Blackie Academic & Professional.

Dimidi, E., S. R. Cox, M. Rossi, and K. Whelan. 2019. Fermented foods: Definitions and characteristics, impact on the gut microbiota and effects on gastrointestinal health and disease. *Nutrients* 11:1806. https://doi.org/10.3390/nu11081806.

Diniz Felipe, A. L., C. Oliveira Souza, L. Ferreira Santos, and A. Cestari. 2019. Synthesis and characterization of mead: From the past to the future and development of a new fermentative route. *Journal of Food Science and Technology* 56:4966–71.

Dubinina, E. V., D. V. Andrievskaya, S. M. Tomgorova, and K. V. Nebezhev. 2020. Innovative technologies of alcoholic beverages based on fruit distillates. *Food Systems* 3:18–23.

Dufossé, L. 2017. Red colourants from filamentous fungi: Are they ready for the food industry? *Journal of Food Composition and Analysis* 69:156–61.

Farnworth, E. 2008. *Handbook of Fermented Functional Foods.* Boca Raton, FL: CRC Press.

Fentie, E. G., S. A. Emire, H. D. Demsash, D. W. Dadi, and J. H. Shin. 2020. Cereal- and fruit-based Ethiopian traditional fermented alcoholic beverages. *Foods* 9:1781. https://doi.org/10.3390/foods9121781.

Ghorai, S., S. Prosad Banik, D. Verma, S. Chowdhury, S. Mukherjee, and S. Khowala. 2009. Fungal biotechnology in food and feed processing. *Food Research International* 42:577–87.

Gorgus, E., M. Hittinger, and D. Schrenk. 2016. Estimates of ethanol exposure in children from food not labeled as alcohol-containing. *Journal of Analytical Toxicology* 40:537–42.

Guiné, R. P. F., M. J. Barroca, T. E. Coldea, E. Bartkiene, and O. Anjos. 2021. Apple fermented products: An overview of technology, properties and health effects. *Processes* 9:223. https://doi.org/10.3390/pr9020223.

Guizani, N., and A. Mothershaw. Fermentation as a method for food preservation. In *Handbook of Food Fermentation*, ed. M. S. Rahman, 215–236. Boca Raton: CRC Press.

Gulitz, A., J. Stadie, M. Wenning, M. A. Ehrmann, and R. F. Vogel. 2011. The microbial diversity of water kefir. *International Journal of Food Microbiology* 151:284–8.

Guo, L., M. Ya, Y. S. Guo, et al. 2019. Study of bacterial and fungal community structures in traditional koumiss from Inner Mongolia. *Journal of Dairy Science* 102:1972–84.

Guzel-Seydim, Z. B., Ç. Gökırmaklı, and A. K. Greene. 2021. A comparison of milk kefir and water kefir: Physical, chemical, microbiological and functional properties. *Trends in Food Science & Technology* 113:42–53.

Guzel-Seydim, Z. B., T. Kok-Tas, A. K. Greene, and A. C. Seydim. 2011. Review: Functional properties of kefir. *Critical Reviews in Food Science and Nutrition* 51:261–8.

Hachmeister, K. A., and D. Y. C. Fung. 1993. Tempeh: A mold-modified indigenous fermented food made from soybeans and/or cereal crains. *Critical Reviews in Microbiology* 19:137–88.

Han, B.-Z., C.-F. Cao, F. M. Rombouts, and M. J. R. Nout. 2004. Microbial changes during the production of Sufu – A Chinese fermented soybean food. *Food Control* 15:265–70.

Hedger, J. 1978. Tempe, oncom and other mycological oddities. *Bulletin of the British Mycological Society* 12:53–5.

Holt, S., V. Mukherjee, B. Lievens, K. J. Verstrepen, and J. M. Thevelein. 2018. Bioflavoring by non-conventional yeasts in sequential beer fermentations. *Food Microbiology* 72:55–66.

Horiuchi, J. I., T. Kanno, and M. Kobayashi. 2000. Effective onion vinegar production by a two-step fermentation system. *Journal of Bioscience and Bioengineering* 90:289–93.

Iglesias, A., A. Pascoal, A. Branco Choupina, C. Alfredo Carvalho, X. Feás and L. M. Estevinho. 2014. Developments in the fermentation process and quality improvement strategies for mead production. *Molecules* 19, 12577–90.

International Food Information Council (IFIC) and U.S. Food and Drug Administration (FDA). 2010. *Overview of Food Ingredients, Additives & Colors.* U.S. Food and Drug Administration. https://www.fda.gov/food/food-ingredients-packaging/overview-food-ingredients-additives-colors (accessed February 12, 2022).

Irlinger, F., S. Layec, S. Helinck, and E. Dugat-Bony. 2015. Cheese rind microbial communities: Diversity, composition and origin. *FEMS Microbiology Letters* 362:1–11.

Ivey, M. L., and T. G. Phister. 2011. Detection and identification of microorganisms in wine: A review of molecular technique. *Journal of Industrial Microbiology and Biotechnology* 38:1619–34.

Jacobson, G., and J. Wasileski. 1994. Production of food colorants by fermentation. In *Bioprocess Production of Flavor, Fragrance, and Color Ingredients*, ed. A. Gabelman, 205–237. New York: John Wiley & Sons.

Jatuwong, K., N. Suwannarach, J. Kumla, W. Penkhrue, P. Kakumyan, and S. Lumyong. 2020. Bioprocess for production, characteristics, and biotechnological applications of fungal phytases. *Frontiers in Microbiology* 11:188. https://doi.org/10.3389/fmicb.2020.00188.

Jeong, S. C., M. J. Yu, Y. K. Cho, and J. S. Lee. 2003. Characteristics of traditional wine-koji and isolation of fungi. *Journal of Natural Sciences* 13:73–82.

Jeyaram, K. 2009. Traditional fermented foods of Manipur. *Indian Journal of Traditional Knowledge* 8:115–21.

Johansson, L., J. Nikulin, R. Juvonen, et al. 2021. Sourdough cultures as reservoirs of maltose-negative yeasts for low-alcohol beer brewing. *Food Microbiology* 94:103629. https://doi.org/10.1016/j.fm.2020.103629.

Jollivet, N., J. M. Belin, and Y. Vayssier. 1993. Comparison of volatile flavor compounds produced by ten strains of *Penicillium camemberti* Thom. *Journal of Dairy Science* 76:1837–44.

Joshi, V. K. 1997. *Fruit Wines. Directorate of Extension Education.* Nauni, Solan: YS Parmar University of Horticulture and Forestry.

Joshi, V. K., and S. Sharma. 2009 Cider vinegar: Microbiology, technology and quality. In *Vinegars of the World*, eds. L. Solieri, and P. Giudici, 197–207. Milano: Springer.

Joshi, V. K., D. Attri, T. K. Singh, and G. S. Abrol. 2011. Fruit wines: production technology. In *Handbook of Enology: Principles, Practices and Recent Innovations*, ed. V. K. Joshi, 1177–1221. New Delhi: Asia Tech Publisher.

Joshi, V. K., P. K. Shah, K. Kumar. 2005. Evaluation of different peach cultivars for wine preparation. *Journal of Food Science and Technology* 42:83–9.

Joshi, V. K., S. Sharma, and K. Kumar. 2006. Technology for production and evaluation of strawberry wine. *Beverage and Food World* 33:77–8.

Joshi, V. K., S. K. Chauhan, and S. Bhushan. 2000. Technology of fruit based alcoholic beverages. In *Postharvest Technology of Fruits and Vegetables*, eds. L. R. Verma, and V. K. Joshi, 1019–101. New Delhi: Indus Publishing.

Joshi, V. K., V. P. Bhutani, and R. C. Sharma. 1990. The effect of dilution and addition of nitrogen source on chemical, mineral and sensory qualities of wild apricot wine. *American Journal of Enology and Viticulture* 41:229–31.

Keshavarz, T. 2014. Control of fermentation conditions. In *Encyclopedia of Food Microbiology*, ed. C. A. Batt, and M. L. Tortorello, 762–768. Amsterdam: Academic Press.

Ko, S. D. 1986. Indonesian fermented foods not based on soybeans. *Mycological Memoir* 11: 67–84.

Kure, C. F., I. Skaar, and J. Brendehaug. 2004. Mould contamination in production of semi-hard cheese. *International Journal of Food Microbiology* 93:41–9.

Kusumoto, K.-I., Y. Yamagata, R. Tazawa, 2021. Japanese traditional miso and koji making. *Journal of Fungi* 7:579. https://doi.org/10.3390/jof70705.

Latorre-Garcia, L., L. del Castillo-Agudo, and J. Polaina. 2007. Taxonomical classification of yeasts isolated from kefir based on the sequence of their ribosomal RNA genes. *World Journal of Microbiology and Biotechnology* 23:785. https://doi.org/10.1007/s11274-006-9298-y.

Laureys, D., M. Cnockaert, L. De Vuyst, and P. Vandamme. 2016. *Bifidobacterium aquikefiri* sp. nov., isolated from water kefir. *International Journal of Systematic and Evolutionary Microbiology* 66:1281–6.

Lee, S. W., J. H. Kwon, S. R. Yoon, et al. 2010. Quality characteristics of brown rice vinegar by different yeasts and fermentation condition. *Journal of the Korean Society of Food Science and Nutrition* 39:1366–72.

Leistner, L. (1986). Schimmelpilz-gereifte Lebensmittel. *Fleischwirtsch* 66:168–73.

Leistner, L., and Z. Bern. 1970. Vorkommen und bedeutung von hefen bei pokelfleischwaren. *Fleischwirtschaft* 50:350–I.

Li, S., P. Li, F. Feng, et al. 2015. Microbial diversity and their roles in the vinegar fermentation process. *Applied Microbiology and Biotechnology* 99:4997–5024.

Lim, G. 1991. Indigenous fermented foods in south east Asia. *Asian Food Journal* 6:83–101.

Liu, T., Y. Li, Y. Yang, H. Yi, L. Zhang, and G. He. 2019. The influence of different lactic acid bacteria on sourdough flavor and a deep insight into sourdough fermentation through RNA sequencing. *Food Chemistry* 307:125529. https://doi.org/10.1016/j.foodchem.2019.125529.

Lonchamp, J., P. S. Clegg, and S. R. Euston. 2019. Foaming, emulsifying and rheological properties of extracts from a co-product of the Quorn fermentation process. *European Food Research and Technology* 245:1825–39.

Lorenzini, M., B. Simonato, and G. Zapparoli. 2018. Yeast species diversity in apple juice for cider production evidenced by culture-based method. *Folia Microbiologica* 63:677–84.

Lücke, F. K. 1994. Fermented meat products. *Food Research International* 27:299–307.

Lucia, S. A 1963. *History of Wine as Therapy*. Philadelphia: Lippincott.

Lund, F., O. Filtenborg, and J. C. Frisvad. 1995. Associated mycoflora of cheese. *Food Microbiology* 12:173–80.

Lynch, K. M., S. Wilkinson, L. Daenen, and E. K. Arendt. 2021. An update on water kefir: Microbiology, composition and production. *International Journal of Food Microbiology* 345:109128. https://doi.org/10.1016/j.ijfoodmicro.2021.109128.

Maheshwari, G., J. Ahlborn, M. Rühl. 2020. Role of fungi in fermented foods. Reference Module in Life Sciences. *Encyclopedia of Mycology* 2:590–600.

Maïworé, J., L. Tatsadjieu Ngoune, I. Piro-Metayer, and D. Montet. 2019. Identification of yeasts present in artisanal yoghurt and traditionally fermented milks consumed in the northern part of Cameroon. *Scientific African* 6:e00159. https://doi.org/10.1016/j.sciaf.2019.e00159.

Mapari, S. A., A. S. Meyer, U. Thrane, and J. C. Frisvad. 2009. Identification of potentially safe promising fungal cell factories for the production of polyketide natural food colorants using chemotaxonomic rationale. *Microbial Cell Factories* 8:24. https://doi.org/10.1186/1475-2859-8-24.

Mapari, S. A., U. Thrane, and A. S. Meyer. 2010. Fungal polyketide azaphilone pigments as future natural food colorants? *Trends in Biotechnology* 28:300–7.

Marcellino, N. O. S. B., and D. R. Benson. 2013. The good, the bad, and the ugly: Tales of mold-ripened cheese. *Microbiology Spectrum* 1. https://doi.org/10.1128/microbiolspec.CM-0005-12.

Marsh, A. J., C. Hill, R. P. Ross, and P. D. Cotter. 2014. Fermented beverages with health-promoting potential: Past and future perspectives. *Trends in Food Science and Technology* 38:113–24.

Martínez-Torres, A., S. Gutierrez-Ambrocio, P. Heredia-del-Orbe, et al. 2017. Inferring the role of microorganisms in water kefir fermentations. *International Journal of Food Science and Technology* 52:559–71.

Mateo, N., N. W. Nader, and G. Tamayo. 2001. Bioprocessing. *Encyclopedia of Biodiversity* 1:471–88.

Mayolo-Deloisa, K., M. González-González, and M. Rito-Palomares. 2020. Laccases in food industry: Bioprocessing, potential industrial and biotechnological applications. *Frontiers in Bioengineering and Biotechnology* 8:222. https://doi.org/10.3389/fbioe.2020.00222.

Medina, K., E. Boido, L. Fariña, et al. 2013. Increased flavour diversity of Chardonnay wines by spontaneous fermentation and co-fermentation with *Hanseniaspora vineae*. *Food Chemistry* 141:2513–21.

Molinet, J., and F. A. Cubillos. 2020. Wild yeast for the future: Exploring the use of wild strains for wine and beer fermentation. *Frontiers in Genetics*. 11:589350. https://doi.org/10.3389/fgene.2020.589350.

Montanari, G., C. Zambonelli, L. Grazia, G. K. Kamesheva, and M. K. Shigaeva. 1996. Saccharomyces unisporus as the principal alcoholic fermentation microorganism of traditional koumiss. *Journal of Dairy Research* 63:327–31.

Moore, D., and S. W. Chiu. 2001. Fungal products as food. In *Bio-Exploitation of Filamentous Fungi*, eds. S. B. Pointing, and K. D. Hyde, 223–251. Hong Kong: Fungal Diversity Press.

Morrissey, W. F., B. Davenport, A. Querol, and A. D. Dobson. 2004. The role of indigenous yeasts in traditional Irish cider fermentations. *Journal of Applied Microbiology* 97:647–55.

Mu, Z., X. Yang, and H. Yuan. 2012. Detection and identification of wild yeast in Koumiss. *Food Microbiology* 31:301–8.

Murooka, Y., and M. Yamshita. 2008. Traditional healthful fermented products of Japan. *Journal of Industrial Microbiology and Biotechnology* 35:791–8.

Nielsen, M. S., J. C. Frisvad, and P. V. Nielsen. 1998. Protection by fungal starters against growth and secondary metabolite production of fungal spoilers of cheese. *International Journal of Food Microbiology* 42:91–9.

Nout, M. J. R., P. K. Sarkar, and L. R. Beuchat. 2007. Indigenous fermented foods. In *Food Microbiology: Fundamentals and Frontiers*, eds. M. P. Doyle, and L. E. Beuchat, 817–835, Washington DC: ASM Press.

Nowak, J. 1992. Oats tempeh. *Engineering in Life Sciences* 12:345–8.

Núñez, F., L. S. María, P. Belén, D. Josué, L. Sánchez-Montero, and M. J. Andrade. 2015. Selection and evaluation of *Debaryomyces hansenii* isolates as potential bioprotective agents against toxigenic penicillia in dry-fermented sausages. *Food Microbiology* 46:114–20.

Panda, B. P., S. Javed, and M. Ali. 2010. Production of angkak through co-culture of *Monascus purpureus* and *Monascus ruber*. *Brazilian Journal of Microbiology* 41:57–64.

Pando Bedriñana, R., A. Querol Simón, and B. Suárez Valles. 2010. Genetic and phenotypic diversity of autochthonous cider yeasts in a cellar from Asturias. *Food Microbiology* 27:503–8.

Patrick, C. 1952. *Alcohol, Culture, and Society*. Durham: Duke U Press.

Pereira, G. V. M., V. T. Soccol, and C. R. Soccol. 2016. Current state of research on cocoa and coffee fermentations. *Current Opinion in Food Science* 7:50–7.

Perez, R. H., T. Zendo, and K. Sonomoto. 2014. Novel bacteriocins from lactic acid bacteria (LAB): Various structures and applications. *Microbial Cell Factories* 13:S3. https://doi.org/10.1186/1475-2859-13-S1-S3.

Perry-O'Keefe, H., H. Stender, A. Broomer, K. Oliveira, J. Coull, and J. J. Hyldig-Nielsen. 2001. Filter-based PNA in situ hybridization for rapid detection, identification and enumeration of specific micro-organisms. *Journal of Applied Microbiology* 90:180–9.

Pitt, J. I., R. H. Cruickshank, and L. Leistner. 1986. *Penicillium commune, P camembertii*, the origin of white cheese moulds, and the production of cyclopiazonic acid. *Food Microbiology* 3:363–71.

Pontes E. K., G. C. G. Cunha, C. Muller, et al. 2019. Advances in yeast alcoholic fermentations for the production of bioethanol, beer and wine. *Advances in Applied Microbiology* 109:61–119.

Punt, P. J., N. van Biezen, A. Conesa, A. Albers, J. Mangnus, and C. van den Hondel. 2002. Filamentous fungi as cell factories for heterologous protein production. *Trends in Biotechnology* 20:200–6.

Rainieri, S., and C. Zambonelli. 2009. Organisms associated with acetic acid bacteria in vinegar production. In *Vinegars of the World*, eds. L. Solieri, and P. Giudici, 73–95. Milano: Springer.

Ramos-Moreno, L., F. Ruiz-Pérez, E. Rodríguez-Castro, and J. Ramos. 2021. *Debaryomyces hansenii* is a real tool to improve a diversity of characteristics in sausages and dry-meat products. *Microorganisms* 9:1512.

Rizal, S., Murhadi, M.E. Kustyawati, and U. Hasanudin. 2020. Growth optimization of *Saccharomyces cerevisiae* and *Rhizopus oligosporus* during fermentation to produce tempeh with high β-glucan content. *Biodiversitas* 21:2667–73.

# Index

Note: **Bold** page numbers refer to tables and *italic* page numbers refer to figures.

Ingram Content Group UK Ltd.
Milton Keynes UK
UKHW050008050523
421200UK00003B/12